An Introduction to
Stochastic Processes and Nonequilibrium Statistical Physics

Revised Edition

SERIES ON ADVANCES IN STATISTICAL MECHANICS*

Editor-in-Chief: M. Rasetti *(Politecnico di Torino, Italy)*

Published

Vol. 7:   The Hubbard Model – Recent Results
          edited by M. Rasetti

Vol. 8:   Statistical Thermodynamics and Stochastic Theory of Nonlinear
          Systems Far From Equilibrium
          by W. Ebeling & L. Schimansky-Geier

Vol. 9:   Disorder and Competition in Soluble Lattice Models
          by W. F. Wreszinski & S. R. A. Salinas

Vol. 10: An Introduction to Stochastic Processes and Nonequilibrium
          Statistical Physics
          by H. S. Wio

Vol. 12: Quantum Many-Body Systems in One Dimension
          by Zachary N. C. Ha

Vol. 13: Exactly Soluble Models in Statistical Mechanics: Historical Perspectives
          and Current Status
          edited by C. King & F. Y. Wu

Vol. 14: Statistical Physics on the Eve of the 21st Century: In Honour of
          J. B. McGuire on the Occasion of his 65th Birthday
          edited by M. T. Batchelor & L. T. Wille

Vol. 15: Lattice Statistics and Mathematical Physics: Festschrift Dedicated to
          Professor Fa-Yueh Wu on the Occasion of his 70th Birthday
          edited by J. H. H. Perk & M.-L. Ge

Vol. 16: Non-Equilibrium Thermodynamics of Heterogeneous Systems
          by S. Kjelstrup & D. Bedeaux

Vol. 17: Chaos: From Simple Models to Complex Systems
          by M. Cencini, F. Cecconi & A. Vulpiani

Vol. 19: An Introduction to Stochastic Processes and Nonequilibrium Statistical Physics
          by H. S. Wio, R. R. Deza & J. M. López

Forthcoming

Vol. 18   Statistical Mechanics of Magnetic Excitations: From Spin Waves to Stripes
          and Checkerboards
          by E. Rastelli

*For the complete list of titles in this series, please go to
http://www.worldscibooks.com/series/sasm_series

Series on Advances in Statistical Mechanics – Vol. 19

An Introduction to
Stochastic Processes and Nonequilibrium Statistical Physics

Revised Edition

Horacio S. Wio
Instituto de Física de Cantabria
Universidad de Cantabria & CSIC, Spain

Roberto R. Deza
Instituto de Física de Mar del Plata
Universidad Nacional de Mar del Plata & CONICET, Argentina

Juan M. López
Instituto de Física de Cantabria
Universidad de Cantabria & CSIC, Spain

 World Scientific

NEW JERSEY • LONDON • SINGAPORE • BEIJING • SHANGHAI • HONG KONG • TAIPEI • CHENNAI

Published by

World Scientific Publishing Co. Pte. Ltd.

5 Toh Tuck Link, Singapore 596224

USA office: 27 Warren Street, Suite 401-402, Hackensack, NJ 07601

UK office: 57 Shelton Street, Covent Garden, London WC2H 9HE

British Library Cataloguing-in-Publication Data
A catalogue record for this book is available from the British Library.

Series on Advances in Statistical Mechanics — Vol. 19
AN INTRODUCTION TO STOCHASTIC PROCESSES AND
NONEQUILIBRIUM STATISTICAL PHYSICS
(Revised Edition)

Copyright © 2012 by World Scientific Publishing Co. Pte. Ltd.

ISBN 978-981-4374-78-1

Printed in Singapore.

Preface

*Philosophy is written in this great book (by which I mean the
Universe) which stands always open to our view, but it cannot
be understood unless one first learns how to comprehend the
language and interpret the symbols in which it is written, and
its symbols are triangles, circles, and other geometric figures,
without which it is not humanly possible to comprehend even
one word of it; without it one wanders in a dark labyrinth ...*
Galileo Galilei

It has been almost twenty years since the first edition of this book emerged.
Perhaps too long for so actively developing field. As then, the primary
guiding principle is still—besides looking forward to introduce Nonequilibrium Statistical Physics as coherently as possible by gluing its many facette
together—to provide a working knowledge of some of the field's relevant de-
velopment directions by the time of the book's appearance (as it was the
case with pattern formation). To have a broader panorama, the author of
the original edition (H.S.Wio) joined efforts with two colleagues (R.R.Deza
and J.M.López) to revise the first edition including some new aspects in order to make it a more updated exposition of nonequilibrium and stochastic
processes and applications.

Clearly, one of the most conspicuous developments during this period
has been on the side of nanoscience, which has transited from being somewhat more than a dream to an everyday reality for the public at large. Of
course, only a few out of the many developments in this field belong into
a book on Nonequilibrium Statistical Physics. Consequently, some of the
new material in this edition deals with Brownian transport and its relatives,
which can be loosely grouped as "constructive effects of noise".

A secondary but natural guiding principle in selecting the material for this edition was to cover mostly those topics that belong to one's expertise. In fact, that was the case in the first edition with the topic of pattern formation. The present one provides an opportunity to advocate for the *nonequilibrium potential*. Together with several people we have endeavored many of these years to this useful description framework of nonequilibrium phenomena.

There are of course many lines in the field of Nonequilibrium Statistical Physics in which the nonequilibrium-potential framework is still not useful (or perhaps never will). We have tried to outline an account of those directions keeping the original spirit of the book. Among them we have included concepts of kinetic roughening and scaling. We have also, in addition to the previous material, provided some references to more recent literature.

The responsibility for what appears in this new edition is clearly ours, but we would like to thank many friends and colleagues for collaborations, comments, suggestions, etc, through which we have gained a better understanding of this field.

We also thank our families for their continuous support and patience.

H.S.W, R.R.D, J.M.L.

Preface to the first edition

This book arose from notes for several diferent courses that I have taught at the Instituto Balseiro, Universidad Nacional de Cuyo, Bariloche and the Departamento de Física, Universidad Nacional de Mar del Plata. Their purpose was to bring in a more or less self-contained introductory form certain material that is scattered in the literature. Such an introduction will offer graduate students the opportunity of becoming acquainted with elements of stochastic, kinetic and nonequilibrium processes. These, though standard techniques by now, are not usually taught in the usual courses on statistical physics. The main audience for the work will, therefore, be undergraduate physics students in their last year and graduate students.

As already indicated, the material covered in this work is scattered in texts, review articles and proceedings of summer schools. However, those exceptional titles that cover most of the material do so either too superficially or at a level far above an introduction. The aim of this presentation is then to arrange the material in such a way as to introduce adequately a set of ideas and techniques that provide the theoretical framework for the description of far from equilibrium phenomena, involving meso- and macroscopic pictures. The relevance of this framework is obvious from its many applications in the most diverse fields, ranging from physics, chemistry and biology to population dynamics, economy and sociology. I hope this book will be found suitable as a basis for such an introductory course, and bring to the student a feeling of the tools and techniques most usually employed in the treatment of far from equilibrium phenomena.

The area of nonequilibrium phenomena, that has received different names (H. Haken called it *Synergetics*, I. Prigogine and collaborators *Self-Organization Systems*, while others know it as *Complex Systems*), covers such a wide spectrum of subjects that it was impossible to include them all. For this reason, it was necessary to make a selection of subjects leaving out some very interesting and important ones such as disordered systems, instabilities in fluids as well as in lasers, neural networks, cellular automata, and so on. However, I hope that the material included will offer at least, a feeling of and attract attention to, the many interesting aspects of this field.

The organization of the book is as follows:

- In the first chapter I briefly introduce the theory of stochastic processes as the most adequate framework to describe the temporal behavior of fluctuations. The discussions and examples aim to make clear why stochastic methods have become so important in so many different branches of science and technology. The common principles and methods that arise in those fields are presented here.
- The second chapter introduces some basic ideas of the kinetic approach. A presentation of the BBGKY hierarchy, within a classical context, together with some examples, gives a hint of how it is possible to derive transport as well as microscopic balance equations. A connection with the quantal problem and the idea of reduced density operators is also given.
- In the third chapter, I describe and discuss the Onsager relations and approach. First, Onsager's ideas about approach to equilibrium are introduced in an elementary way making clear the role of fluctuations, followed by a more general presentation of the Onsager relations, and some examples of application in simple systems. A discussion of the *minimum entropy production theorem*, shows that, for linear systems, steady states out of equilibrium play a similar role to that of equilibrium states in equilibrium thermodynamics.
- In chapter four, I start reviewing the definitions and deriving some properties of self-correlation functions. Next, the framework of the *linear response theory* is introduced, and the well-known *fluctuation-dissipation theorem* is presented.
- The next chapter introduces some basic tools needed for an adequate analysis of a system in a far from equilibrium situation. These nonequilibrium phenomena that lead usually to space-time or *dissipative structures*, have a widespread interest due to their implications for the understanding of cooperative phenomena in physics, chemistry, biology, etc. Next I turn to discuss a type of behavior we can expect at the macroscopic level when an external

control parameter is varied. Through some examples, such notions as *attractors*, *limit cycles*, *bifurcations* and *symmetry breaking*, are introduced. Next, the effect of external fluctuations on macroscopic behavior is analyzed. It is shown that near an instability point, they can give rise to completely new behavior.

- In the sixth and last chapter I present, studying some simple model examples, some of the general underlying principles that exist in most of the nonequilibrium phenomena usually leading to *dissipative structures*. Among these model examples, I focus on the *active* (or *excitable*) *media* picture, which has become very useful in the description of pattern formation and propagation. I also discuss the *reaction-diffusion model*, for the one- and two-component cases. We study not only the formation of static patterns, but also a few principles governing their propagation.

Responsibility for what appears here is, of course, my own, but I would like to acknowledge the assistance and help I have received while writing this book from many colleagues, graduate students and friends. To name a few, I want to thank Damián H. Zanette, Veronica Grünfeld, Marcelo Kuperman, Guillermo Abramson, Roberto Deza and Carlos Borzi. I also extend my thanks to Maxi San Miguel, Miguel A. Rodriguez, Lázaro D. Salem and Luis Pesquera, with whom—through long standing collaborations and interminable discussions—I have gained in my understanding not only of stochastic processes and statistical physics, but other aspects of physics as well. Most important, with all of them I have enjoyed the pleasure of friendship. Also, I thank the many students that have endured with stoicism the courses on stochastic processes, nonequilibrium statistical physics and instabilities, that I have taught during these years and have made possible this textbook.

Ms. Mirta Rangone has transformed the original childlike sketches into the nice figures that appear in the book. Damián H. Zanette has reviewed the entire manuscript, and I have greatly appreciated his many detailed comments and suggestions. I would like particularly to give my thanks to Veronica Grünfeld for her continuous encouragement and support. She undertook the heavy task of correcting the English version of the whole manuscript. If the book has attained a readable English level it is her sole merit; however, all remaining errors are only my fault. My daughter Mayra also helped me with some parts of the English version. Finally, the warmest thanks are to my wife María Luz who encouraged me along the whole period of this enterprise and, patiently, accepted to be supplanted by drafts, papers, textbooks, and word processors through far too many evenings and long weekends.

Horacio S. Wio

Contents

Preface   vii

Preface to the first edition   ix

The stochastic approach   1

1. Stochastic processes and the master equation   3

   1.1 Introduction   3
   1.2 Stochastic processes   4
        1.2.1 Distribution functions and mean values   4
        1.2.2 Joint and conditional probabilities   7
   1.3 Markovian processes   9
        1.3.1 Wiener–Lévy process   10
        1.3.2 Ornstein–Uhlenbeck process   10
        1.3.3 Poisson processes   11
   1.4 Master equation   11
        1.4.1 Decay processes (radioactive decay of nuclei,
               excited atoms, etc)   13
        1.4.2 Kinetics of the Ising model   15
   1.5 Approach to equilibrium: Ehrenfest urn model
        (dog–flea model)   16

2. The Fokker–Planck equation   21

   2.1 Kramers–Moyal expansion   21
   2.2 Brownian motion, Langevin and Fokker–Planck
        equations   23

2.3   Stochastic differential equations and Fokker–Planck equations 27

2.4   Path integral for Markov processes 30

2.5   A note on numerical simulation 35

3. The Ω–expansion   37

3.1   Van Kampen's Ω–expansion 37

    3.1.1   Effusion of a dilute gas 37

    3.1.2   Dissociation of a diatomic gas 45

3.2   A case with many variables 48

    3.2.1   Description of the model 49

    3.2.2   The expansion 50

    3.2.3   Behavior of fluctuations 51

    3.2.4   Some results 54

3.3   Limitations of the Ω–expansion 56

Thermodynamics and kinetics near equilibrium   59

4. Distributions, BBGKY–hierarchy, balance equations, and the density operator   61

4.1   Introduction 61

4.2   Probability density as a fluid 62

4.3   BBGKY hierarchy 67

4.4   Simple kinetic equations: Vlasov and Boltzmann 70

4.5   Microscopic balance equations 74

    4.5.1   Balance equation for the particle density 75

    4.5.2   Balance equation for the momentum density . . . 75

    4.5.3   Balance equation for the energy density 77

    4.5.4   Balance equation for the entropy density 77

4.6   Density operator 78

4.7   Reduced density operator 80

4.8 $\mathcal{H}$–theorem 82

4.9   Transport coefficients 84

5. Linear nonequilibrium thermodynamics and Onsager relations   89

5.1   Introduction 89

5.2   Onsager's regression-to-equilibrium hypothesis 90

        5.2.1   Chemical kinetics   94
        5.2.2   Self-diffusion   95
    5.3   Onsager's relations   97
    5.4   Examples of Onsager's relations 100
        5.4.1   Mechano-caloric effect 103
        5.4.2   Thermo-molecular pressure effect 104
    5.5   Minimum production of entropy 108
    5.6   Some concepts of fluctuations around equilibrium 113
    5.7   Fluctuation theorems 114
        5.7.1   Nonequilibrium partition identity 116
        5.7.2   The fluctuation theorem and Loschmidt's
                paradox 117
        5.7.3   Jarzynski's equality 118
        5.7.4   Crooks' fluctuation theorem 119
        5.7.5   Final comments 120

6. Linear response theory, fluctuation–dissipation theorem   123

    6.1   Introduction 123
    6.2   Correlation functions: Definitions and properties 124
        6.2.1   Properties 128
        6.2.2   Sum rules 132
    6.3   Linear response theory 133
        6.3.1   Inelastic scattering cross section 134
        6.3.2   Response functions 139
    6.4   Fluctuation–dissipation theorem, and properties of
          response functions 142
    6.5   Some examples 146
        6.5.1   Dielectric susceptibility 146
        6.5.2   Transport coefficients 147

The mesoscopic approach far from equilibrium: Non-extended systems   **151**

7. Stability of dissipative dynamical systems   153

    7.1   Introduction 153
    7.2   Phase plane analysis and linear stability theory 154
    7.3   Limit cycles, bifurcations, symmetry breaking 159
    7.4   Notions about the "nonequilibrium potential" 168

7.4.1   Lyapunov function and global stability   169

7.4.2   Nonequilibrium potential   172

8. Noise-induced phenomena in non-extended dynamical systems   175

8.1   Introduction   175

8.2   Bistability, escape times, critical phenomena   176

8.3   Noise-induced transitions   184

8.4   Notions of stochastic resonance   195

8.5   Brownian motors   199

The mesoscopic approach far from equilibrium: Extended systems   207

9. Formation and propagation of patterns in far-from-
    equilibrium systems   209

9.1   Introduction   209

9.2   Reaction–diffusion descriptions and pattern formation . .   210

9.3   Example: An electrothermal instability   217

9.4   Pattern propagation: (a) one component systems   226

9.5   Pattern propagation: (b) two component systems   232

9.6   The genesis of spirals   235

9.7   Examples of NEP in extended systems   239

9.7.1   Scalar reaction–diffusion model   239

9.7.2   Activator–inhibitor systems   241

9.7.3   Three-component activator–inhibitor model   244

10. Noise-induced phenomena in extended systems   247

10.1   Introduction   247

10.2   Stochastic resonance in spatially extended media   247

10.2.1   Scalar reaction–diffusion model   247

10.2.2   Role of the NEP symmetry   251

10.2.3   Enhancement due to selective coupling   252

10.2.4   Nonlocal coupling   253

10.2.5   Other cases   254

10.3   Noise-induced phase transitions   254

10.3.1   Initial idea: Short-time instability   254

10.3.2   NIPT due to a dynamical instability   256

10.3.3   Effect of colored noise   259

10.3.4 Another form of NIPT 260
10.3.5 Coupled ratchets: anomalous hysteresis 263

11. Surface growth and kinetic roughening 265

11.1 Introduction 265
11.2 Scaling of the surface width and critical exponents 266
11.3 Scale-invariant surfaces, models, and universality classes . 268
11.3.1 Random growth 271
11.3.2 Edwards–Wilkinson equation 271
11.3.3 Kardar–Parisi–Zhang equation 272
11.3.4 Mullins–Herring equation 273
11.3.5 Lai–Das Sarma–Villain equation 274
11.4 Anomalous roughening: Global versus local surface
fluctuations 274
11.4.1 Family–Vicsek scaling 278
11.4.2 Super-roughening 279
11.4.3 Intrinsic anomalous roughening 280
11.5 Generic scaling Ansatz 283
11.6 Dynamics of the surface slope and the origin of
anomalous scaling 288
11.6.1 Kardar–Parisi–Zhang equation 290
11.6.2 Surface growth with conservation law 291
11.6.3 Linear MBE model 292
11.6.4 Lai–Das Sarma–Villain equation 293

12. Final Comments 295

Bibliography suggested in the first edition 299

Bibliography 305

Index 311

About the Authors 315

PART 1

The stochastic approach

*The longest period of time for which a modern painting has
hung upside down in a public gallery unnoticed is 47 days.
This occurred to* Le Bateau *by Matisse
in the Museum of Modern Art, New York City.
In this time 116 000 people had passed through the gallery.*
Ross McWhirter and Norris McWhirter

*To the Wise there is no wealth. For Virtuos there is no power.
And to the Powerful there is neither wise nor virtuous.*
R. Fontanarrosa

Chapter 1

Stochastic processes and the master equation

God moves the player, and he, in turn, each piece.
Which god behind God the web begins,
of agonies and time and dust and dreams?
Jorge Luis Borges

1.1 Introduction

The original framework of equilibrium thermodynamics, considers only relations among quantities that correspond to measured macroscopic equilibrium values of some physical variables. From the point of view of (equilibrium) statistical mechanics, those values are typically given by ensemble averages. As it is known, the Gibbs statistical approach leads us to
consider fluctuations around such average values, corresponding to the instantaneous values of the physical quantities, and behaving according to a
well known theory. Some examples, however, like Brownian motion, have
shown that there are some phenomena that cannot be described by (equilibrium) statistical thermodynamics, and require a more detailed analysis
of the behavior of fluctuations. Within such kind of framework the interest
arises in the behavior of the fluctuations as functions of time. This implies
a more difficult study, that involves not only the equilibrium distribution,
but also its temporal evolution. At least in principle, this requires solving
the equations of motion of the physical system under consideration.

There are many reasons that justify the increasing interest in the study
of such fluctuations. Firstly, because they present serious impediments to
accurate measurements in very sensitive experiments, demanding some very
specific techniques in order to reduce their effects. Besides, the fluctuations
might be used as a source of additional information about the system. An-

other important aspect is that fluctuations can produce macroscopic effects such as the appearance of *spatio-temporal patterns*—or, according to Prigogine: *dissipative structures*—in physical, chemical, or biological systems. The most adequate framework to describe the temporal behavior of fluctuations is the theory of stochastic processes. It is then clear why stochastic methods have become so important in different branches of physics, chemistry, biology, technology, population dynamics, economics, and sociology. In spite of the large number of different problems that arise in all these fields, there are some common principles and methods. This first chapter aims to present a brief introduction to such techniques.

1.2 Stochastic processes

1.2.1 *Distribution functions and mean values*

A *stochastic or random variable* is a quantity X, defined by a set of possible values $\{x\}$, and a probability distribution on this set. Consider the usual example of a dice: after each throw, the number in the upper face corresponds to the variable X, with possible outcomes $x = \{1, 2, 3, 4, 5, 6\}$, and probabilities of $p = 1/6$ (in a honest dice) for each value of x. The set of possible outcomes (called *range* or *set of states*) could be discrete or continuous, finite or infinite. If the range is discrete and denumerable (as for the case of the dice), the *probability distribution* will be given by a set of nonnegative numbers $\{p_n\}$ such that

$$\sum_n p_n = 1 \qquad (1.1)$$

When the range corresponds to an interval $[a, b]$ over the x-axis, the probability distribution is determined by a nonnegative function $P(x)$, with $P(x)dx$ the probability of $X \in [x, x + dx]$, and such that

$$\int_a^b dx P(x) = 1 \qquad (1.2)$$

This function is usually called *probability density*, and the possibility that it contains one or more delta-like contributions should not be discarded. As a matter of fact, a discrete distribution may be written as a continuous one, but only composed of delta contributions. Figure 1.1 shows typical examples of both discrete and continuous distributions. The previous definition corresponds to a one-dimensional variable, but could be immediately extended to higher dimensional cases.

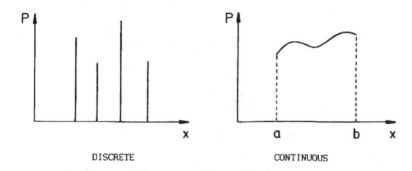

P, P, DISCRETE, CONTINUOUS, x, a, b, x

Fig. 1.1

When a stochastic variable is given, every related quantity $Y = f(X)$ is again a stochastic variable. The latter could be any kind of mathematical object, and particularly also a function of an auxiliary variable t, i.e. $Y = f(X, t)$, where t could be the time or some other parameter. Such $Y(t)$ is called a *stochastic process*. It could be considered as a set of *sample functions* or *realizations* $y(t) = f(x, t)$, each one obtained when we fix X in one of its possible values. In Fig. 1.2 we have depicted, in the $y - t$ plane, different forms for the trajectories for the process $Y(t)$ (that is, the set of points (y_n, t_n)), corresponding to choosing different values of X.

Let X be a stochastic variable defined on the range $(-\infty, +\infty)$ and with distribution $P(x)$. We define the *average* of some function $f(x)$ over the distribution as

$$\langle f(x) \rangle = \int_{-\infty}^{\infty} dx f(x) P(x) \tag{1.3a}$$

The *moments* of the variable X, are quantities of particular relevance in the whole theory and are defined as

$$\mu_m = \langle X^m \rangle = \int_{-\infty}^{\infty} dx\, x^m P(x) \tag{1.3b}$$

It is also useful to introduce the *characteristic function*

$$G(k) = \langle e^{ikx} \rangle = \int_{-\infty}^{\infty} dx\, e^{ikx} P(x) \tag{1.4}$$

as it results to be the *moment generating function*

$$G(k) = \sum_m \frac{(ik)^m}{m!} \mu_m \quad ; \quad \mu_m = (-i)^m \frac{\partial^m}{\partial k^m} G(k = 0) \tag{1.5}$$

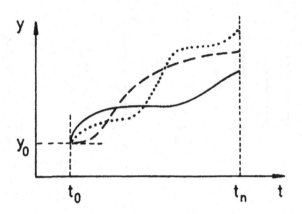

Fig. 1.2

Other useful quantities are the *cumulants* κ_m, defined by

$$\ln[G(k)] = \sum_m \frac{(ik)^m}{m!} \kappa_m = \ln\left[\sum_m \frac{(ik)^m}{m!} \mu_m\right] \tag{1.6}$$

The first cumulant is coincident with the first moment (*mean*) of the stochastic variable: $\kappa_1 = \mu_1 = \langle x \rangle$. The second cumulant, called *variance*, is related to the second moment through $\kappa_2 = \mu_2 - \mu_1^2 = \sigma^2$.

All these notions can be extended to several variables: consider $\mathbf{X} = (x_1, x_2, \ldots, x_n)$, with a probability distribution $P(x_1, x_2, \ldots, x_n)$, also called the *joint probability distribution*, that gives the probability that the set of variables have their values within $(x_1, x_1 + dx_1)$ and $(x_2, x_2 + dx_2)$, etc. This quantity allows us to define the moments

$$\langle X_1^\mu X_2^\nu \ldots X_n^\eta \rangle = \tag{1.7}$$

$$\int_{-\infty}^{\infty} \int_{-\infty}^{\infty} \ldots \int_{-\infty}^{\infty} dx_1 dx_2 \ldots dx_n x_1^\mu x_2^\nu \ldots x_n^\eta P(x_1, x_2, \ldots, x_n)$$

In terms of these moments the generating function is given by

$$G(\mathbf{k}) = \langle e^{i\mathbf{k}\cdot\mathbf{x}} \rangle = \sum_\mu \sum_\nu \ldots \sum_\eta \frac{(ik)^\mu}{\mu!} \frac{(ik)^\nu}{\nu!} \ldots \frac{(ik)^\eta}{\eta!} \langle X_1^\mu X_2^\nu \ldots X_n^\eta \rangle \tag{1.8}$$

Correspondingly, in terms of the generalized cumulants it is

$$G(\mathbf{k}) = \exp\left\{\sum_\mu \sum_\nu \ldots \sum_\eta \frac{(ik)^\mu}{\mu!} \frac{(ik)^\nu}{\nu!} \ldots \frac{(ik)^\eta}{\eta!} \kappa_\mu \kappa_\nu \ldots \kappa_\eta\right\} \tag{1.9}$$

For the particular case of a *Gaussian* distribution, in the one dimensional case we have

$$P(x) = (2\pi\sigma^2)^{-1/2} \exp\left\{ -\frac{[x - \mu_1]^2}{2\sigma^2} \right\} \tag{1.10}$$

and all cumulants with $m > 2$ are zero. The multivariable Gaussian distribution turns out to be

$$P(\mathbf{x}) = \left[\frac{\det \mathbb{A}}{(2\pi)^n} \right]^{-1/2} \exp\left\{ -(\mathbf{x} - \bar{\mu}_1)\mathbb{A}(\mathbf{x} - \bar{\mu}_1)^t \right\} \tag{1.11}$$

where $\bar{\mu}_1$ is the constant vector of the first moments $(\mu_{1,j} = \langle x_j \rangle)$ and $\mathbb{A}$ is the *correlation matrix*

$$[\mathbb{A}^{-1}]_{ij} = \langle (x_i - \mu_{1,i})(x_j - \mu_{1,j}) \rangle \tag{1.12}$$

1.2.2 *Joint and conditional probabilities*

A stochastic process $Y(t)$, defined from a stochastic variable X as indicated before, leads us to a *hierarchy* of probability densities. We write

$$P_n(y_1, t_1; y_2, t_2; \ldots; y_n, t_n)\, dy_1 dy_2 \ldots dy_n \tag{1.13}$$

for the probability that $Y(t_1)$ is within the interval $(y_1, y_1 + dy_1)$, $Y(t_2)$ in $(y_2, y_2 + dy_2)$, and so on. These P_n may be defined for $n = 1, 2, \ldots$ and only for different times. This hierarchy has the following properties:

(i) $P_n \geq 0$,
(ii) P_n is invariant under permutations of pairs (y_i, t_i) and (y_j, t_j),
(iii)

$$\int dy_n P_n = P_{n-1} \quad \text{and} \quad \int dy_1 P_1 = 1. \tag{1.14}$$

According to a theorem due to Kolmogorov, it is possible to prove that the inverse is also true. That is, a set of functions fulfilling the above conditions defines a stochastic process. An alternative characterization of a stochastic process is also possible through the whole hierarchy of moments

$$\mu_n(t_1, t_2, \ldots, t_n) = \langle Y(t_1) Y(t_2) \ldots Y(t_n) \rangle = \tag{1.15}$$

$$\int_{-\infty}^{\infty} \ldots \int_{-\infty}^{\infty} dy_1 dy_2 \ldots dy_n \, y_1 y_2 \ldots y_n \, P_n(y_1, t_1; y_2, t_2; \ldots; y_n, t_n)$$

Another very important quantity is the *conditional probability density* $P_{n/m}$ (according to van Kampen's notation) that corresponds to the probability of having the value y_1 at time t_1, y_2 at $t_2, \ldots, y_n$ at t_n, given that we have $Y(t_{n+1}) = y_{n+1}$, $Y(t_{n+2}) = y_{n+2}, \ldots, Y(t_{n+m}) = y_{n+m}$. Its definition is

$$P_{n/m}(y_1, t_1; ..; y_n, t_n | y_{n+1}, t_{n+1}; \ldots; y_{n+m}, t_{n+m}) =$$
$$\frac{P_{n+m}(y_1, t_1; \ldots; y_n, t_n; y_{n+1}, t_{n+1}; \ldots; y_{n+m}, t_{n+m})}{P_m(y_{n+1}, t_{n+1}; \ldots; y_{n+m}, t_{n+m})} \qquad (1.16)$$

The kind of trajectories contributing to the *two-time joint probability* and to the *two-time conditional probability* distributions are schematically depicted in Fig. 1.3. In the first case, for the joint probability, we consider

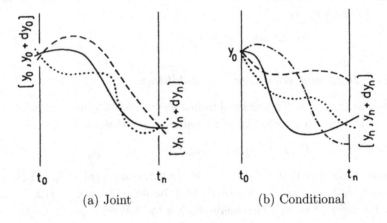

(a) Joint   (b) Conditional

Fig. 1.3

trajectories starting at a given interval $(y_0, y_0 + dy_0)$ and reaching another given interval $(y_n, y_n + dy_n)$. On the other hand, for the case of the conditional distribution, we consider out of all the trajectories starting at fixed point y_0, only those reaching the interval $(y_n, y_n + dy_n)$. Coming back to the example of the dice, the joint probability will inform us about the probability of obtaining a given sequence of outcomes in successive throws, while the conditional probability corresponds to the probability of obtaining a certain outcome (or series of outcomes) if the result of the last throw (or set of throws) is known.

Another concept of great importance is the *statistical independence of variables*. We say that the values of a stochastic process at different times are *independent* if

$$P_2(y_1, t_1; y_2, t_2) = P_1(y_1, t_1) P_1(y_2, t_2)$$

More generally, we will say that two stochastic variables are independent when one of the following equivalent properties is fulfilled:

(i) all moments factorize: $\langle X_1^m X_2^n \rangle = \langle X_1^m \rangle \langle X_2^n \rangle$,
(ii) the characteristic function, as given by Eq. (1.9), factorizes: $G(k_1, k_2) = G(k_1)G(k_2)$,
(iii) all cumulants $\langle\langle X_1^m X_2^n \rangle\rangle$ vanish when both m and n differ from zero simultaneously.

For instance, if we have a pair of dice, the probability that the second dice has some outcome, is independent of the result obtained with the first one. However, the possibility of obtaining a given combination is not. All this can be adequately generalized for more than two stochastic variables.

1.3 Markovian processes

Among the many possible classes of stochastic processes, there is one that merits a special treatment: we refer to the *Markovian processes*. We will discuss here this class of processes with some detail.

For a stochastic process $Y(t)$, $P(y_2, t_2 | y_1, t_1)$ is the *conditional probability* (also called the *transition probability*) that $Y(t_2)$ takes the value y_2, knowing that $Y(t_1)$ has taken the value y_1. From this definition and Eq. (1.16) results the following identity for the *joint probability* $P_2(y_1, t_1; y_2, t_2)$ (Bayes' rule):

$$P_2(y_1, t_1; y_2, t_2) = P_1(y_1, t_1)P(y_2, t_2 | y_1, t_1) \tag{1.17}$$

Such a process $Y(t)$ is called Markovian if for *every* set of successive times $t_1 < t_2 < \ldots < t_n$, the following condition holds

$$P_n(y_1, t_1, \ldots, y_n, t_n) = P_1(y_1, t_1)P_{n-1}(y_2, t_2, \ldots, y_n, t_n | y_1, t_1) =$$

$$= \ldots = P_1(y_1, t_1)P(y_n, t_n | y_{n-1}, t_{n-1}) \ldots P(y_2, t_2 | y_1, t_1) \tag{1.18}$$

From this definition, it results that a Markovian process is completely determined if we know $P_1(y_1, t_1)$ and $P(y_2, t_2 | y_1, t_1)$. It is easy to find a relevant condition to be fulfilled for Markovian processes: specifying the previous equation for the case $n = 3$ and integrating over y_2, we obtain

$$\int dy_2 P_3(y_1, t_1, y_2, t_2, y_3, t_3) = P_2(y_1, t_1, y_3, t_3) = \tag{1.19a}$$

$$P_1(y_1, t_1)\, P(y_3, t_3 | y_1, t_1) = \int dy_2 P_1(y_1, t_1)P(y_3, t_3 | y_2, t_2)P(y_2, t_2 | y_1, t_1)$$

For $t_1 < t_2 < t_3$ we find the identity

$$P(y_3, t_3 | y_1, t_1) = \int dy_2 P(y_3, t_3 | y_2, t_2) P(y_2, t_2 | y_1, t_1) \qquad (1.19b)$$

which is the *Chapman–Kolmogorov equation* for Markovian processes. Every pair of non-negative functions $P_1(y_1, t_1)$ and $P(y_2, t_2 | y_1, t_1)$, adequately normalized and satisfying not only Eq. (1.19b) but also

$$P_1(y_2, t_2) = \int dy_1 P_1(y_1, t_1) P(y_2, t_2 | y_1, t_1) \qquad (1.19c)$$

defines a Markovian process.

Let us analyze a few relevant examples of Markovian processes:

1.3.1 *Wiener–Lévy process*

It is defined in the range $-\infty < y < \infty$ and $t \geq 0$, through

$$P_1(y, t) = [2\pi t]^{-1/2} \exp\{-y^2/2t\} \qquad (1.20a)$$

$$P(y_2, t_2 | y_1, t_1) = [2\pi(t_2 - t_1)]^{-1/2} \exp\left\{-\frac{[y_2 - y_1]^2}{2(t_2 - t_1)}\right\}$$

It is easy to prove that these functions fulfill the Chapman–Kolmogorov equation. It is also possible to show that the *self-correlation function* (which is defined as a particular kind of moment, i.e. $\langle y(t_1) y(t_2) \rangle$) is given by

$$\langle y(t_1) y(t_2) \rangle = \min(t_1, t_2) \qquad (1.20b)$$

This process describes the position of a Brownian particle (a problem to be discussed in Sec. 2.2) in one dimension. It is referred to as a *Gaussian process* in order to indicate that all P_n are Gaussian (multivariate) distributions.

1.3.2 *Ornstein–Uhlenbeck process*

This process is defined in the range $-\infty < y < \infty, -\infty < t < \infty$ and $(t_2 - t_1) = \tau > 0$ through

$$P_1(y, t) = [2\pi]^{-1/2} \exp\{-y^2/2\} \qquad (1.21a)$$

$$P(y_2, t_2 | y_1, t_1) = [2\pi(1 - e^{-2\tau})]^{-1/2} \exp\left\{-\frac{[y_2 - y_1 e^{-\tau}]^2}{2(1 - e^{-2\tau})}\right\}$$

This process describes the *velocity* of a Brownian particle and is also Gaussian. Furthermore, it is *stationary*, meaning that

$$P_n(y_1, t_1, \ldots, y_n, t_n) = P_n(y_1, t_1 + \tau, \ldots, y_n, t_n + \tau) \qquad (1.21b)$$

According to a theorem due to Doob, it is (essentially) the only simultaneously Markovian, Gaussian and stationary process.

The *self-correlation function* of this process is given by

$$\langle y(t_1)y(t_2)\rangle = \exp\{-|t_2 - t_1|\}.$$

Writing $Y(t) = aL(t)$, $t = b$, and taking the limit $b \to \infty$ and $a \to \infty$, but in such a way that $2a^2/b \simeq 1$, we have

$$\langle L(t_1)L(t_2)\rangle = \delta(t_1 - t_2) \tag{1.22}$$

corresponding to the (so called) *white noise limit* or *Langevin process* (see Sec. 2.3). Even though $L(t)$ is not a *true* stochastic process, its integral corresponds to the previously defined Wiener process.

1.3.3 *Poisson processes*

Assume that y takes only discrete integer values: $n = 1, 2, \ldots$, and $t \geq 0$. We can define a Markovian process (for $t_2 > t_1 > 0$) through

$$P(n_2, t_2|n_1, t_1) = \frac{(t_2 - t_1)^{n_2 - n_1}}{(n_2 - n_1)!} \exp\{-(t_2 - t_1)\}$$

$$P_1(n, 0) = \delta_{n,0} \tag{1.23}$$

with the understanding that for $n_2 < n_1$, $P(n_2, t_2|n_1, t_1) = 0$. Thus, each sample function $y(t)$ is a succession of steps of unit height, and randomly distributed in time according to the last equation, which is called the Poisson distribution.

1.4 Master equation

Actually, the Chapman–Kolmogorov equation for Markovian processes is not of much help when we are looking for solutions of a given problem, because it is essentially a property of the solution. However, it can be recast in a more useful form. Returning to Eq. (1.19b), we take $t_3 = t_2 + \delta t$ and consider the limit $\delta t \to 0$. It is clear that we have $P(y_3, t_3|y_2, t_2) = \delta(y_3 - y_2)$, and it is intuitive to assume that, if $t_3 - t_2 \simeq \delta t$ (very small), the probability that a transition happens must be proportional to δt. According to this we adopt

$$P(y_3, t_2 + \delta t|y_2, t_2) = \delta(y_3 - y_2)\left[1 - A(y_2)\delta t\right] + \delta t W(y_3|y_2) + \mathcal{O}(\delta t^2) \tag{1.24a}$$

where $W(y_3|y_2)$ is the *transition probability per unit time* from y_2 to y_3 (which in general could be also a function of t_2), and the probability normalization tells us that

$$A(y_2) = \int dy_3 W(y_3|y_2) \qquad (1.24\text{b})$$

Substitution of the form for $P(y_3, t_2 + \delta t|y_2, t_2)$ into the Chapman–Kolmogorov equation (1.19b) gives

$$P(y_3, t_2 + \delta t|y_1, t_1) = \int dy_2 P(y_3, t_2 + \delta t|y_2, t_2) P(y_2, t_2|y_1, t_1) =$$

$$[1 - A(y_3)\delta t] P(y_3, t_2|y_1, t_1) + \delta t \int dy_2 W(y_3|y_2) P(y_2, t_2|y_1, t_1) =$$

$$P(y_3, t_2|y_1, t_1) - \delta t \int W(y_2|y_3) P(y_3, t_2|y_1, t_1) dy_2 +$$

$$\delta t \int dy_2 W(y_3|y_2) P(y_2, t_2|y_1, t_1) \qquad (1.25)$$

This can be rearranged as

$$\frac{P(y_3, t_2 + \delta t|y_1, t_1) - P(y_3, t_2|y_1, t_1)}{\delta t} = \qquad (1.26)$$

$$\int dy_2 \left[W(y_3|y_2) P(y_2, t_2|y_1, t_1) - W(y_2|y_3) P(y_3, t_2|y_1, t_1) \right]$$

and in the limit $\delta t \to 0$, we find

$$\frac{\partial}{\partial t} P(y, t|y_0, t_0) = \int dy' \left[W(y|y') P(y', t'|y_0, t_0) - W(y'|y) P(y, t|y_0, t_0) \right]$$
$$(1.27\text{a})$$

which is the celebrated *master equation*. When the range of the variables is discrete instead of continuous, we find

$$\frac{\partial}{\partial t} p_\nu(t) = \sum_\nu \left[W_{\nu\nu'} p_{\nu'}(t) - W_{\nu'\nu} p_\nu(t) \right] \qquad (1.27\text{b})$$

Here the usual interpretation of the master equation as a *balance equation* becomes apparent, and it is easy to identify the *gain* and *loss* terms for each state ν.

The master equation is a differential form of the Chapman–Kolmogorov equation. It is an equation for the transition probability $P(y, t|y_0, t_0)$, but not for $P_1(x, t)$. However, when we fix the point (x_0, t_0), we may assume that it becomes an equation for $P_1(x, t) \simeq P(y, t|y_0, t_0)$. This equation is more adequate for mathematical manipulations than the Chapman–Kolmogorov equation, and has a direct physical interpretation as a balance equation. At the same time, $W(y|y')\delta t$ and $W_{\nu\nu'}\delta t$, are the transition probabilities during a very short time (δt). They could be evaluated by approximate methods, for instance by time dependent perturbation theory (i.e. the *Fermi golden rule*) as

$$W_{\nu\nu'} = [2\pi/\hbar]\,|H^*_{\nu\nu'}|^2\rho(\epsilon_\nu) \tag{1.28}$$

where H^* is the perturbation Hamiltonian, and $\rho(\epsilon_\nu)$ the density of states of the unperturbed states. It is clear that the master equation allows us to determine the evolution of the system for times longer than δt: both time scales (times larger or shorter than δt) can be treated separately assuming that the Markovian property holds. We note the different role played by the master equation. The Chapman–Kolmogorov equation, besides being nonlinear, is only a manifestation of the Markovian character of the process under study, but contains no specific information regarding a particular Markovian process. In contrast, the master equation considers the transition probabilities for a specific process. Furthermore, it is a linear equation for the probabilities determining the macroscopic state of the system.

1.4.1 *Decay processes (radioactive decay of nuclei, excited atoms, etc)*

Let γ be the decay probability per unit time (it is a property of the excited atom). The transition probability for $n' \to n$ in an interval δt is given by

$$W_{nn'}\,\delta t = \begin{cases} 0 & n \geq n' \\ n'\gamma\delta t & n = n' - 1 \\ \mathcal{O}(\delta t^2) & n < n' - 1 \end{cases}$$

where n' is the number of excited atoms at time t. Let $p_n(t)$ be the probability density of n excited atoms at t (given that there were n_0 at $t_0 < t$), then

$$W_{nn'} = n'\gamma\delta_{n,n'-1} \qquad (n \neq n')$$

Substituting this form into the master equation Eq. (1.27b), we obtain

$$\dot{p}_n(t) = (n+1)\gamma p_{n+1}(t) - n\gamma p_n(t) \tag{1.29}$$

with the initial condition ($t_0 = 0$) $p_n(0) = \delta_{n,n_0}$. It is possible to obtain partial results without deriving the explicit form of the solution. Calling

$$N(t) = \langle n \rangle = \sum_n np_n(t)$$

we have

$$\frac{d}{dt}\langle n \rangle = \sum_n n\dot{p}_n(t) = \gamma \sum_n n(n+1)p_{n+1}(t) - \gamma \sum_n n^2 p_n(t)$$

and, after rearranging indices

$$= \gamma \sum_n (n-1)np_n(t) - \gamma \sum_n n^2 p_n(t) = -\gamma \sum_n np_n(t)$$

finally rendering

$$\frac{d}{dt}N(t) = -\gamma N(t).$$

Considering the initial condition $N(0) = n_0$, we obtain the well known solution

$$N(t) = n_0 e^{-\gamma t}$$

Also, by defining the second moment

$$N_2(t) = \langle n^2 \rangle = \sum_n n^2 p_n(t)$$

we obtain

$$\frac{d}{dt}\langle n^2 \rangle = \sum_n n^2 \dot{p}_n(t) = \gamma \sum_n n^2(n+1)p_{n+1}(t) - \gamma \sum_n n^3 p_n(t)$$

$$= \gamma \sum_n (n-1)^2 np_n(t) - \gamma \sum_n n^3 p_n(t)$$

$$= \gamma \sum_n (-2n^2 + n)p_n(t)$$

yielding

$$\frac{d}{dt}N_2(t) = -2\gamma N_2(t) + \gamma N(t)$$

and substituting the previous solution for $N(t)$ this reduces to

$$= -2\gamma N_2(t) + \gamma n_0 e^{-\gamma t}.$$

If we consider the initial condition $N_2(0) = n_0^2$, we find the solution

$$N_2(t) = n_0(n_0 - 1)e^{-2\gamma t} + n_0 e^{-\gamma t}.$$

We shall discuss here an alternative way of solving Eq. (1.29) through the use of a different and convenient form of the generating function, defined as

$$G(s,t) = \sum_{\nu=0}^{\infty} s^\nu p_\nu(t).$$

Replacing this form in Eq. (1.29), and after some rearrangements, we obtain an equation for $G(s,t)$:

$$\frac{d}{dt}G(s,t) = -\gamma(s-1)\frac{\partial}{\partial s}G(s,t).$$

The substitution $s - 1 = e^z$ $(G(s,t) \equiv G(z,t))$ leads us to

$$\frac{d}{dt}G(z,t) = -\gamma\frac{\partial}{\partial z}G(z,t),$$

with a general solution of the form $G(z,t) = F[\exp(-\gamma t + z)] = F[(s - 1)e^{-\gamma t}]$, F being an arbitrary function of the variable $\eta = \exp[z - \gamma t]$. However, normalization requires $G(1,t) = 1$, and hence $F(0) = 1$. Using again the initial condition $p(n,0|n_0,0) = \delta_{n,n_0}$, meaning

$$G(s,0) = s^{n_0} = F[s-1] = (1+e^z)^{n_0}$$

we finally obtain

$$G(s,t) = \left[1 + (s-1)e^{-\gamma t}\right]^{n_0}.$$

From the last expression the whole hierarchy of moments can be obtained, and by inverting it (for instance by its expansion in powers of s) it is possible to obtain directly the complete solution p_ν (which has a complicated form that we do not include here).

1.4.2 *Kinetics of the Ising model*

We are interested in the probability $P(N_+, N_-, t)$, of finding N_+ spins *"up"* and N_- spins *"down"* at time t $(N_+ + N_- = N$, the total number of spins, fixed), given a certain initial configuration at (the initial) time $t_0 = 0$. We will assume that only one spin may change (*flip*) at each step (*small jumps in the variable*). We then have

$$
\begin{aligned}
\frac{\partial}{\partial t}P(N_+, N_-, t) = &- [W_{+-}(N_+, N_- \to N_+ - 1, N_- + 1) \\
&+ W_{-+}(N_+, N_- \to N_+ + 1, N_- - 1)]\, P(N_+, N_-, t) \\
&+ W_{+-}(N_+ + 1, N_- - 1 \to N_+, N_-)\, P(N_+ + 1, N_- - 1, t) \\
&+ W_{-+}(N_+ - 1, N_- + 1 \to N_+, N_-)\, P(N_+ - 1, N_- + 1, t) \quad (1.30)
\end{aligned}
$$

where

$$W_{+-}(N_+, N_- \to N_+ - 1, N_- + 1) = N_+ \exp\left\{-\mu - \frac{\alpha}{N}(N_+ - N_-)\right\}$$

$$W_{-+}(N_+, N_- \to N_+ + 1, N_- - 1) = N_- \exp\left\{\mu + \frac{\alpha}{N}(N_+ - N_-)\right\}$$

are the transition probabilities for a *spin-flip* process. The coefficients in the exponential (Boltzmann factors) are

$$\mu = \frac{\mu_0 H}{kT} \qquad \alpha = \frac{J}{kT},$$

corresponding to the *external* and the *molecular* magnetic field. In general the transition probabilities could be more complex, but the assumption of independence of the spin configurations is the simplest one that allows us to obtain the Weiss limit, corresponding to the system being in contact with a thermal bath. These transition probabilities give the equilibrium distribution

$$P_{\rm st}(N_+, N_-) = \frac{N!}{N_+!N_-!} \exp\left[\mu(N_+ - N_-) + \frac{\alpha}{2N}(N_+ - N_-)^2\right].$$

1.5   Approach to equilibrium: Ehrenfest urn model (dog–flea model)

This simple model was analyzed by Ehrenfest at the beginning of the XXth century in order to study the approach to equilibrium and the problem of irreversibility. The model is as follows: we consider $2N$ balls, numbered from 1 to $2N$, distributed between two urns I and II (similarly, and coincident with Ehrenfest's original version, we could also consider two dogs and $2N$ fleas distributed between them). We pick at random a number η such that $1 \leq \eta \leq 2N$, and change the ball labelled by this number from the urn where it is (say I) to the other one (say II). We repeat this procedure several times. It is clear that we have $N_{\rm I} + N_{\rm II} = 2N$, and we call $N_{\rm I} - N_{\rm II} = 2n$. A typical representation of n after repeating the indicated procedure is sketched in Fig. 1.4.

What we could expect is to reach *the naive equilibrium condition* (as verified in the figure), corresponding to having N balls in each urn (that is $n = 0$). But this idea is not strictly realistic due to the existence of *fluctuations*. The concept of equilibrium is more *subtle*, and is connected with the probability of departure from such a *naive* equilibrium picture, introducing two problems:

(i) To find the equilibrium distribution (*static problem*). In a real physical situation, the answer to this problem is known and is given through the *canonical* or *microcanonical* distributions.

(ii) To determine the time that elapses until the decay of a departure from the *naive* equilibrium (*dynamical problem*). For real physical

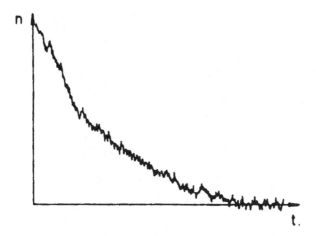

Fig. 1.4

situations, this problem is not solved in general. For *small* departures, the answer is provided by *linear irreversible thermodynamics* (also by the *Onsager relations*, or *Linear Response Theory*, that we will discuss in Chapters 3 and 4 respectively).

Within the framework of this Markovian model we can solve both problems, shedding light on their solution for more general cases. As the problem is Markovian, we have that

$$P(m_1, s_1; m_2, s_2; \ldots; m_r, s_r | m_0) = \prod_{i=0}^{r-1} P(m_{i+1}, s_{i+1} - s_i | m_i) \qquad (1.31)$$

where $P(m, s | m_0)$ corresponds to the conditional probability of having $N_I = m$, after s steps, if originally we had $N_I = m_0$. Let us try to write a recurrence relation for this conditional probability (having the form of a master equation). The transition probabilities of *one-step*, results of considering:

$$W(m|m') = \text{Prob}\{s + 1 : m | s : m'\}$$

$$= \frac{m'}{2N} \delta_{m',m-1} + \delta_{m',m+1} \qquad (1.32)$$

Hence, the conditional probability of originally having m_0 balls inside the urn I (i.e. $N_I = m_0$), and m after $s + 1$ steps (adopting the notation

$P(m, s|m_0) = P(m, s))$, fulfills the relation

$$P(m, s + 1) = W(m|m - 1)P(m - 1, s) + W(m|m + 1)P(m + 1, s)$$

$$= \left(1 - \frac{m - 1}{2N}\right) P(m - 1, s) + \frac{m + 1}{2N} P(m + 1, s) \quad (1.33)$$

This equation corresponds to the Chapman–Kolmogorov equation for the present model. The stationary solution is given by

$$\lim_{s \to \infty} P(m, s + 1) = P_0(m) = \frac{[2N]!}{m![2N - m]!} \frac{1}{2}^{2N} \quad (1.34)$$

which, remembering that $P(m, 0|m_0) = \delta_{m,m_0}$, results from

$$P_0(m) = \sum_m P(m_0)P(m, 1|m_0)$$

By means of Eq. (1.33) we are able to calculate moments and averages such as

$$\langle m(s)^p \rangle = \sum_m m^p P(m, s). \quad (1.35)$$

For instance, it is easy to prove that

$$\langle m(s) \rangle = \sum_m m P(m, s) = 1 + \left(1 - \frac{2}{2N}\right) \langle m(s - 1) \rangle$$

$$= N + [m(0) - N]\left(1 - \frac{2}{2N}\right)^s \quad (1.36)$$

with $m(0) = m$. Hence, $\langle m \rangle = m_0$ when $s = 0$, and for s large enough this average goes to $N_I = N = N_{II}$. The same result is obtained from $P_0(m)$. The expression for the stationary probability Eq. (1.34), has a very sharp maximum for N large enough when $N_I = N_{II} = N$, that corresponds to the *naive equilibrium*. In fact, for very large N, we obtain

$$P(m) \cong (\pi N)^{-1/2} e^{-m^2/N}. \quad (1.37)$$

The average number of balls inside urn I, given by Eq. (1.36), could be written as:

$$\langle m(s + 1) \rangle = N + [m_0 - N]\left(1 - \frac{1}{N}\right)^{s+1}$$

Calling $\eta = m - N$ (indicating the departure from the *naive equilibrium*), we rewrite the last equation as

$$\langle \eta(s + 1) \rangle = \eta_0 \left(1 - \frac{1}{N}\right)^{s+1} \quad (1.38)$$

with $\eta_0 = \langle \eta(0) \rangle = m_0 - N = \langle m(0) \rangle - N$. In the very large N limit, and for τ (the time between jumps) very small ($\tau \to 0$), but such that $(1/N\tau) \to \gamma$ (with γ finite) and $s\tau = t$, we obtain

$$\langle \eta(t) \rangle = \eta_0 \exp\{-\gamma t\}, \tag{1.39}$$

indicating a monotonic approach to equilibrium, characteristic of linear laws.

Chapter 2

The Fokker–Planck equation

Eccentric, intervolved, yet regular
Then most, when most irregular, they seem,
And in their motions harmony divine ...
John Milton

2.1 Kramers–Moyal expansion

We assume now that y is a continuous variable, and that its changes correspond to *small jumps* (or variations). In this case it is possible to derive, starting from the master equation, a differential equation. The transition probability $W(y|y')$ will decay very fast as a function of $|y - y'|$. We could then write $W(y|y') = W(y', \xi)$, where $\xi = y - y'$ corresponds to the size of the jump. The master equation will take the form

$$\frac{\partial}{\partial t} P(y, t|y_0, t_0) = \tag{2.1}$$

$$\int d\xi \, W(y - \xi, \xi) \, P(y - \xi, t|y_0, t_0) \; - \; P(y, t|y_0, t_0) \int d\xi \, W(y, -\xi)$$

According to the assumption of small jumps, and adding the argument that P varies slowly with y, we make a Taylor expansion in ξ that gives

$$\frac{\partial}{\partial t} P(y, t|y_0, t_0) = \tag{2.2}$$

$$\int d\xi \left[W(y, \xi) \, P(y - \xi, t|y_0, t_0) - \xi \frac{\partial}{\partial y} P(y, t|y_0, t_0) + \right.$$

$$\left. \frac{1}{2} \xi^2 \frac{\partial^2}{\partial y^2} P(y, t|y_0, t_0) - \ldots \right] - P(y, t|y_0, t_0) \int d\xi \, W(y, -\xi)$$

21

As the first and the last terms are equal (just change $-\xi$ by ξ and the integration limits in the latter), we get

$$\frac{\partial}{\partial t} P(y,t|y_0,t_0) = \sum_{\nu=1}^{\infty} \frac{(-1)^\nu}{\nu!} \frac{\partial^\nu}{\partial y}\nu \alpha_\nu(y) P(y,t|y_0,t_0),$$

with

$$\alpha_\nu(y) = \int d\xi \xi^\nu \, W(y,\xi).$$

This result corresponds to the *Kramers–Moyal expansion* of the master equation. Up to this point we have gained nothing. However, there could be situations where for $\nu > 2$, the α_ν are either zero or very small (even though there are no a priori criteria about the relative size of the terms). If this is the case, we have

$$\frac{\partial}{\partial t} P(y,t|y_0,t_0) = -\frac{\partial}{\partial y}\alpha_1(y) P(y,t|y_0,t_0) + \frac{1}{2}\frac{\partial^2}{\partial y^2}\alpha_2(y) P(y,t|y_0,t_0), \quad (2.3)$$

which has the form of the (also celebrated) *Fokker–Planck equation*.

Let us see a couple of examples. For the Wiener–Lévy process we find that $\alpha_\nu = 0$ ($\nu > 2$), and then

$$\frac{\partial}{\partial t} P(y,t|y_0,t_0) = \frac{\partial^2}{\partial y^2} P(y,t|y_0,t_0).$$

The case of the Ornstein–Uhlenbeck process is similar, giving

$$\frac{\partial}{\partial t} P(y,t|y_0,t_0) = -\frac{\partial}{\partial y} y P(y,t|y_0,t_0) + \frac{\partial^2}{\partial y^2} P(y,t|y_0,t_0).$$

A couple of comments are in order:

(i) It has become customary to call Eq. (2.3)—with an arbitrary dependence of α_1 and α_2 on y (which is the result of not well grounded assumptions, like the criteria on where to cut the expansion, etc.)— a "nonlinear" Fokker–Planck equation, even when it is clearly a *linear* equation in $P(y,t|y_0,t_0)$. We warn the reader not to confuse these with the true *nonlinear Fokker–Planck equations*—resulting from mean-field-like approximations to many-body problems—which have come recently into consideration [Frank (2005)].

(ii) Note that the Kramers–Moyal expansion is **not a systematic** approximation to the master equation. After discussing other forms of describing stochastic processes and their connection with Fokker–Planck equations, we will see how is it possible to build up such a systematic procedure.

2.2 Brownian motion, Langevin and Fokker–Planck equations

Probably the oldest and best known physical example of a Markov process is the so called *Brownian motion*. This phenomenon corresponds to the motion of a heavy test particle, immersed in a fluid composed of light particles in random motion. Due to the (random) collisions of these against the test particle, the velocity of the latter varies in a (large) sequence of small, uncorrelated jumps. To simplify the presentation we restrict the description to a one dimensional system.

If the particle has a velocity v, in average there are more front than back collisions. Then the possibility of a certain velocity change δv within the next time interval Δt, depends on v, but not on the previous velocity values. We then have that the velocity of a Brownian particle is described by a Markov process. When the system as a whole is in equilibrium, the process is stationary and the *self-correlation time* [defined through Eq. (1.20b)] is given by the time that elapses till the information about the initial velocity is lost. However, this scheme is not in complete agreement with the experimental observations. This phenomenon was better understood after the contributions of Einstein and Smoluchowski. They were the first to recognize that what was experimentally observed was not the above described motion. What happens is that, between two successive observations of the test particle position, its velocity has increased and decreased several times; implying that the observational time is longer than the *velocity correlation time*. Assume that a set of measurements on the same Brownian particle gives us a sequence of positions: $\mathbf{x}_1, \mathbf{x}_2, \ldots, \mathbf{x}_n$ (Fig. 2.1). Each displacement $\mathbf{x}_{l+1} - \mathbf{x}_l$ is random, and its probability distribution is independent of $\mathbf{x}_{l-1}, \mathbf{x}_{l-2}$, etc. Hence, not only is the velocity of the Brownian particle a Markov process in itself, but in a *coarse-grained* time scale (imposed by the experimental situation) its position is also a Markov process.

On the basis of these ideas, we will try now to give a quantitative picture of Brownian motion. We start by writing the Newton equation as:

$$m\dot{v} = \mathcal{F}(t) + \mathcal{F}'(t) \tag{2.4}$$

where m is the mass of the Brownian particle, v its velocity, $\mathcal{F}(t)$ the force due to some external field (i.e. gravitational, electrical for charged particles, etc), and $\mathcal{F}'(t)$ is the force produced by the collisions of fluid particles against the test particle. Due to the above indicated rapid fluctuations in v, we have two effects. On one hand a *systematic* one, i.e., a kind of *friction*

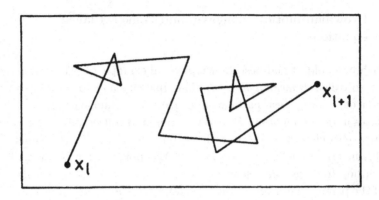

Fig. 2.1

that tends to slow down the particle, and on the other hand, a *random* contribution originated in the random hits of the fluid particle. If the mass of the test particle is much larger than the mass of the fluid particles (implying that the fluid *relaxes* faster than the test particle, allowing us to assume that it is in equilibrium), we can write

$$\frac{1}{m}\mathcal{F}'(t) = -\gamma v + \xi(t) \qquad (2.5)$$

In the r.h.s., γ is the friction coefficient, and the minus sign in the first term indicates that this contribution (as a good friction term) opposes to the motion. The second term corresponds to the stochastic or random contribution, since we have separated the systematic contribution in the first term, and this random contribution averages to zero: $\langle \xi(t) \rangle = 0$ (where the average is over an *ensemble* of noninteracting Brownian particles). In order to define the so called *Langevin force* it is required that

$$\langle \xi(t)\xi(t') \rangle = D\delta(t - t') \qquad (2.6)$$

We will not consider higher order moments, but it is clear that to fully characterize the fluctuating force, we need the whole hierarchy of moments.

With the above indicated arguments, and without an external field, Eq. (2.4) takes the form

$$m\dot{v} = -\gamma v + \xi(t) \qquad (2.7)$$

which is known as the *Langevin equation*. This is the simplest example of a *stochastic differential equation* (that is, a differential equation whose coefficients are random functions with known stochastic properties, see Sec.

2.3). Hence $v(t)$ is a stochastic process, with a given initial condition. Let us consider an ensemble of Brownian particles with initial velocity $v(t = 0) = v_0$. Hence, for $t \geq 0$, the formal solution of Eq. (2.7) is

$$v(t) = v_0 e^{-\gamma t} + e^{-\gamma t} \int_0^t dt' e^{\gamma t'} \xi(t') \tag{2.8a}$$

and after averaging over ξ, we get

$$\langle v(t) \rangle = v_0 e^{-\gamma t} \tag{2.8b}$$

Moreover, considering the square of Eq. (2.8a) and averaging

$$\langle v(t)^2 \rangle = v_0^2 e^{-2\gamma t} + e^{-2\gamma t} \int_0^t dt' \int_0^t dt'' e^{\gamma(t'+t'')} \langle \xi(t')\xi(t'') \rangle =$$

$$v_0^2 e^{-2\gamma t} + \frac{D}{2} \left[1 - e^{-2\gamma t} \right] \tag{2.8c}$$

For $t \to \infty$, the last equation gives

$$\langle v(t)^2 \rangle = \frac{D}{2} = \frac{kT}{m} \quad ; \quad D = \frac{2kT}{m} \gamma \tag{2.9}$$

where the *equipartition theorem* has been used (because we expect that at $t \to \infty$, equilibrium must be reached). This relation between the parameter D (diffusion coefficient), that measures the *size of the fluctuations*, and the constant γ, that measures the *friction*, is a simple form of the *fluctuation–dissipation theorem* to be discussed latter within the context of the *linear response theory*.

We now evaluate the r.m.s. displacement. Multiplying Eq. (2.7) by x, we have

$$x \frac{d}{dt} \dot{x} = \frac{d}{dt}(x\dot{x}) - \dot{x}^2 = -\gamma x \dot{x} + x \xi(t) \tag{2.10}$$

At this point, Langevin's original argument was to assume that $\xi(t)$ and $x(t)$ were uncorrelated

$$\langle x(t)\xi(t) \rangle \equiv 0 \tag{2.11}$$

In a pedagogical article, [Manoliu and Kittel (1979)] have shown that the property (2.11) is neither evident nor necessary. However, in order to simplify the presentation, we will use it. Hence, after averaging Eq. (2.10)

$$\left\langle \frac{d}{dt}(x\dot{x}) \right\rangle = \frac{d}{dt}\langle x\dot{x} \rangle = \frac{kT}{m} - \gamma \langle x\dot{x} \rangle \tag{2.12}$$

we obtain

$$\langle x(t)\dot{x}(t) \rangle = C e^{-\gamma t} + \frac{kT}{\gamma m} \tag{2.13a}$$

If x measures the displacement from the origin (where we have put all the Brownian particles at $t = 0$), we find the condition $0 = C + \frac{kT}{\gamma m}$, that gives

$$\langle x(t)\dot{x}(t)\rangle = \frac{d}{dt}\langle x^2\rangle = \frac{kT}{\gamma m}\left(1 - e^{-\gamma t}\right) \tag{2.13b}$$

Integrating once more we obtain

$$\langle x(t)^2\rangle = \frac{2kT}{\gamma m}\left[t - \frac{1}{\gamma}\left(1 - e^{-\gamma t}\right)\right] \tag{2.13c}$$

Now, we consider two limit cases

(a) Initial transient regime: $t \ll \frac{1}{\gamma}$, where we can expand $e^{-\gamma t} \cong 1 - \gamma t + \frac{1}{2}(\gamma t)^2 - \ldots$, and obtain

$$\langle x(t)^2\rangle \cong \frac{kT}{m}t^2 \tag{2.14a}$$

that corresponds to the particle *inertial* motion during the initial transient (with *thermal velocity* $\bar{v} = \sqrt{\frac{kT}{m}}$).

(b) Asymptotic regime: $t \gg \frac{1}{\gamma}$, where we can approximate $e^{-\gamma t} \cong 0$, and then

$$\langle x(t)^2\rangle \cong \frac{2kT}{m}t \tag{2.14b}$$

which is characteristic of a diffusive motion, as is discussed, for instance, in the framework of *random-walk* schemes, in most statistical physics textbooks.

An alternative way to analyze the problem of Brownian motion is to consider the probability distribution of finding the system within a given velocity range $(v, v + dv)$ rather than the process itself, knowing that at some initial time t_0, its velocity was v_0

$$P\,dv = P(v, t|v_0, t_0)\,dv \tag{2.15a}$$

We know that

$$\lim_{\delta t \to 0} P(v, t + \delta t|v', t) = \delta(v - v') \tag{2.15b}$$

and it is possible to prove (based on equilibrium arguments, i.e. *detailed balance*) that

$$\lim_{\delta t \to \infty} P(v, t + \delta t|v', t) = \left(\frac{m}{2\pi kT}\right)^{1/2} e^{-mv^2/2kT} \tag{2.15c}$$

As was indicated in the previous section, whenever in the Kramers–Moyal expansion of the master equation Eq. (2.3) the moments of order higher than two are zero, we get a Fokker–Planck equation (FPE). In this case, the master equation is written for the gain and loss contributions within the interval $(v, v + dv)$. According to the average values obtained for $\langle v \rangle$ and $\langle v^2 \rangle$, we have

$$\frac{\partial}{\partial t}P(v, t|v_0, t_0) = -\frac{\partial}{\partial v}\gamma v P(v, t|v_0, t_0) + \frac{D}{2}\frac{\partial^2}{\partial v^2}P(y, t|y_0, t_0) \qquad (2.16)$$

In order to get the same results with this equation as with the Langevin approach, we need to impose an extra condition on $\xi(t)$: that this process be Gaussian. This means that all odd moments are zero and that even moments can be written in terms of the second moment, for instance as

$$\langle \xi(t_1)\xi(t_2)\xi(t_3)\xi(t_4) \rangle =$$
$$\langle \xi(t_1)\xi(t_2) \rangle \langle \xi(t_3)\xi(t_4) \rangle + \langle \xi(t_1)\xi(t_3) \rangle \langle \xi(t_2)\xi(t_4) \rangle + \ldots =$$
$$D^2\{\delta(t_1 - t_2)\delta(t_3 - t_4) + \ldots\} \qquad (2.17)$$

When studying the linear response theory we will come back to discuss other ideas related with Brownian motion. It is worth remarking that the picture of a Brownian particle immersed in a fluid is typical of a variety of problems, even when there are no real particles. For instance, it is the case if there is only a certain *degree of freedom* that interacts, in a more or less random way, with other (*irrelevant*) *degrees of freedom* playing the role of the bath. This indicates the importance of analyzing and understanding such an *archetypical* situation.

2.3 Stochastic differential equations and Fokker–Planck equations

Here we want to give a more formal presentation (but still not completely rigorous from a mathematical point of view) of the relation between *stochastic differential equations* (SDE) of the *Langevin type*, and *Fokker–Planck equations* (FPE). We start considering a very general form for the one-dimensional SDE:

$$\dot{x}(t) = \frac{d}{dt}x(t) = f[x(t), t] + g[x(t), t]\xi(t), \qquad (2.18)$$

where $\xi(t)$ is the so called *white noise* with $\langle \xi(t) \rangle = 0$ and $\langle \xi(t)\xi(t') \rangle = \delta(t - t')$, as in Eqs. (2.5) and (2.6), with $D = 1$. We do not consider higher moments, but the usual assumption is that the process is Gaussian. However, as we indicated in Sec. 2.2, $\xi(t)$ is not a well defined stochastic process. In a loose way, it could be considered as the "derivative" of the well defined *Wiener process*, but such a derivative does not exist at all. We now integrate Eq. (2.18) over a short time interval δt

$$x(t + \delta t) - x(t) = f[x(t), t]\delta t + g[x(t), t]\xi(t)\delta t. \tag{2.19}$$

If $x(t)$ is a Markov process (which is true), it is well defined if we are able to determine its probability distribution $P_1(x, t)$ as well as its conditional probability distribution $P(x, t|x', t')$ for $t > t'$. In order to obtain an equation for the latter quantity, we define now a *conditional average*, corresponding to the average of a function of the stochastic variable x (say $F(x)$), given that x has the value y at $t' < t$:

$$\langle F[x(t)]|x(t') = y\rangle = \langle\langle F[x(t)]\rangle\rangle = \int dx' F(x') P(x', t|y, t'). \tag{2.20}$$

Due to the property $P(x, t|x', t) = \delta(x - x')$, we have

$$\langle F(x(t))|x(t) = y\rangle = \int dx' F(x') P(x', t|y, t) = \int dx' F(x') \delta(x - x').$$

We use now this definition in order to obtain the first few *conditional moments* of $x(t)$

$$\langle\langle F[x(t)]\rangle\rangle = \langle x(t + \delta t)|x(t) = x\rangle = \int dx' F(x') \delta(x - x') P(x', t + \delta t|x, t)$$

$$= \langle\langle f[x(t), t]\delta t\rangle\rangle + \langle\langle g[x(t), t]\xi(t)\delta t\rangle\rangle \tag{2.21}$$

It is clear that for the first term on the r.h.s. we have

$$\langle\langle f[x(t), t]\delta t\rangle\rangle = f[x(t), t]\delta t. \tag{2.22}$$

Meanwhile, for the second term we have

$$\langle\langle g[x(t), t]\xi(t)\delta t\rangle\rangle = g[x(t), t]\langle\langle\xi(t)\rangle\rangle\delta t \tag{2.23}$$

(remember that, according to Langevin's argument, $\langle x\xi\rangle = 0$) which yields

$$\langle\langle\Delta x(t)\rangle\rangle = \int dx' (x - x') P(x', t + \delta t|x, t) = f[x(t), t]\delta t \tag{2.24}$$

For the second moment we have

$$\langle\langle\Delta x(t)^2\rangle\rangle = \int dx' (x - x')^2 P(x', t + \delta t|x, t)$$

$$= \langle\langle[f[x(t), t]\delta t + g[x(t), t]\xi(t)\delta t]^2\rangle\rangle$$

$$= \langle\langle[f[x(t), t]\delta t]^2\rangle\rangle + \langle\langle 2f[x(t), t]g[x(t), t]\xi(t)\delta t^2\rangle\rangle$$

$$+ \langle\langle[g[x(t), t]\xi(t)\delta t]^2\rangle\rangle$$

$$= [f[x(t), t]\delta t]^2 + 2f[x(t), t]g[x(t), t]\langle\langle\xi(t)\rangle\rangle\delta t^2$$

$$+ g[x(t), t]^2\langle\langle[\xi(t)\delta t]^2\rangle\rangle \tag{2.25}$$

Here we resort to properties of the Wiener process. Using that

$$\xi(t)\delta t = \int_t^{t+\delta t} dt'\xi(t') = \Delta W(t),$$

where $W(t)$ is the Wiener process, and according to Eq. (1.20b), $\langle[\xi(t)\delta t]^2\rangle \simeq \langle\Delta W(t)^2\rangle = \Delta t$, renders

$$\langle\langle\Delta x(t)^2\rangle\rangle = \int dx'(x-x')^2 P(x',t+\delta t|x,t) = g[x(t),t]^2\delta t + \mathcal{O}(\delta t^2). \quad (2.26)$$

It is possible to show that in general

$$\langle\langle\Delta x(t)^\nu\rangle\rangle \simeq \mathcal{O}(\delta t^\nu), \quad \nu \geq 2.$$

Let us consider now an arbitrary function $R(x)$, and evaluate its conditional average. Using the Chapman–Kolmogorov equation [Eq. (1.19b)]

$$\int dx R(x)P(x,t+\delta t|y,s) = \int dx R(x)\int dz P(x,t+\delta t|z,t)P(z,t|y,s)$$

$$= \int dz P(z,t|y,s)\int dx R(x)P(x,t+\delta t|z,t) \quad (2.27)$$

Expanding $R(x)$ in a Taylor series around z, as for $\delta t \simeq 0$ we know that $P(x,t+\delta t|z,t) \simeq \delta(x-z)$, and only a neighbourhood of z will be relevant,

$$\int dx R(x)P(x,t+\delta t|y,s) = \int dz P(z,t|y,s) \quad (2.28)$$

$$\int dx \left[R(z) + (x-z)R'(z) + \frac{1}{2}R''(z)(x-z)^2 + \ldots\right] P(x,t+\delta t|z,t).$$

Remembering the normalization condition for $P(z,t|y,s)$,

$$= \int dz P(z,t|y,s)R(z) + \int dz R'(z)P(z,t|y,s)\int dx(x-z)$$

$$+ \int dz \frac{1}{2}R''(z)P(z,t|y,s)\int dx(x-z)^2 P(x,t+\delta t|z,t)$$

$$+ + \ldots \quad (2.29)$$

Integrating by parts and using Eqs. (2.24), (2.26) we obtain

$$\int dx R(x)P(x,t+\delta t|y,s) = \int dx R(x) + [P(x,t|y,s) - \quad (2.30)$$

$$- \frac{\partial}{\partial x}[f(x,t)P(x,t|y,s)]\delta t + \frac{1}{2}\frac{\partial^2}{\partial x^2}[g(x,t)^2 P(x,t|y,s)]\delta t + \mathcal{O}(\delta t^2)]$$

Arranging terms and taking the limit $\delta t \to 0$, gives

$$0 = \int dx R(x) \left[\frac{\partial}{\partial t} P(x,t|y,s) - \right. \tag{2.31}$$

$$\left. - \left(-\frac{\partial}{\partial x}[f(x,t)P(x,t|y,s)] + \frac{1}{2}\frac{\partial^2}{\partial x^2}[g(x,t)^2 P(x,t|y,s)] \right) \right].$$

Due to the arbitrariness of the function $R(x)$, we arrive at the condition

$$\frac{\partial}{\partial t} P(x,t|y,s) = \tag{2.32}$$

$$= -\frac{\partial}{\partial x}\left[f(x,t)P(x,t|y,s) \right] + \frac{1}{2}\frac{\partial^2}{\partial x^2}\left[g(x,t)^2 P(x,t|y,s) \right],$$

which is the desired Fokker–Planck equation for the transition probability $P(x,t|y,s)$ associated with the stochastic process driven by the SDE Eq. (2.18).

2.4   Path integral for Markov processes

The aim of this section is to offer an introduction to the *path-integral* approach for Markovian stochastic processes. The description of non Markovian processes by means of path-integrals is also possible, but too involved for an introductory presentation.

The path-integral technique has proved to be a very powerful tool in various areas of physics, both computationally and conceptually. It often provides an alternative route for the derivation of perturbation expansions as well as an excellent framework for non-perturbative analysis. However, the applications of path-integrals that are usually found are related with problems in quantum mechanics, field theory and statistical physics, with only a very few exceptions where a presentation of the path-integral technique within the realm of stochastic processes is done. As a matter of fact, and from a historical point of view, the latter was the context where path-integrals were firstly discussed, when Wiener introduced such an approach to describe diffusion processes.

We will focus our discussion on one-dimensional Markovian processes describable through Langevin or Fokker–Planck equations. The form of the Langevin equation is

$$\dot{q}(t) = \frac{d}{dt}q(t) = f[q(t),t] + g[q(t),t]\,\xi(t), \tag{2.33}$$

where $\xi(t)$ is a *white noise* with $\langle \xi(t) \rangle = 0$ and $\langle \xi(t)\xi(t') \rangle = D\,\delta(t-t')$, as in Eqs. (2.5) and (2.6). As we have indicated in Sec. 2.3, $\xi(t)$ is not a well-defined stochastic process, but it could be considered as the "derivative" of the well-defined *Wiener process*.

The form of the Fokker–Planck equation related with a Langevin equation like Eq. (2.33), is (see Sec. 2.3)

$$\frac{\partial}{\partial t}P(q,t|q',s) = -\frac{\partial}{\partial q}\left[f(q,t)P(q,t|q',s)\right] +$$

$$+ \frac{D}{2}\frac{\partial^2}{\partial q^2}\left[g(q,t)^2 P(q,t|q',s)\right], \qquad (2.34)$$

This is an equation for the transition probability $P(q,t|q',s)$ $(t > s)$, this transition probability being also the propagator of this Markov process. As is well known, $P(q,t|q_0,t_0)$ fulfills the Chapman–Kolmogorov equation $(t_0 < t' < t)$

$$P(q,t|q_0,t_0) = \int_{-\infty}^{\infty} dz\, P(q,t|z,t')P(z,t'|q_0,t_0). \qquad (2.35)$$

By making a partition of the time interval in N steps $t_0 < t_1 < \ldots < t_f$ with $t_j = t_0 + (t_f - t_0)/N$, this equation allows to obtain a path-integral representation of the propagator. With the referred partition, we reiterate Eq. (2.35) and write

$$P(q_f,t_f|q_0,t_0) = \int_{-\infty}^{\infty}\cdots\int_{-\infty}^{\infty} dq_1 dq_2 \ldots dq_{N-1} P(q_f,t_f|q_{N-1},t_{N-1})\cdots$$

$$P(q_2,t_2|q_1,t_1)P(q_1,t_1|q_0,t_0). \qquad (2.36)$$

Now, the probability that at a given time t, the process takes a value between a and b is given by

$$\int_a^b dq\, P(q,t|q_0,t_0).$$

In an analogous way, the probability that the process, starting at $q = q_0$ at $t = t_0$, has a value between a_1 and b_1 at t_1, between a_2 and b_2 at $t_2,\ldots$, between a_{N-1} and b_{N-1} at t_{N-1} (with $a_j < b_j$ and $t_j < t_{j+1}$), and reaching q_N at t_N, will be given by

$$\int_{a_1}^{b_1}\int_{a_2}^{b_2}\cdots\int_{a_{N-1}}^{b_{N-1}} dq_1 dq_2 \ldots dq_{N-1} P(q_1,t_1|q_0,t_0)$$

$$P(q_2,t_2|q_1,t_1)\ldots P(q_N,t_N|q_{N-1},t_{N-1}). \qquad (2.37)$$

If we increase the number of time slices within the time partition where the intervals (a_j, b_j) are specified, and at the same time we take the limit $|a_j - b_j| \to 0$, the trajectory is defined with higher and higher precision. Clearly, a requisite is that the trajectories be continuous. This happens in particular for the Wiener process. With all this in mind, Eq. (2.36) can be interpreted as an integration over all the possible paths that the process could follow (corresponding to the different values of the sequence: $q_0, q_1, q_2, \ldots, q_{N-1}, q_N = q_f$). As was discussed in Sec. 1.3.1, for the Wiener process we have that

$$P(W_2, t_2 | W_1, t_1) = [2\pi D(t_2 - t_1)]^{-1/2} \exp\left\{ -\frac{[W_2 - W_1]^2}{2D(t_2 - t_1)} \right\}. \qquad (2.38)$$

For $N \to \infty$ we can define a *measure* in the path space, known as the *Wiener measure*. By substituting Eq. (2.38) into Eq. (2.37) we get

$$\prod_{j=1}^{N} \frac{dW_j}{(4\pi\epsilon D)^{1/2}} \exp\left(-\frac{1}{4D\epsilon} \sum_j (W_j - W_{j-1})^2 \right), \qquad (2.39)$$

which is the desired probability of following a given path.

In the limit of $\epsilon \to 0$ and $N \to \infty$, we can write the exponential in Eq. (2.39) in the continuous limit as

$$\exp\left[-\frac{1}{4D} \int_{t_0}^{t} d\tau \left(\frac{dW}{d\tau} \right)^2 \right]. \qquad (2.40)$$

If we integrate the expression in Eq. (2.39) over all the intermediate points (which is equivalent to sum over all the possible paths), as all the integrands are Gaussian, and the convolution of two Gaussian is again a Gaussian, we recover the result of Eq. (2.38) for the probability density of the Wiener process. Hence, we have expressed the probability density as a path-integral (*Wiener integral*)

$$P(W, t | W_0, t_0) = \int \mathcal{D}[W(\tau)] \exp\left[-\frac{1}{4D} \int_{t_0}^{t} d\tau \left(\frac{dW}{d\tau} \right)^2 \right], \qquad (2.41)$$

where the expression inside the integral represents the continuous version of the integral of Eq. (2.39), over all possible values of the intermediate points $\{W_j\}$.

Let us go now back to the general SDE in Eq. (2.33). We start by writing the discretized version of the Langevin equation given by Eq. (2.33) (in order to simplify the notation we adopt $g(q, t) = 1$ and $f(q, t)$ to be independent of t):

$$q_{j+1} - q_j \simeq \{\alpha f(q_{j+1}) + (1 - \alpha)f(q_j)\}\epsilon + [W_{j+1} - W_j], \qquad (2.42)$$

where $\epsilon = (t_f - t_0)/N$, and $W_j = W(t_j)$ is the Wiener process (as indicated in Sec. 2.3, formally, $dW(t) \simeq \xi(t)dt$). The parameter α ($0 \leq \alpha \leq 1$) is arbitrary. The most usual choices are $\alpha = 0$ (Itô scheme) and $\alpha = 1/2$ (Stratonovich scheme). We do not want to come into the usual difficulties related with this problem, but will keep this parameter in order to show the dependence of the final *Lagrangian* on it. The probability that

$$W(t_0) = 0; W_1 < W(t_1) < W_1 + dW_1; \ldots; W_N < W(t_N) < W_N + dW_N$$

is given, according to the previous results, by

$$P(\{W_j\}) = \prod_{j=1}^{N} \frac{dW_j}{(4\pi\epsilon D)^{1/2}} \exp\left(-\frac{1}{4D\epsilon}\sum_j (W_j - W_{j-1})^2\right). \qquad (2.43)$$

As our interest is to have the corresponding probability in the q-space, we need to transform the probability given in the last equation. As is well known, to do the transformation we need $\mathbb{J}$, which is the Jacobian of the transformation connecting both sets of stochastic variables ($\{W_j\} \to \{q_j\}$). To find it we write Eq. (2.37) as

$$W_j = q_j - q_{j-1} \simeq \{\alpha f(q_j) + (1 - \alpha)f(q_{j-1})\}\epsilon + [W_{j+1} - W_{j-1}]. \quad (2.44)$$

The above indicated Jacobian is given by

$$\mathbb{J} = \det\left(\frac{dW_j}{dq_k}\right) = \prod_{j=1}^{N}\left(1 - \epsilon\alpha\sum_j \frac{df(q_j)}{dq_j}\right). \qquad (2.45)$$

For $\epsilon \to 0$, it can be approximated by

$$\mathbb{J} = \exp\left(-\epsilon\alpha\sum_j \frac{df(q_j)}{dq_j}\right). \qquad (2.46)$$

Now, recalling that $P(\{q_j\}) = \mathbb{J}P(\{W_j\})$, and taking into account that the conditional probability $P(q, t|q_0, t_0)$ is given as a sum over all the possible paths, we get

$$
P(q, t|q_0, t_0) = \lim_{N \to \infty} \left(\frac{1}{4\pi\epsilon D} \right)^{N/2} \int_{-\infty}^{\infty} \cdots \int_{-\infty}^{\infty} dW_1 dW_2 \ldots dW_N
$$

$$
\delta(q_f - q_N) \exp \left[-\frac{1}{4D\epsilon} \sum_j (W_j - W_{j-1})^2 \right] = \lim_{N \to \infty} \left(\frac{1}{4\pi\epsilon D} \right)^{N/2}
$$

$$
\int_{-\infty}^{\infty} \cdots \int_{-\infty}^{\infty} \prod_{j=1}^{N} \frac{dq_j}{(4\pi\epsilon D)^{N/2}} dq_N \delta(q_N - q) \exp \left(-\epsilon\alpha \sum_j \frac{df(q_j)}{dq_j} \right)
$$

$$
\exp \left[-\frac{\epsilon}{4D} \sum_j \left(\frac{q_{j+1} - q_j + \epsilon\{\alpha f(q_{j+1}) + (1 - \alpha)f(q_j)\}}{\epsilon} \right)^2 \right] . \tag{2.47}
$$

In the continuous limit, the different terms in the exponentials yield

$$
\lim_{\substack{N \to \infty \\ \epsilon \to 0}} \epsilon\alpha \sum_j \frac{df(q_j)}{dq_j} \to \alpha \int_{t_0}^{t} ds \frac{df[q(s)]}{dq} \tag{2.48}
$$

$$
\lim_{\substack{N \to \infty \\ \epsilon \to 0}} \frac{\epsilon}{2} \sum_j [\alpha f(q_{j+1}) + (-\alpha)f(q_j)]^2 \to \frac{1}{2} \int_{t_0}^{t} ds f[q(s)]^2 \tag{2.49}
$$

$$
\lim_{\substack{N \to \infty \\ \epsilon \to 0}} \frac{\epsilon}{2} \sum_j (\frac{q_{j+1} - q_j}{\epsilon})^2 \to \frac{1}{2} \int_{t_0}^{t} ds \dot{q}(s)^2 \tag{2.50}
$$

$$
\lim_{\substack{N \to \infty \\ \epsilon \to 0}} \sum_j [\alpha f(q_{j+1}) + (-\alpha)f(q_j)] \to \int_{t_0}^{t} ds f[q(s)] \tag{2.51}
$$

Hence, the *path-integral representation* of $P(q, t|q_0, t_0)$ turns out to be

$$
P(q, t|q_0, t_0) = \int \mathcal{D}[q(t)] \exp \left(\int_{t_0}^{t} ds \mathcal{L}[q(s), \dot{q}(s)] \right), \tag{2.52}
$$

where

$$
\mathcal{S}[q(t)] = \int_{t_0}^{t} ds \mathcal{L}[q(s), \dot{q}(s)] \tag{2.53}
$$

is the *stochastic action*, and

$$
\mathcal{L}[q(s), \dot{q}(s)] = \frac{1}{4D} \left(\dot{q}(s) + f[q(s)] \right)^2 - \alpha \frac{df[q(s)]}{dq} \tag{2.54}
$$

is the *stochastic Lagrangian* (also called the *Onsager–Machlup functional*). The dependence of the Lagrangian on α is clearly seen in the last expression.

In Eq. (2.52), $\mathcal{D}[q]$ corresponds to the differential of the path, that is, the continuous expression of the discrete product of differentials at the end of Eq. (2.46).

We can also consider to start from the FPE in Eq. (2.34), using an operator formalism similarly as for the quantal case (i.e. via *Trotter's formula*). In this context the discretization problem associated with the different possibilities for the parameter α, transforms into a problem of operator ordering. However the result must be the same. Notwithstanding, for the several variable case, if the diffusion matrix is singular (null determinant), this is the only way to obtain a path-integral representation, but now in a phase-space like picture.

Clearly, almost all of the techniques developed within the other fields of application to evaluate path-integrals, can be adequately translated to the present context. For instance, it is possible to choose a *reference path* (that within a quantal context is the *classical path*, and in the stochastic context will the *most probable path*), and to expand the action in terms of the departure of the *actual path* from the *reference path*. As in the quantal case, this procedure gives exact results as far as the stochastic Lagrangian is at most quadratic in q and $\dot{q}$.

We stop this discussion at this point, and refer the reader to a few textbooks related with path-integration techniques [Langouche *et al.* (1982); Schulman (1981); Wiegel (1986); Wio (1990)].

2.5   A note on numerical simulation

As pointed out at the end of Sec. 2.1, the FPE is a *linear* equation in $P(y, t | y_0, t_0)$. On one hand, we saw that it could be derived—by means of a Kramers–Moyal expansion—from a master equation, which is *linear* by virtue of introducing phenomenological "transition rates" into a differential Chapman–Kolmogorov equation. On the other hand, we have just seen the FPE to be equivalent as a description to a path-integral formulation, much as a Schrödinger equation is. Thus, a FPE can in principle be solved (even for non-stationary regimes) by finding its eigenvalues and eigenvectors. However—as is the case for Schrödinger's equation—this may be a formidable and hopeless task.

With the massive availability of computing capabilities, it has become customary to solve the FPE by following the Langevin evolution of an ensemble of systems submitted to different realizations of the driving noise.

Whereas it is strongly advisable for the reader to get acquainted with the techniques involved in such a numerical simulation, we feel that a detailed description of those techniques lies outside the scope of this book. As a fully operative introduction, we recommend [San Miguel and Toral (2000)]. In this brief section, we limit ourselves to stress a couple of basic facts:

(a) The numerical integration of a (system of) Langevin equation(s) requires *discretizing time*, namely to assign a point to an interval. Whereas the reader may be familiar with the choices involved in the integration (in the sense of Riemann) of smooth functions like $f[q(t), t]$ in Eq. (2.18) or (2.33)—as well as with stability issues regarding the numerical scheme (namely, Runge–Kutta *vs.* Euler)—he/she is likely to be unaware of the special meaning this choice acquires when it comes to a "wild" term like $g[x(t), t]\xi(t)$. The issue between Itô's ($\alpha = 0$) and Stratonovich's ($\alpha = 1/2$) choices (noted just in passing in Sec. 2.4) is far from being a simple matter of preference. It does lead to different rules of calculus and different expressions of the FPE. Whereas mathematicians side usually with Itô's interpretation because of its advantages for the mathematical analysis (despite this choice leading them to a non-standard stochastic calculus), physicists recognize in Stratonovich's interpretation (besides the advantage of using standard stochastic calculus) the natural $\tau \to 0$ limit of "colored" noise (as the Ornstein–Uhlenbeck and other self-correlated processes are known).

(b) The other issue worth commenting is the fact that—from Eq. (1.20b) and the result $\langle [\xi(t)\delta t]^2 \rangle \simeq \langle \Delta W(t)^2 \rangle = \Delta t$ stated just before Eq. (2.26)—the term $g[x(t), t]\xi(t)$ is of order $\mathcal{O}(\Delta t)^{1/2}$ if $\xi(t)$ is a "white noise", as opposed to regular terms which are $\mathcal{O}(\Delta t)$. Instead, if $\xi(t)$ were a "colored noise", $g[x(t), t]\xi(t)$ would be $\mathcal{O}(\Delta t)$.

We close here and refer to [San Miguel and Toral (2000)] for further details, after recommending to use (whenever possible) Runge–Kutta-like methods instead of those inspired in Euler's algorithm. They are certainly more cumbersome to program but they are numerically more stable.

Chapter 3

The Ω–expansion

. . . we are like the wizard, who weaves a labyrinth and
is forced to wander inside it till the end of his days . . .
Jorge Luis Borges

3.1   Van Kampen's Ω–expansion

In the previous sections we have discussed about the master equation. In general, solving it is not a trivial task, and requires adequate approximate methods. The Kramers–Moyal expansion is a useful approach but it is not a systematic way of obtaining a FPE that approximates the master equation. However, such a systematic procedure does exist: it was introduced by van Kampen, and turns out to be valid for a wide class of systems. The extent of its applications fully justifies its presentation in this course. Within this scheme, one is able to show how to extract the macroscopic equation that drives the process, as well as the FPE for the fluctuations around such macroscopic behavior. This approach has become a standard technique and is a very efficient method to extract a FPE from a master equation (if the latter fulfills some conditions to be discussed later).

In order to introduce this procedure, instead of a formal presentation, we will discuss a couple of examples of physical and chemical origin.

3.1.1   *Effusion of a dilute gas*

The first (physical) problem to be discussed corresponds to the *effusion of a dilute gas*. This problem configures a neat and very clear classroom example to introduce van Kampen's Ω–expansion method for the master equation, as well as a transparent, but not completely trivial application of the FPE.

The system we consider (Fig. 3.1) is an isothermal container divided in two equal volumes, V_A and V_B, connected through a small hole of area s. We will call the total volume $V = V_A + V_B$, and N_A and N_B the number of particles within each volume, with $N = N_A + N_B$, the total number of particles. The respective densities are $n_A = N_A/V_A$ and $n_B = N_B/V_B$, and nonequilibrium situations are characterized by $n_A \neq n_B$. In this situation we have effusion, that is the passage through the communication hole, till the equilibrium density ($n_A = n_B = N/V$) is reached. A natural question to be asked is: what is the temporal behavior of this simple diffusion process? We restrict ourselves to the case in which the molecular densities are so low that the particle *mean-free-path* is (much) larger than the linear dimensions of the container. This effusion process is isothermal as in the Joule-Thompson experiment, that is, as in the expansion of a dilute gas against the vacuum (remember that the internal energy of a dilute perfect gas is independent of the volume). In a (very) short time interval Δt, the

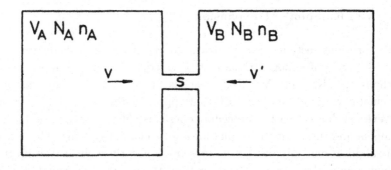

Fig. 3.1

average number of molecules transferred from V_A to V_B, through the hole, is

$$\Delta N_A = n_A s v \Delta t \qquad (3.1\text{a})$$

where v is the average component of the particle velocity, normal to s, and pointing outward from V_A. Analogously, for the passage from V_B to V_A we have

$$\Delta N_B = n_B s v' \Delta t \qquad (3.1\text{b})$$

The assumption of having an isothermal process implies $v = v'$, that remains constant in time. These numbers, ΔN_A and ΔN_B, must obey a

Poisson distribution with averages ΔN_A and ΔN_B, that is

$$p(\Delta N_{A,B}) = \frac{\Delta N_{A,B}^{N_{A,B}} e^{-\Delta N_{A,B}}}{(N_{A,B}!)} \tag{3.2}$$

When we consider the limit $\Delta t \to 0$, the only relevant cases are

$$p(0) = 1 - \Delta N_{A,B} + \mathcal{O}(\Delta t^2)$$
$$p(1) = \Delta N_{A,B} + \mathcal{O}(\Delta t^2) \tag{3.3}$$

corresponding to the *cross* and *non-cross* probabilities, respectively. Other values ($\Delta N_{A,B} \geq 2$) are at least quadratic in Δt. From Eqs. (3.1a) and (3.1b) we have that

$$\Delta N_{A,B} = n_{A,B} s v \Delta t = \frac{N_{A,B}}{V_{A,B}} s v \Delta t$$

and one is tempted to consider that $N_{A,B}$ is so large compared with 1 that we can neglect the difference between $N_{A,B} \pm 1$ and $N_{A,B}$. However, the conservation of the particle number prevents the use of this approximation.

We could then write the master equation for the probability density $P(N_A, t)$ of having N_A molecules in V_A at time t (provided there were N_{A0} particles at the initial time $t_0 < t$), with the normalization condition:

$$\sum_{N_A=0}^{N} P(N_A, t) = 1 \tag{3.4}$$

Calling $W_A(\delta N_A, N_A)$ the transition probability (per unit time) of transferring δN_A molecules from V_A to V_B, if there are N_A in V_A, and similarly for the transfer from V_B to V_A, we have for the master equation

$$P(N_A, t + \delta t) = W_A(0, N_A) W_B(0, N_B) P(N_A, t)$$
$$+ W_A(1, N_A + 1) W_B(0, N_A - 1) P(N_A + 1, t)$$
$$+ W_A(0, N_A - 1) W_B(1, N_A + 1) P(N_A - 1, t) + \mathcal{O}(\delta t^2) \tag{3.5}$$

Introducing now the parameter a, defined through the relation $V_A = aV$—or the equivalent one $V_B = (1-a)V$—in the limit of $\delta t \to 0$, we obtain the following form for the desired master equation

$$\frac{d}{dt} P(N_A, t) = \frac{sv}{a(1-a)V} \{(1-a)[(N_A+1)P(N_A+1,t) - N_A P(N_A,t)]$$
$$+ a[(N - N_A + 1)P(N_A - 1, t) - (N - N_A)P(N_A, t)]\} \tag{3.6}$$

where we have used $N_B = N - N_A$. At this point, it is useful to introduce the *step operator* $\mathbb{E}$, which is defined by the relations

$$\mathbb{E}\varphi(N_A) = \varphi(N_A + 1)$$

$$\mathbb{E}^{-1}\varphi(N_A) = \varphi(N_A - 1) \tag{3.7}$$

where $\varphi(N_A)$ is an arbitrary function of N_A. In terms of these operators, Eq. (3.6) may be written as

$$\frac{d}{dt}P(N_A, t) = (sv/a(1-a)V)\{(1-a)[\mathbb{E} - 1]N_A P(N_A, t)$$

$$+ a[\mathbb{E}^{-1} - 1](N_A - N)P(N_A, t)\} \tag{3.8}$$

As indicated earlier, in order to illustrate the Ω–expansion method we will work with the above obtained master equation, applying the method directly and making the relevant comments at each stage. Ω refers to a parameter that is *large* (compared with others in the system), and such that its inverse, or more correctly $\Omega^{-1/2}$, being a *small* quantity, may be used as an expansion parameter (for instance, Ω could be the system volume, the mass of the system, a population number, etc). The first point to consider is that the transition rates $W(n|m)$ from the state m to the state n appearing in the master equation, must have the *canonical form*. This essentially means that $W(n|m)$ must have the form of a Taylor-like expansion in terms of Ω^{-1}, with each coefficient being a function of Ω only through the intensive variable m/Ω; that is:

$$W(m|n) = f(\Omega)\{\phi_0(m/\Omega, r) + \Omega^{-1}\phi_1(m/\Omega, r) + \ldots\} \tag{3.9}$$

where $r = n - m$ is the step size, and $f(\Omega)$ is some arbitrary function of Ω. As one can see from Eqs. (3.6) or (3.8), adopting $\Omega = V$, this condition is fulfilled in our master equation, the arbitrary function f being $f(V) = sv/a(1 - a)$. A second point is the following (Fig. 3.2): we can assume that $P(N_A, t)$ must be sharply peaked at values of $N_A \simeq \langle N_A \rangle$, which is of *macroscopic order* V (or Ω) and, according to the *central limit theorem*, has a width of order $V^{1/2}$ (or, correspondingly, $\Omega^{1/2}$). Then, it is reasonable to make the Ansatz of separating N_A into a macroscopic part of order V, and fluctuations around it of order $V^{1/2}$:

$$N_A = V\Psi_A(t) + V^{1/2}\xi \tag{3.10}$$

Next we rewrite the master equation in terms of the variable ξ instead of N_A. We want to know how to relate the probability of finding the system with a value of the stochastic variable N_A in the range $(N_A, N_A + 1)$ with

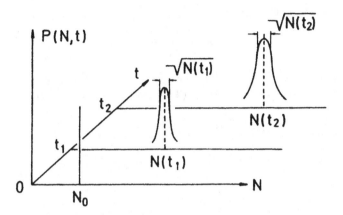

Fig. 3.2

the corresponding probability in terms of the new stochastic variable ξ. The relation is

$$P(N_A, t)\Delta N_A = \Pi(\xi, t)\Delta\xi \tag{3.11}$$

where

$$P(V\Psi_A(t) + V^{1/2}\xi, t) = \Pi(\xi, t) \tag{3.12a}$$

It is clear that

$$\frac{\partial}{\partial\xi}\Pi(\xi, t) = V^{1/2}\frac{\partial}{\partial N_A}P(N_A, t) \tag{3.12b}$$

and also

$$\frac{\partial}{\partial t}\Pi(\xi, t) = \frac{\partial}{\partial t}P(N_A, t) + V^{1/2}\frac{\partial}{\partial\xi}\Pi(\xi, t)\frac{d}{dt}\Psi_A(t)$$

$$= \frac{\partial}{\partial t}P(N_A, t) + V\frac{\partial}{\partial N_A}P(N_A, t)\frac{d}{dt}\Psi_A(t) \tag{3.12c}$$

The step operator $\mathbb{E}$ defined by Eqs. (3.7) can be written in terms of the variable ξ as a *shift operator*

$$\mathbb{E}^{\pm 1} = \exp\{\pm V^{-1/2}\frac{\partial}{\partial\xi}\} = 1 \pm V^{-1/2}\frac{\partial}{\partial\xi} + 1/2V^{-1}\frac{\partial^2}{\partial\xi^2} + \dots \tag{3.13}$$

We can also scale the time variable as:

$$t \to \tau = \frac{sv}{a(1-a)V}t \tag{3.14}$$

After the change of variables and the scaling and expansion indicated above, the master equation in Eq. (3.8) has the expanded form

$$\frac{\partial}{\partial \tau}\Pi(\xi,\tau) - V^{1/2}\frac{\partial}{\partial \xi}\Pi(\xi,\tau)\frac{d}{d\tau}\Psi_A(\tau) =$$

$$= \left\{V^{-1/2}\frac{\partial}{\partial \xi} + V^{-1}\frac{\partial^2}{\partial \xi^2} + \dots\right\}(1-a)[V\Psi_A + V^{1/2}\xi]\Pi(\xi,\tau)$$

$$\left\{V^{-1/2}\frac{\partial}{\partial \xi} + V^{-1}\frac{\partial^2}{\partial \xi^2} + \dots\right\}a[V(n-\Psi_A) + V^{1/2}\xi]\Pi(\xi,\tau) \tag{3.15}$$

We now collect and equate powers of V. In order that the large terms proportional to $V^{1/2}$ disappear from Eq. (3.15), we demand that Ψ_A fulfill the following equation

$$\frac{d}{dt}\Psi_A(t) = na - \Psi_A(t) \tag{3.16}$$

which corresponds to the *macroscopic equation*, that is, the equation driving the macroscopic evolution of the system (i.e. Ψ_A). Its solution will have the form

$$\Psi_A(\tau) = \Psi_A(0)e^{-\tau} + na[1 - e^{-\tau}] \tag{3.17}$$

where $\Psi_A(0)$ corresponds to the initial condition. It is clear that for $\tau \to \infty$, $\Psi_A(\infty) \to na$ corresponding to the stationary equilibrium value. In the above solution, we see that $t_d = [sv/a(1-a)V]^{-1}$ plays the role of a relaxation time. For instance, if we start with the volume V_A empty (i.e.: $N_A = 0$), from Eq. (3.17) we see that the stationary value $\Psi_A(\infty) = na$, will be reached only after a period of time of order t_d has elapsed.

For the next order in V (i.e.: $V^0 = 1$), Eq. (3.15) gives

$$\frac{\partial}{\partial \tau}\Pi(\xi,\tau) = \frac{\partial}{\partial \xi}[\xi\Pi(\xi,\tau)] + [(1-a)\Psi_A(\tau) + na]\frac{\partial^2}{\partial \xi^2}\Pi(\xi,\tau) \tag{3.18}$$

which is the desired *Fokker–Planck equation*. It is worth remarking that this is the equation that governs the time behavior of the fluctuations, described by the stochastic variable ξ, *around* the deterministic or macroscopic one, described by the variable Ψ_A. As it has the same form as the

FPE corresponding to the Ornstein–Uhlenbeck process [see after Eq. (2.3), however, with coefficients that are functions of time through its dependence on the solution of the macroscopic equation $\Psi_A(\tau)$], its solution *must be Gaussian*. Such equations are called *linear*, due to the linear dependence of the drift coefficient on ξ. In order to have the explicit form of the solution, it is enough to solve the equations for the first two moments, which can be obtained multiplying Eq. (3.18) by ξ and ξ^2, respectively, and integrating over ξ. The resulting equations are

$$\frac{d}{d\tau}\langle\xi\rangle = -\langle\xi\rangle \tag{3.19a}$$

$$\frac{d}{d\tau}\langle\xi^2\rangle = (1-2a)\Psi_A(\tau) + na - 2\langle\xi\rangle \tag{3.19b}$$

The solution of Eq. (3.19a) is clearly a decaying exponential

$$\langle\xi(\tau)\rangle = \langle\xi\rangle_0 e^{-\tau}$$

while the solution of Eq. (3.19b) has the form

$$\langle\xi^2(\tau)\rangle = \langle\xi^2\rangle_0 e^{-2\tau} + [(1-2a)\Psi_A(0)e^{-\tau} + na](1-e^{-\tau})$$

where $\langle\xi\rangle_0$ and $\langle\xi^2\rangle_0$ are the corresponding initial conditions. Taking those initial values to be zero, we have depicted in Fig. 3.3 the solutions to Eq.

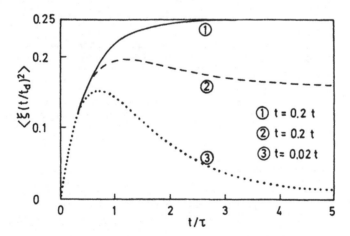

Fig. 3.3 $\langle\xi^2\rangle$ as a function of t (in units of t_d), for different values of a, indicated in the figure. Remaining parameters are $V = 5$, $n_A(0) = 0.6$, $n = 1$, $\langle\xi^2(0)\rangle = 0$.

(3.19b), as given above, for different values of the parameters. There we see that, depending on the situation under study, we can find an increase of the fluctuations beyond the stationary value before the fluctuations decay reaching the stationary regime.

With the above results we can write the solution to the FPE Eq. (3.18) as

$$\Pi(\xi, \tau) = [2\pi\sigma(\tau)^2]^{-1/2} \exp\left\{ -\frac{[\xi - \xi(\tau)]^2}{2\sigma(\tau)^2} \right\}, \qquad (3.20)$$

where $\sigma(\tau)^2 = \langle\xi(\tau)^2\rangle - \langle\xi(\tau)\rangle^2$. In Fig. 3.4 we present some results for this probability distribution, for three different times. The increase in the distribution width before reaching the smaller value corresponding to the stationary regime, is clearly seen.

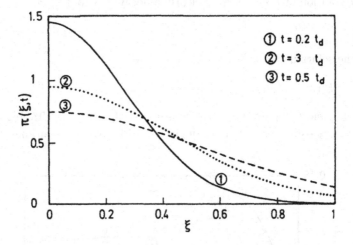

Fig. 3.4   Probability density $\Pi(\xi, \tau)$ as function of τ, for different times. We consider, for the fluctuations, zero initial conditions and the other parameters are: $V = 5$, $n_A(0) = 0.6$, $n = 1$, $a = 0.1$.

The procedure developed above makes it clear that van Kampen's Ω–expansion is a systematic method, which allows us to extract from the master equation, at different orders in $\Omega^{-1/2}$, the macroscopic equation which drives the system towards the stationary equilibrium condition and the FPE for the fluctuations around such macroscopic behavior.

In the next example we will show that it is also possible to obtain the *stationary correlation function for the fluctuations*.

3.1.2 Dissociation of a diatomic gas

The second example is related to a chemical-like problem: namely the *dissociation of a diatomic gas*

$$AB \underset{\beta}{\overset{\alpha}{\rightleftarrows}} A + B$$

We will assume that the total number of atoms A and B (N_A and N_B) is fixed. Calling n the number of molecules AB, the dissociation probability per unit time will be αn. For an association to occur, one of the $(N_A - n)$ free A atoms must collide with one of the $(N_B - n)$ free B atoms. Hence, the corresponding probability of reaction is $\beta(N_A - n)(N_B - n)/\Omega$, where Ω is the volume of the container where the reaction takes place. The master equation, in terms of p_n (the probability of having n molecules AB), is

$$\dot{p}_n = \alpha(n + 1)p_{n+1} - \alpha n p_n$$

$$+ \frac{\beta}{\Omega}\{(N_A - n + 1)(N_B - n + 1)p_{n-1} - (N_A - n)(N_B - n)p_n\}$$

$$= \alpha[\mathbb{E} - 1]n p_n + \frac{\beta}{\Omega}[\mathbb{E}^{-1} - 1](N_A - n)(N_B - n)p_n \qquad (3.21)$$

The stationary solution ($\dot{p}_n = 0$) can be found, and related with the known equilibrium distribution, giving

$$\frac{\alpha}{\beta} = \left[2\pi kT \frac{m_A + m_B}{m_A m_B}\right]^{3/2} e^{\kappa/kT}$$

where κ is the bonding energy. However, the time-dependent solutions cannot be found, due to the nonlinearity of the second coefficient. It is clear that we are dealing with a Markov process (there are no *age* effects: α, β are fixed, etc). A point worth stressing is that *whether a physical phenomenon is or not Markovian depends on which variables we choose.*

In order to obtain a solvable equation for this process, we repeat the procedure done for the previous example. We consider again that we have a large parameter Ω (associated with the volume), and write $N_A = \Omega\rho_A$ and $N_B = \Omega\rho_B$, where ρ_A and ρ_B are the respective densities which, for simplicity, we assume equal: $\rho_A = \rho_B = \rho$. We can again argue about p_n having a marked maximum centered around a value n, of order Ω, and a width of order $\Omega^{1/2}$, yielding

$$n = \Omega\phi(t) + \Omega^{1/2}\xi$$

$$p_n = \Pi(\xi, t) \qquad (3.22)$$

As before, we consider the expansion of the creation and annihilation operators, Eq. (3.13), and substituting into the master equation, we obtain

$$\frac{\partial}{\partial t}\Pi(\xi,t) - \Omega^{1/2}\frac{\partial}{\partial \xi}\Pi(\xi,t)\frac{d}{dt}\phi(t) =$$

$$= \alpha\left\{\Omega^{1/2}\frac{\partial}{\partial \xi} + \frac{1}{2}\frac{\partial^2}{\partial \xi^2} + \dots\right\}\left[\phi + \Omega^{-1/2}\xi\right]\Pi(\xi,t)$$

$$+ \beta\left\{-\Omega^{1/2}\frac{\partial}{\partial \xi} + \frac{1}{2}\frac{\partial^2}{\partial \xi^2} + \dots\right\}\left[\rho - \phi - \Omega^{-1/2}\xi\right]^2\Pi(\xi,t) \quad (3.23)$$

Again, separating terms of order $\Omega^{1/2}$, and asking for this contribution to be zero, we obtain the macroscopic equation for ϕ (or $n \simeq \Omega\phi$):

$$\dot{\phi} = -\alpha\phi + \beta[\rho - \phi]^2, \quad (3.24)$$

which is the macroscopic equation governing the evolution of the AB concentration. Considering the next order (Ω^0), we find

$$\frac{\partial}{\partial t}\Pi(\xi,t) = \{\alpha + 2\beta[\rho - \phi]\}\frac{\partial}{\partial \xi}\xi\Pi(\xi,t)$$

$$+ \frac{1}{2}\left\{\alpha\phi + \beta[\rho - \phi]^2\right\}\frac{\partial^2}{\partial \xi^2}\Pi(\xi,t) \quad (3.25)$$

that again results to be a linear FPE, with time-dependent coefficients (through the dependence of $\phi(t)$). As we said before, the solution is Gaussian, and it is enough to calculate the first two moments of ξ, with equations given by

$$\frac{d}{dt}\langle\xi\rangle = -2\{\alpha + 2\beta[\rho - \phi]\}\langle\xi\rangle$$

$$\frac{d}{dt}\langle\xi^2\rangle = -2\{\alpha + 2\beta[\rho - \phi]\}\langle\xi^2\rangle + \{\alpha\phi + \beta[\rho - \phi]^2\} \quad (3.26)$$

After solving these equations we will have all the necessary elements to build up the complete solution $\Pi(\xi,t)$, for the behavior of the fluctuations around the macroscopic trajectory.

An interesting point, that we have not discussed so far, is the possibility of obtaining the correlation of fluctuations when the stationary state is reached. The stationary value of ϕ is obtained taking $\dot{\phi} = 0$ in Eq. (3.24). This results in

$$0 = -\alpha\phi + \beta[\rho - \phi]^2. \quad (3.27)$$

Being a quadratic equation in ϕ, it has two roots:

$$\phi_{\text{st}} = \{\alpha/2\beta + \rho\} \pm \left[[\alpha/2\beta]^2 + 2[\alpha/2\beta]\rho\right]^{1/2}. \quad (3.28)$$

As easily seen, one of the roots is larger than ρ (corresponding to the plus sign), without physical meaning and to be disregarded, leaving the other (corresponding to the minus sign) as the only one yielding ϕ_{st}. Substituting this value in the FPE for $\Pi(\xi, t)$, the coefficients turn out to be independent of t, and ξ becomes a pure Ornstein–Uhlenbeck process. However, we will not use this last possibility as we can directly obtain

$$\langle\xi\rangle_{st} = 0$$

$$\langle\xi^2\rangle_{st} = \frac{-\alpha\phi_{st} + \beta[\rho - \phi_{st}]^2}{2\{\alpha + 2\beta[\rho - \phi_{st}]\}} \tag{3.29}$$

Let us assume that at t_1 we have the value ξ_1; then the average of the variable ξ at $t = t_2$, subject to the previous condition at $t = t_1$, is

$$\langle\xi(t_2)\rangle_{\xi_1} = \xi_1 \exp\{-[\alpha + 2\beta\rho - 2\beta\phi_{st}][t_2 - t_1]\}$$

$$= \xi_1 \exp\{-[t_2 - t_1]/\tau\} \tag{3.30}$$

The correlation function will be

$$\langle\xi(t_2)\xi(t_1)\rangle = \int\int d\xi_1 d\xi_2 \xi_1 \xi_2 P(\xi_1, t_1; \xi_2, t_2)$$

$$= \int\int d\xi_1 d\xi_2 \xi_1 \xi_2 P(\xi_1, t_1) P(\xi_2, t_2|\xi_1, t_1)$$

$$= \int d\xi_1 \xi_1 P(\xi_1)^{st} \langle\xi(t_2)\rangle_{\xi_1}$$

$$= \langle\xi_1^2\rangle_{st} \exp\{-[\alpha + 2\beta\rho - 2\beta\phi_{st}][t_2 - t_1]\}$$

$$= \left[\frac{\alpha\phi_{st}}{\{\alpha + 2\beta[\rho - \phi_{st}]\}}\right] \exp\{-[t_2 - t_1]/\tau\} \tag{3.31}$$

Multiplying this equation by Ω^2, we get the self correlation function for n. As we will see latter, the *Wiener–Khintchine theorem* states that the Fourier transform of this correlation function is the spectral density of the fluctuations, which is a directly measurable quantity.

The above presented Ω-expansion procedure is also known as *the linear noise approximation for the fluctuations.* If we include higher order terms in $\Omega^{-1/2}$, we will have additional terms in Eqs. (3.18) and (3.25), with the effect that *the first coefficient will loose its linear character in ξ, and the second one will become dependent on ξ,* higher order derivatives arising as well. The solution will not be Gaussian anymore, but it will still be possible to obtain the successive moments till the desired order. As a consequence, in the self-correlation function, Eq. (3.31), an additional exponential term arises, decaying twice as faster as the previous one.

On the other hand, a theorem due to Pawula tells us that, within a Kramers–Moyal-like expansion, if the order of derivatives included in the equation is greater than 2 and finite, it is not possible to guarantee the positivity of the solution.

3.2   A case with many variables

We present now a third example where van Kampen's method is applied to a many component system. The last few years have witnessed a growing interest among theoretical physicists in complex phenomena in fields departing from the classical mainstream of physics research, particularly the application of statistical physics methods to social phenomena [Weidlich (2002); Stauffer *et al.* (2006)] we will consider an example within this realm. Among those sociological problems, one that has attracted much attention was the building (or the lack) of consensus in social groups. There are many different models that simulate and analyze the dynamics of such processes in opinion formation, cultural dynamics, etc [Castellano *et al.* (2009)].

We will analyze a simple *opinion formation model* consisting of two parties, A and B, and an "intermediate" group I, called *undecided agents* [de la Lama *et al.* (2006, 2007)]. It is worth noting that these three groups are not in the same step. We consider that members of groups A and B have well established positions about a given subject and I constitutes a group of undecided agents that would probably be converted to one of the dominant positions. We assume that the supporters of parties A and B do not interact among them, but only through their interaction with the group I, convincing one of its members using a simple rule that is within a mean-field treatment. However, we don't consider that members of I can convince those of A or B, mainly because they do not have a definite opinion, but instead we assume that there is a nonzero probability of a spontaneous change of opinion from I to the other two parties and viceversa $I \leftrightarrows A$ and $I \leftrightarrows B$. We will see that this probability of spontaneous change of opinion (that implies the existence of a *social temperature* inhibits the possibility of reaching a consensus. Instead of consensus, we find that each party has some statistical density of supporters, and there is also a statistical stationary number of undecided (I) agents.

3.2.1 *Description of the model*

Hence, we consider a system composed of three different groups of agents

- supporters of the A party, indicated by N_A,
- supporters of the B party, indicated by N_B,
- undecided ones, indicated by N_I.

As indicated, the interactions we are going to consider are only between A and I, and B and I. That means that we do not consider direct interactions among A and B. The different contributions that we include are

- spontaneous transitions $A \to I$, occurring with a rate $\alpha_1 N_A$;
- spontaneous transitions $I \to A$, occurring with a rate $\alpha_2 N_I$;
- spontaneous transitions $B \to I$, occurring with a rate $\alpha_3 N_B$;
- spontaneous transitions $I \to B$, occurring with a rate $\alpha_4 N_I$;
- convincing rule $A + I \to 2\,A$, occurring with rate $\frac{\beta_1}{N} N_A N_I$;
- convincing rule $B + I \to 2\,B$, occurring with rate $\frac{\beta_2}{N} N_B N_I$.

As indicated above, here N_i is the number of agents supporting the party or opinion "i" (with $i = A, B, I$). We have the constraint $N_A + N_B + N_I = N$, where N is the total number of agents. Such a constraint implies that, for fixed N, there are only two independent variables N_A and N_B. By using this constraint, the rates indicated above associated to processes involving N_I, could be written replacing $N_I = (N - N_A - N_B)$.

With the above indicated interactions and rates, the master equation for the probability $P(N_A, N_B, t)$ of having populations N_A and N_B at time t (due we have had populations N_A^o and N_B^o at an initial time $t_o < t$), may be written as

$$
\begin{aligned}
\frac{\partial}{\partial t} P(N_A, N_B, t) =\ & \alpha_1 (N_A + 1) P(N_A + 1, N_B, t) \\
& + \alpha_3 (N_B + 1) P(N_A, N_B + 1, t) \\
& + \alpha_2 (N - N_A - N_B + 1) P(N_A - 1, N_B, t) \\
& + \alpha_4 (N - N_A - N_B + 1) P(N_A, N_B - 1, t) \\
& + \frac{\beta_1}{\Omega} (N_A - 1)(N - N_A - N_B + 1) P(N_A - 1, N_B, t) \\
& + \frac{\beta_2}{\Omega} (N_B - 1)(N - N_A - N_B + 1) P(N_A, N_B - 1, t) \\
& - \Big[\alpha_1 N_A + \alpha_3 N_B + (\alpha_2 + \alpha_4)(N - N_A - N_B) \\
& \quad + \left(\frac{\beta_1 N_A + \beta_2 N_B}{\Omega} \right)(N - N_A - N_B) \Big] P(N_A, N_B, t).
\end{aligned}
\tag{3.32}
$$

This is the model master equation to which we will apply van Kampen's approach.

3.2.2   The expansion

In order to apply van Kampen's approach we identify the large parameter Ω with N (assuming $N \gg 1$); and define—as usual—the following separation of the N_i's into a macroscopic part of size N, and a fluctuational part of size $N^{\frac{1}{2}}$,

$$N_A = N\Psi_A(t) + N^{\frac{1}{2}}\xi_A(t),$$

$$N_B = N\Psi_B(t) + N^{\frac{1}{2}}\xi_B(t), \tag{3.33}$$

and define a "reference" density $\rho = \frac{N}{\Omega}$, that in our case is $\rho = 1$. We define again the "step operators"

$$\mathbb{E}_i^1 f(N_i) = f(N_i + 1),$$

$$\mathbb{E}_i^{-1} f(N_i) = f(N_i - 1),$$

with $f(N_i)$ an arbitrary function. Using the forms indicated in Eqs. (3.33), in the limit of $N \gg 1$, the step operators adopt the differential form

$$\mathbb{E}_i^{\pm 1} = 1 \pm \left(\frac{1}{N}\right)^{\frac{1}{2}} \frac{\partial}{\partial \xi_i} + \frac{1}{2}\left(\frac{1}{N}\right) \frac{\partial^2}{\partial \xi_i^2} \pm \cdots, \tag{3.34}$$

with $i = A, B$. Transforming from the original variables (N_A, N_B) to the new ones (ξ_A, ξ_B), we have the relations

$$P(N_A, N_B, t) \to \Pi(\xi_A, \xi_B, t), \tag{3.35a}$$

$$N^{\frac{1}{2}} \frac{\partial}{\partial N_i} P(N_A, N_B, t) = \frac{\partial}{\partial \xi_i} \Pi(\xi_A, \xi_B, t). \tag{3.35b}$$

Putting everything together, and considering contributions up to order $N^{\frac{1}{2}}$, yields the following two coupled differential equations for the macroscopic behavior

$$\frac{d}{dt}\Psi_A(t) = -\alpha_1 \Psi_A + \left[\alpha_2 + \beta_1 \Psi_A\right]\left(\rho - \Psi_A - \Psi_B\right), \tag{3.36a}$$

$$\frac{d}{dt}\Psi_B(t) = -\alpha_3 \Psi_B + \left[\alpha_4 + \beta_2 \Psi_B\right]\left(\rho - \Psi_A - \Psi_B\right). \tag{3.36b}$$

It can be proved that the last set of equations has a unique (physically sound) stationary solution, i.e. a unique attractor

$$\Psi_A(t \to \infty) = \Psi_A^{\text{st}}$$

$$\Psi_B(t \to \infty) = \Psi_B^{\text{st}}.$$

This is the main condition to validate the application of van Kampen's Ω-expansion approach.

The following order, that is N^0, yields the Fokker–Planck equation (FPE) governing the fluctuations around the macroscopic behavior. It is given by

$$\frac{\partial}{\partial t} \Pi(\xi_A, \xi_B, t) =$$

$$\frac{\partial}{\partial \xi_A} \Big[(\alpha_1 \xi_A + (\alpha_2 + \beta_1 \Psi_A)(\xi_A + \xi_B) - \beta_1 \xi_A (\rho - \Psi_A - \Psi_B)) \Pi(\xi_A, \xi_B, t) \Big]$$

$$+ \frac{\partial}{\partial \xi_B} \Big[(\alpha_3 \xi_B + (\alpha_4 + \beta_2 \Psi_B)(\xi_A + \xi_B) - \beta_2 \xi_B (\rho - \Psi_A - \Psi_B)) \Pi(\xi_A, \xi_B, t) \Big]$$

$$+ \frac{1}{2} \Big[\alpha_1 \Psi_A + (\alpha_2 + \beta_1 \Psi_A)(\rho - \Psi_A - \Psi_B) \Big] \frac{\partial^2}{\partial \xi_A^2} \Pi(\xi_A, \xi_B, t)$$

$$+ \frac{1}{2} \Big[\alpha_3 \Psi_B + (\alpha_4 + \beta_2 \Psi_B)(\rho - \Psi_A - \Psi_B) \Big] \frac{\partial^2}{\partial \xi_B^2} \Pi(\xi_A, \xi_B, t). \tag{3.37}$$

As is well known for this approach, the solution of this FPE will have a Gaussian form determined by the first and second moments of the fluctuations.

3.2.3 *Behavior of fluctuations*

From the FPE indicated above (Eq. (3.37)), it is possible to obtain equations for the mean value of the fluctuations as well as for the correlations of those fluctuations. For the fluctuations, $\langle \xi_A(t) \rangle = \eta_A$ and $\langle \xi_B(t) \rangle = \eta_B$, we have

$$\frac{d}{dt} \eta_A(t) = - \Big[\alpha_1 + \alpha_2 + \beta_1(2\Psi_A + \Psi_B) - \beta_1 \rho \Big] \eta_A$$

$$- (\alpha_2 + \beta_1 \Psi_A) \eta_B \tag{3.38a}$$

$$\frac{d}{dt} \eta_B(t) = - \Big[\alpha_3 + \alpha_4 + \beta_2(\Psi_A + 2\Psi_B) - \beta_2 \rho \Big] \eta_B$$

$$- (\alpha_4 + \beta_2 \Psi_B) \eta_A. \tag{3.38b}$$

Calling $\sigma_A = \langle \xi_A(t)^2 \rangle$, $\sigma_B = \langle \xi_B(t)^2 \rangle$, and $\sigma_{AB} = \langle \xi_A(t)\xi_B(t) \rangle$, we obtain for the correlation of fluctuations

$$\frac{d}{dt}\sigma_A(t) = -2\alpha_1\sigma_A - 2[\alpha_2 + \beta_1\Psi_A][\sigma_A + \sigma_{AB}] + 2\beta_1\sigma_A[\rho - \Psi_A - \Psi_B]$$

$$+[\alpha_1\Psi_A + (\alpha_2 + \beta_1\Psi_A)(\rho - \Psi_A - \Psi_B)], \qquad (3.39a)$$

$$\frac{d}{dt}\sigma_B(t) = -2\alpha_3\sigma_B - 2[\alpha_4 + \beta_2\Psi_B][\sigma_{AB} + \sigma_B] + 2\beta_2\sigma_B[\rho - \Psi_A - \Psi_B]$$

$$+[\alpha_3\Psi_B + (\alpha_4 + \beta_2\Psi_B)(\rho - \Psi_A - \Psi_B)], \qquad (3.39b)$$

$$\frac{d}{dt}\sigma_{AB}(t) = -[\alpha_1 + \alpha_3]\sigma_{AB} - [\alpha_2 + \beta_1\Psi_A][\sigma_{AB} + \sigma_B]$$

$$-[\alpha_4 + \beta_2\Psi_B][\sigma_A + \sigma_{AB}] + [\rho - \Psi_A - \Psi_B][\beta_1 + \beta_2]\sigma_{AB}. \qquad (3.39c)$$

3.2.3.1   *Reference state: Symmetric case*

Now we particularize the above indicated equations to the symmetric case i.e., the case when $\Psi_A^{st} = \Psi_B^{st}$. Hence, we adopt

$$\alpha_1 = \alpha_3 = \alpha, \quad \alpha_2 = \alpha_4 = \alpha', \text{ and } \beta_1 = \beta_2 = \beta.$$

The macroscopic equations (3.36) reduce to

$$\frac{d}{dt}\Psi_A(t) = -[\alpha + \alpha' - \beta]\Psi_A - \beta\Psi_A^2 - \beta\Psi_A\Psi_B - \alpha'\Psi_B + \alpha' \quad (3.40a)$$

$$\frac{d}{dt}\Psi_B(t) = -[\alpha + \alpha' - \beta]\Psi_B - \beta\Psi_B^2 - \beta\Psi_A\Psi_B - \alpha'\Psi_A + \alpha'. \quad (3.40b)$$

In order to make more explicit the solution of these equations, we introduce the auxiliary variables $\Sigma = \Psi_A + \Psi_B$ and $\Delta = \Psi_A - \Psi_B$ (and use that $\rho = 1$). The last equations now transform into

$$\frac{d}{dt}\Sigma(t) = -\left[\alpha + 2\alpha' - \beta\right]\Sigma - \beta\Sigma^2 + 2\alpha', \qquad (3.41a)$$

$$\frac{d}{dt}\Delta(t) = -\left[\alpha - \beta\right]\Delta - \beta\Delta\Sigma. \qquad (3.41b)$$

In the long time limit, $t \to \infty$, we found on one hand that $\Delta^{st} = 0$, implying $\Psi_A^{st} = \Psi_B^{st}$, while on the other hand

$$0 = \beta\,\Sigma^2 + \left[\alpha + 2\alpha' - \beta\right]\Sigma - 2\alpha'.$$

This polynomial has two roots, but only one is physically sound, namely

$$\Sigma^{\text{st}} = \frac{\alpha + 2\alpha' - \beta}{2\beta} \left(-1 + \sqrt{1 + \frac{8\alpha'\beta}{[\alpha + 2\alpha' - \beta]^2}} \right), \tag{3.42}$$

yielding $\Psi_A^{\text{st}} = \Psi_B^{\text{st}} = \Psi_o^{\text{st}} = \frac{1}{2}\Sigma^{\text{st}}$.

In a similar way, we can also simplify the equations for η_A and η_B, calling $S(t) = \eta_A + \eta_B$ and $D(t) = \eta_A - \eta_B$. The corresponding equations are then rewritten as

$$\frac{d}{dt}S(t) = -\left[\alpha + 2\alpha' + 2\beta(\Psi_A + \Psi_B) - \beta\right] S, \tag{3.43a}$$

$$\frac{d}{dt}D(t) = -\left[\alpha + \beta(\Psi_A + \Psi_B) - \beta\right] D - \beta\left[\Psi_a - \Psi_B\right] S,, \tag{3.43b}$$

while for the correlation of the fluctuations we have

$$\frac{d}{dt}\sigma_A(t) = -2\alpha\sigma_A - 2[\alpha' + \beta\Psi_A][\sigma_A + \sigma_{AB}] + 2\beta[1 - \Psi_A - \Psi_B]\sigma_A$$
$$+ [\alpha\Psi_A + (\alpha' + \beta\Psi_A)(1 - \Psi_A - \Psi_B)], \tag{3.44a}$$

$$\frac{d}{dt}\sigma_B(t) = -2\alpha\sigma_B - 2[\alpha' + \beta\Psi_B][\sigma_{AB} + \sigma_B] + 2\beta[1 - \Psi_A - \Psi_B]\sigma_B$$
$$+ [\alpha\Psi_B + (\alpha' + \beta\Psi_B)(1 - \Psi_A - \Psi_B)], \tag{3.44b}$$

$$\frac{d}{dt}\sigma_{AB}(t) = -2\alpha\sigma_{AB} - [\alpha' + \beta\Psi_A][\sigma_{AB} + \sigma_B]$$
$$- [\alpha' + \beta\Psi_B][\sigma_{AB} + \sigma_A] + 2\beta[1 - \Psi_A - \Psi_B]\sigma_{AB}. \tag{3.44c}$$

Equations (3.43) show that both $S = 0$ and $D = 0$ in the asymptotic limit $t \to \infty$, implying that $\eta_A^{\text{st}} = \eta_B^{\text{st}} = 0$. However, also in the general (non symmetric) case we expect to find $\eta_A^{\text{st}} = \eta_B^{\text{st}} = 0$. In addition, it is clear from Eqs. (3.44) that for $t \to \infty$ we generally obtain $\sigma_i^{\text{st}} \neq 0$ ($i = A, B, AB$).

As we have seen, in the symmetric case we have $\Psi_A^{\text{st}} = \Psi_B^{\text{st}} = \Psi_o^{\text{st}}$, hence it is clear that $\sigma_A(t)$ and $\sigma_B(t)$ behave in a similar way. And in particular $\sigma_A^{\text{st}} = \sigma_B^{\text{st}} = \sigma_o^{\text{st}}$. In order to analyze the typical time for return to the stationary situation under small perturbations, we assume small perturbations of the form $\sigma_i^{\text{st}} \approx \sigma_o^{\text{st}} + \delta\sigma_i(t)$ ($i = A, B$) and $\sigma_{AB}^{\text{st}} \approx \sigma_{AB,o}^{\text{st}} + \delta\sigma_i(t)$, and fix $\Psi_A^{\text{st}} = \Psi_B^{\text{st}} = \Psi_o^{\text{st}}$. We find again that both $\delta\sigma_A(t)$ and $\delta\sigma_B(t)$ behave in the same way, and this help us to reduce the number of equations for the decay of correlations. Hence, we can put $\delta\sigma_A(t) = \delta\sigma_B(t) = \delta\sigma_o(t)$.

The system driving the correlations becomes

$$\frac{d}{dt}\delta\sigma_o(t) = -2\Big[\alpha + \alpha' - \beta + 3\beta\Psi_o^{\text{st}}\Big]\delta\sigma_o - 2\Big[\alpha' + \beta\Psi_o^{\text{st}}\Big]\delta\sigma_{AB}, \qquad (3.45\text{a})$$

$$\frac{d}{dt}\delta\sigma_{AB}(t) = -2\Big[\alpha + \alpha' - \beta + 3\beta\Psi_o^{\text{st}}\Big]\delta\sigma_{AB} - 2\Big[\alpha' + \beta\Psi_o^{\text{st}}\Big]\delta\sigma_o. \quad (3.45\text{b})$$

Clearly, $\delta\sigma_o^{\text{st}} = \delta\sigma_{ab}^{\text{st}} \equiv 0$. After some algebraic steps we obtain

$$\delta\sigma_o(t) \simeq \delta\sigma_o(0) \exp\Big[-2[\alpha + 2\beta\Psi_o^{\text{st}} - \beta]\,t\Big] \qquad (3.46\text{a})$$

$$\delta\sigma_{AB}(t) \simeq \delta\sigma_{AB}(0) \exp\Big[-2[\alpha + 2\beta\Psi_o^{\text{st}} - \beta]\,t\Big]. \qquad (3.46\text{b})$$

These results indicate that, for the symmetrical case, the typical relaxation time is given by

$$\tau_{\text{relax}} = \frac{1}{2}[\alpha + 2\beta\Psi_o^{\text{st}} - \beta]^{-1}. \qquad (3.47)$$

3.2.4 *Some results*

In order to show how the scheme works, here we present a few results that show some typical behavior to be expected from the model and the approximation method. In what follows, all parameters are measured in arbitrary units.

In Fig. 3.5 we show the evolution of $\Psi_A(t)$ and $\Psi_B(t)$, the macroscopic solutions, indicating some trajectories towards the attractor: (a)

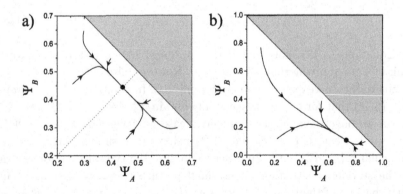

Fig. 3.5   Evolution of the macroscopic solutions, Eqs. (3.36). Case (a) corresponds to trajectories towards a symmetric solution. Case (b) corresponds to trajectories towards an asymmetric solution.

for a symmetric (i.e. with $\Psi_A^{st} = \Psi_B^{st}$), and (b) an asymmetric case (i.e. with $\Psi_A^{st} \neq \Psi_B^{st}$). In the first case we have used that $\alpha_1 = \alpha_3 = 1$, $\alpha_2 = \alpha_4 = 3$, and $\beta_1 = \beta_2 = 2$. In the second case we adopted $\alpha_1 = 1$, $\alpha_3 = 5$, $\alpha_2 = \alpha_4 = 3$, and $\beta_1 = \beta_2 = 2$. It is worth recalling that Ψ_A and Ψ_B are the density of supporters of party A and party B, respectively. During the evolution towards the attractor, starting from arbitrary initial conditions, we observe the possibility of a marked initial increase of the macroscopic density for one of the parties, follow by a marked reduction, or other situations showing only a decrease of an initial high density. Such cases indicate the need of taking with care the results of surveys and polls during, say, an electoral process. It is possible that an impressive initial increase in the support of a party can be followed for an also impressive decay of such a support.

We remark that, due to the symmetry of the problem, it is equivalent to varying the set of parameters $(\alpha_3, \alpha_4, \beta_2)$ or the set $(\alpha_1, \alpha_2, \beta_1)$. Also worth remarking is that in both panels of Fig. 3.5 the sum of Ψ_A and Ψ_B is always $\Psi_A + \Psi_B < 1$, so verifying that there is always a finite fraction of undecided agents.

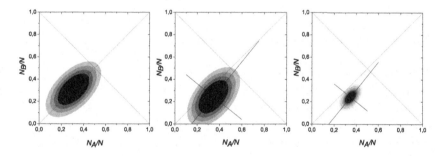

Fig. 3.6   Stationary, Gaussian, probability distribution $\Pi(\xi_A, \xi_B)^{st}$ projected on the original (N_A, N_B) plane. On the left we have a symmetric case, the central plot shows an asymmetric case, while on the right we have the same asymmetric case as before, but now $N = 1000$, showing the dispersion reduction of the Gaussian distribution.

In Fig. 3.6 we depict the stationary (Gaussian) probability distribution (pdf) $\Pi(\xi_A, \xi_B)^{st}$ projected on the original (N_A, N_B) plane. We show three cases: on the left a symmetric case ($\alpha_1 = \alpha_3 = 2$, $\alpha_2 = \alpha_4 = 1$, $\beta_1 = \beta_2 = 2$, and $N = 100$), the central one corresponds to an asymmetric situation ($\alpha_1 = 2$ and $\alpha_3 = 2.5$, while $\alpha_2 = \alpha_4 = 1$, $\beta_1 = \beta_2 = 2$) with a population of $N = 100$, and on the right the same asymmetric situation but with a population of $N = 1000$. This last case clearly shows the influence of the population number in reducing the dispersion (as the population increases). We can use this pdf in order to estimate the probability p_i $(i = A, B)$, of winning for one or the other party. It corresponds to the volume of the distribution remaining above, or below, the bisectrix $N_A/N = N_B/N$. In the symmetric case, as is obvious, we obtain $p_A = p_B = 0.5$ (or 50%), while in the asymmetric case we found $p_B = 0.257$ (or 25.7%) and $p_B = 0.015$ (or 1.5%) for $N = 100$ and $N = 1000$, respectively. These results indicate that, for an asymmetric situation like the one indicated here, we have a nonzero probability that the minority party could, due to a fluctuation during the voting day, win a close election. However, in agreement with intuition, as far as $N \gg 1$, and the stationary macroscopic solution departs from the symmetric case, such a probability p_i reduces proportionally to N^{-1}.

The simple *toy model* for opinion formation we have studied, includes a nonzero probability of a spontaneous change of opinion from I to the other two parties and viceversa. It shows that it is spontaneous change of opinion that inhibits the possibility of reaching a consensus, and yields that each party has some statistical density of supporters, while a statistical stationary number of undecided agents remains.

3.3   Limitations of the Ω–expansion

The Ω–expansion procedure sketched in the previous paragraph, is based on the argument (or hope) that the fluctuations will remain *small*. We need to check a posteriori that its order is $\Omega^{1/2}$, and that other contributions will behave as $\Omega^{-1/2}$, at least. However, if $\langle \xi \rangle$ and $\langle \xi^2 \rangle$ increase with time, such powers of Ω will be an adequate measure of the size of the fluctuations only during a certain limited period of time. In the second of the two examples discussed in the previous paragraph, it is possible to see from the equations for $\langle \xi \rangle$ and $\langle \xi^2 \rangle$ [Eqs. (3.26)], that those quantities remain finite or will increase depending on

$$\alpha - 2\beta\rho + 2\beta\phi \lesseqgtr 0 \qquad (3.48)$$

What does this mean? Let us consider the macroscopic equation for ϕ, Eq. (3.27), and assume that ϕ_1 and $\phi_1 + \delta\phi$ are two *neighbouring* solutions at $t = 0$, we then have

$$\frac{d}{dt}\delta\phi(t) = -\{\alpha - 2\beta\rho + 2\beta\phi\}\delta\phi + \mathcal{O}(\delta\phi^2) = -\lambda\delta\phi + \mathcal{O}(\delta\phi^2) \qquad (3.49)$$

When the coefficient λ is such that $\lambda > 0$, both solutions converge for $t \to \infty$; meanwhile, for $\lambda < 0$ they diverge, implying that $\phi_1(t)$ will be stable or unstable against small perturbations. This point will be discussed again in the last chapter.

We could conclude that on one hand, the fluctuations around the stable solutions of the macroscopic equation will remain small and could be *controlled* by means of the Ω–expansion, and on the other hand, for unstable solutions, the fluctuations increase and the Ω–expansion becomes *spurious* after a transient period.

For a more detailed and general discussion on the Ω–expansion and its validity, we refer the reader to [van Kampen (1982)].

PART 2
Thermodynamics and kinetics near equilibrium

Those are my principles, and if you don't like them ...
well, I have others.
Groucho Marx

Provando e riprovando
Galileo Galilei

Chapter 4

Distributions, BBGKY–hierarchy, balance equations, and the density operator

Time, that mathematical abstraction, that twister of fools' minds,
fools who flaunt the badge of learning. Baby, time is real.
Roberto Arlt

4.1 Introduction

In the previous chapters we studied the temporal evolution of probability distributions for Markov processes, where the dynamics is determined by a transition probability (that very often is phenomenologically determined). The resulting equations, master or Fokker–Planck equations, might be considered to be justified semi-phenomenologically, and to show the kind of behavior needed to describe the irreversible decay toward equilibrium states. We can ask ourselves if it is possible to derive similar (irreversible) *kinetic* equations but starting from a (reversible) rigorous microscopic theoretical point of view (i.e. Newton's or Schrödinger equations).

In order to answer such a question we will concentrate on a classical situation with a large number of degrees of freedom, for instance composed of a large number, say N, of interacting particles in a box, or interacting objects in a lattice. The behavior of such objects will be described by Newton's laws or by Hamiltonian dynamics. In a tridimensional system there are $3N$ degrees of freedom (assuming there are no internal degrees like spin) and classically the state of the system is determined when we specify $6N$ independent coordinates ($3N$ corresponding to position and $3N$ to the -conjugate- momentum variables). That is specifying *a point in the $6N$-dimensional phase space.* Such a point will evolve according to Hamiltonian dynamics, corresponding to the exact knowledge of the state of the system. But it is clear that in general we only know a certain probability of being at a given point (or rather, in a small region). Therefore, we assign a probability distribution on the phase space.

The *N-body probability density* for a classical (or quantal) system contains far more information than what we really need. In practical situations the main use of such a density is to obtain *expectation values*, or *correlation functions* for diverse observables, as these are the quantities measured experimentally. In general the observables that are usually treated correspond to *one-* and *two-body operators*, and then it is necessary to use some *reduced distributions* instead of the complete one. We will see how it is possible to obtain evolution equations for these reduced distributions through the famous *BBGKY hierarchy*. We will also present microscopic balance equations, some methods to calculate a (few) transport coefficients, and a short comment on the extension of these results to the quantum-mechanical case.

4.2   Probability density as a fluid

We will consider a tridimensional closed classical system composed of N particles. The state of such a system will be completely determined by specifying a set of $2N$ independent variables $(\mathbf{p}^N, \mathbf{q}^N)$ (where $\mathbf{p}^N = (\mathbf{p}_1, \mathbf{p}_2, \ldots, \mathbf{p}_N)$ and $\mathbf{q}^N = (\mathbf{q}_1, \mathbf{q}_2, \ldots, \mathbf{q}_N)$), $\mathbf{p}_j$ and $\mathbf{q}_j$ being the vector momentum and coordinate of the j-th particle. If the *state vector* $\mathbf{X}^N = \mathbf{X}^N(\mathbf{p}^N, \mathbf{q}^N)$ is known at a given time it will be determined at all later times through Newton or Hamilton equations of motion. Let us call $\mathcal{H}(\mathbf{X}^N, t)$ the system Hamiltonian, then

$$\dot{\mathbf{p}}_k \equiv \frac{d}{dt}\mathbf{p}_k = -\frac{\partial \mathcal{H}}{\partial \mathbf{q}_k} \tag{4.1a}$$

$$\dot{\mathbf{q}}_k \equiv \frac{d}{dt}\mathbf{q}_k = -\frac{\partial \mathcal{H}}{\partial \mathbf{p}_k} \tag{4.1b}$$

When $\mathcal{H}$ does not depend on time explicitly, it is a constant of motion

$$\mathcal{H}(\mathbf{X}^N) = E, \tag{4.2}$$

where E is the total energy of the system. In such a case the system is called *conservative*.

We associate to the system a $6N$ dimensional phase space denoted by Γ. A point in this Γ space is specified by $\mathbf{X}^N(\mathbf{p}^N, \mathbf{q}^N)$. When the system evolves in time, the point $\mathbf{X}^N$ describes a trajectory in Γ space as schematically indicated in Fig. 4.1.

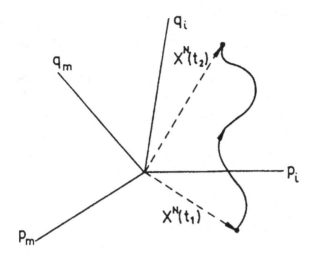

Fig. 4.1

In general, for real physical systems, due to the uncertainty in the knowledge of the initial conditions, the state of the system is not exactly specified. We will thus assume that $\mathbf{X}^N$ is a kind of stochastic variable, and introduce a probability density $\rho(\mathbf{X}^N, t)$ within that Γ space: $\rho(\mathbf{X}^N, t)d\mathbf{X}^N$ being the probability that the point $\mathbf{X}^N$ is inside the phase space volume $\mathbf{X}^N \Leftrightarrow \mathbf{X}^N + d\mathbf{X}^N$ at time t ($d\mathbf{X}^N = d\mathbf{q}_1 d\mathbf{p}_1 d\mathbf{q}_2 d\mathbf{p}_2 \dots d\mathbf{q}_N d\mathbf{p}_N$). We will then introduce a representation of the phase space of the system as a (continuous) fluid composed of the *state points*. We have the normalization condition

$$\int_\Gamma \rho(\mathbf{X}^N, t)d\mathbf{X}^N = 1. \tag{4.3}$$

The probability of finding the system in a given, finite, region $\mathbf{R}$ of Γ space at time t will then be

$$\mathcal{P}(\mathbf{R}, t) = \int_\mathbf{R} \rho(\mathbf{X}^N, t)d\mathbf{X}^N. \tag{4.4}$$

In such a scheme, the probability behaves like a fluid in Γ space. We will hence use fluid mechanical arguments in order to obtain the equation of motion for the probability density. Let us call $\dot{\mathbf{X}}^N = (\dot{\mathbf{q}}^N, \dot{\mathbf{p}}^N)$ the velocity of the point $\mathbf{X}^N$ in Γ space. We consider a small volume V_0, and analyze how the probability that the system is inside this volume varies with time

$$\frac{d}{dt}\mathcal{P}(V_0, t) = \frac{\partial}{\partial t}\int_{V_0}\rho(\mathbf{X}^N, t)d\mathbf{X}^N = -\oint_{S_0}\rho(\mathbf{X}^N, t)\dot{\mathbf{X}}^N \cdot d\mathbf{S}^N \qquad (4.5)$$

where S_0 indicates the surface of the volume V_0. Here $\dot{\mathbf{X}}^N \cdot d\mathbf{S}^N$ indicates the scalar product of the $6N$ dimensional velocity $\dot{\mathbf{X}}^N$, and the normal to the surface S_0, $d\mathbf{S}^N$. We can use Gauss's theorem in order to transform the surface integral into a volume integral

$$\oint_{S_0}\rho(\mathbf{X}^N, t)\dot{\mathbf{X}}^N \cdot d\mathbf{S}^N = -\int_{V_0}\nabla_{\mathbf{X}^N} \cdot [\rho(\mathbf{X}^N, t)\dot{\mathbf{X}}^N]d\mathbf{X}^N \qquad (4.6)$$

$$\nabla_{\mathbf{X}^N} = (\partial_{\mathbf{q}_1}, \partial_{\mathbf{q}_2}, \dots, \partial_{\mathbf{q}_n}, \partial_{\mathbf{p}_1}, \partial_{\mathbf{p}_2}, \dots, \partial_{\mathbf{p}_n}).$$

Bringing the time derivative in Eq. (4.5) inside the integral and rearranging the expression we get

$$\int_{V_0}\left\{\frac{\partial}{\partial t}\rho(\mathbf{X}^N, t) + \nabla_{\mathbf{X}^N} \cdot [\rho(\mathbf{X}^N, t)\dot{\mathbf{X}}^N]\right\}d\mathbf{X}^N = 0 \qquad (4.7a)$$

indicating that the integral is zero. This corresponds to the balance equation for the probability density, i.e.

$$\frac{\partial}{\partial t}\rho(\mathbf{X}^N, t) + \nabla_{\mathbf{X}^N} \cdot [\rho(\mathbf{X}^N, t)\dot{\mathbf{X}}^N] = 0 \qquad (4.7b)$$

From here, and using Hamilton's equations of motion, we will show that the probability behaves like an incompressible fluid. A volume element in Γ space changes with time according to

$$d\mathbf{X}^N(t) = \mathcal{J}^N(t, t_0)d\mathbf{X}^N(t_0) \qquad (4.8)$$

where $\mathcal{J}^N$ is the Jacobian associated to the transformation from $\mathbf{X}^N(t_0)$ to $\mathbf{X}^N(t)$, given by

$$\mathcal{J}^N(t, t_0) = \det\begin{bmatrix} \frac{\partial \mathbf{p}^N(t)}{\partial \mathbf{p}^N(t_0)} & \frac{\partial \mathbf{p}^N(t)}{\partial \mathbf{q}^N(t_0)} \\ \frac{\partial \mathbf{q}^N(t)}{\partial \mathbf{p}^N(t_0)} & \frac{\partial \mathbf{q}^N(t)}{\partial \mathbf{q}^N(t_0)} \end{bmatrix} \qquad (4.9)$$

The Jacobian satisfies the relation

$$\mathcal{J}^N(t, t_0) = \mathcal{J}^N(t, t_1)\mathcal{J}^N(t_1, t_0) \qquad (4.10)$$

$(t_0 < t_1 < t)$. We consider now a very short time interval: $\Delta t = t - t_0$, and write

$$\mathbf{p}^N(t) = \mathbf{p}^N(t_0) + \Delta t\dot{\mathbf{p}}^N(t_0) + \mathcal{O}(\Delta t^2)$$

$$\mathbf{q}^N(t) = \mathbf{q}^N(t_0) + \Delta t\dot{\mathbf{q}}^N(t_0) + \mathcal{O}(\Delta t^2) \qquad (4.11)$$

Replacing this into Eq. (4.9) we get

$$\mathcal{J}^N(t, t_0) = \det \begin{bmatrix} 1 + \Delta t \frac{\partial \dot{\mathbf{p}}^N(t)}{\partial \mathbf{p}^N(t_0)} & \Delta t \frac{\partial \dot{\mathbf{p}}^N(t)}{\partial \mathbf{q}^N(t_0)} \\ \Delta t \frac{\partial \dot{\mathbf{q}}^N(t)}{\partial \mathbf{p}^N(t_0)} & 1 + \Delta t \frac{\partial \dot{\mathbf{q}}^N(t)}{\partial \mathbf{q}^N(t_0)} \end{bmatrix} \tag{4.12}$$

$$= 1 + \Delta t \left(\frac{\partial \dot{\mathbf{q}}^N(t)}{\partial \mathbf{q}^N(t_0)} + \frac{\partial \dot{\mathbf{p}}^N(t)}{\partial \mathbf{p}^N(t_0)} \right) + \mathcal{O}(\Delta t^2).$$

But from Hamilton's equations, Eqs. (4.1), we have

$$\frac{\partial \dot{\mathbf{q}}^N(t)}{\partial \mathbf{q}^N(t_0)} + \frac{\partial \dot{\mathbf{p}}^N(t)}{\partial \mathbf{p}^N(t_0)} = 0, \tag{4.13}$$

leading to

$$\mathcal{J}^N(t, t_0) = 1 + \mathcal{O}(\Delta t^2). \tag{4.14}$$

Hence, from Eq. (4.10) we have

$$\mathcal{J}^N(t, 0) = \mathcal{J}^N(t, t_0) \mathcal{J}^N(t_0, 0) = \mathcal{J}^N(t_0, 0)[1 + \mathcal{O}(\Delta t^2)], \tag{4.15}$$

that yields

$$\frac{d}{dt} \mathcal{J}^N(t_0, 0) = \lim_{\Delta t \to 0} \frac{\mathcal{J}^N(t_0 + \Delta t, 0) - \mathcal{J}^N(t_0, 0)}{\Delta t} = 0 \tag{4.16}$$

Therefore, the Jacobian $\mathcal{J}^N$ does not change with time and then

$$\mathcal{J}^N(t, 0) = \mathcal{J}^N(0, 0) = 1. \tag{4.17}$$

This fact leads to

$$d\mathbf{X}^N(t) = d\mathbf{X}^N(t_0) \tag{4.18}$$

that, along with

$$\nabla_{\mathbf{X}^N} \cdot \dot{\mathbf{X}}^N = 0 \tag{4.19}$$

implies that the probability behaves as an incompressible fluid [see Eqs. (4.6), (4.13)]. According to these results we can rewrite Eq. (4.7b) as

$$\frac{\partial}{\partial t} \rho(\mathbf{X}^N, t) + \dot{\mathbf{X}}^N \cdot [\nabla_{\mathbf{X}^N} \rho(\mathbf{X}^N, t)] + \rho(\mathbf{X}^N, t)[\nabla_{\mathbf{X}^N} \cdot \dot{\mathbf{X}}^N] = 0, \tag{4.20a}$$

that leads to the equation of motion for $\rho(\mathbf{X}^N, t)$

$$\frac{\partial}{\partial t} \rho(\mathbf{X}^N, t) = -\dot{\mathbf{X}}^N \cdot [\nabla_{\mathbf{X}^N} \rho(\mathbf{X}^N, t)]. \tag{4.20b}$$

The last result can be written in terms of the total time derivative, also called the *convective* derivative,

$$\frac{d}{dt} = \frac{\partial}{\partial t} + \dot{\mathbf{X}}^N \cdot \nabla_{\mathbf{X}^N}$$

yielding

$$\frac{d}{dt}\rho(\mathbf{X}^N, t) = 0. \tag{4.20c}$$

The last equation indicates that the probability density remains constant in the neighborhood of a point that moves together with the *probability fluid*.

Now, using Hamilton's equations, we can write Eq. (4.20b) in the form known as *Liouville's equation*

$$\frac{\partial}{\partial t}\rho(\mathbf{X}^N, t) = -\hat{\mathcal{H}}\rho(\mathbf{X}^N, t) = \{\rho(\mathbf{X}^N, t), \hat{\mathcal{H}}\} \tag{4.21a}$$

where the operator $\hat{\mathcal{H}}$ indicates the *Poisson bracket*

$$\hat{\mathcal{H}} = \sum_i \left(\frac{\partial \mathcal{H}}{\partial \mathbf{q}_i} \frac{\partial}{\partial \mathbf{p}_i} - \frac{\partial \mathcal{H}}{\partial \mathbf{p}_i} \frac{\partial}{\partial \mathbf{q}_i} \right) \tag{4.21b}$$

In shorthand notation, Eq. (4.21a) reads

$$i\frac{\partial}{\partial t}\rho(\mathbf{X}^N, t) = \hat{\mathcal{L}}\rho(\mathbf{X}^N, t) \tag{4.22}$$

with $\hat{\mathcal{L}} = -i\hat{\mathcal{H}}$. This is the so called *Liouville equation* ($\hat{\mathcal{L}}$ being an Hermitian operator). It is clear that if we know the probability density at $t = 0$, we can also know the density at $t = t' > 0$

$$\rho(\mathbf{q}^N, \mathbf{p}^N, t) = \exp\{-i\hat{\mathcal{L}}t\}\rho(\mathbf{q}^N, \mathbf{p}^N, 0) \tag{4.23}$$

If the density remains constant in time we have

$$\hat{\mathcal{L}}\rho(\mathbf{X}^N, t) = 0 \tag{4.24}$$

that corresponds to a stationary solution of the Liouville equation. Since $\hat{\mathcal{L}}$ is Hermitian, all its eigenvalues must be real. Hence Eq. (4.23) indicates that the temporal behavior will be oscillatory, and will not decay to a unique state. Moreover, if we make the change $t \to -t$, as the equation is invariant against time inversion, nothing changes, i.e. there is no decay to an equilibrium state. This clearly shows a behavior different from the one observed for the Master or Fokker–Planck equations. In the next section we shall see a way to tackle this problem (which occupies a central place in statistical physics) through a systematic method of deriving equations of motion for few body densities.

4.3   BBGKY hierarchy

As indicated in the last paragraph, the procedure we are going to study in order to overcome the nonexistence of decaying solutions for the Liouville equation, is a systematic one, called the *BBGKY hierarchy*, after the name of its authors (Bogoliubov, Born, Green, Kirkwood and Yvon).

As mentioned earlier, the N-body probability density we have discussed so far, contains more information than what is really needed: most of the quantities measured experimentally correspond to mean values of few body operators (or phase space functions). This implies that we only need to resort to one and two body reduced density probabilities. Let us investigate how we could write such reduced densities. According to what has been discussed in chapter 1, the one body density may be written as ($\rho(\mathbf{X}^N, t) = \rho(\mathbf{X}_1, \mathbf{X}_2, \ldots, \mathbf{X}_N, t)$):

$$\rho_1(\mathbf{X}_1, t) = \int \ldots \int d\mathbf{X}_2 d\mathbf{X}_3 \ldots d\mathbf{X}_N \rho(\mathbf{X}_1, \mathbf{X}_2, \ldots, \mathbf{X}_N, t) \qquad (4.25a)$$

whereas the s ($< N$) body density is

$$\rho_s(\mathbf{X}_1, \ldots, \mathbf{X}_s, t) = \int \ldots \int d\mathbf{X}_{s+1} \ldots d\mathbf{X}_N \rho(\mathbf{X}_1, \ldots, \mathbf{X}_N, t) \qquad (4.25b)$$

Also, as is well known from standard classical and quantum mechanics, the one body (phase space) operators are written as

$$\hat{O}^{(1)}(\mathbf{p}^N, \mathbf{q}^N) = \sum_{i=1}^{N} \hat{O}(\mathbf{q}_i) \qquad (4.26a)$$

while for two-body operators

$$\hat{O}^{(2)}(\mathbf{p}^N, \mathbf{q}^N) = \sum_{i<j}^{N(N-1)/2} \hat{O}(\mathbf{q}_i, \mathbf{q}_j) \qquad (4.26b)$$

Hence, we can write the expectation value of these operators as:

$$\langle \hat{O}^{(1)} \rangle \equiv \int \ldots \int d\mathbf{X}_1 d\mathbf{X}_2 \ldots d\mathbf{X}_N \sum_i \hat{O}^{(1)}(\mathbf{X}_i) \rho(\mathbf{X}_1, \mathbf{X}_2, \ldots, \mathbf{X}_N, t)$$

$$= N \int d\mathbf{X}_1 \hat{O}^{(1)}(\mathbf{X}_1) \rho_1(\mathbf{X}_1, t) = \mathrm{Tr}\left(\hat{O}^{(1)} \rho \right) \qquad (4.27a)$$

$$\langle \hat{O}^{(2)} \rangle \equiv \int \ldots \int d\mathbf{X}_1 d\mathbf{X}_2 \ldots d\mathbf{X}_N \sum_{i<j} \hat{O}^{(2)}(\mathbf{X}_i, \mathbf{X}_j) \rho(\mathbf{X}_1, \mathbf{X}_2, \ldots, \mathbf{X}_N, t)$$

$$= \frac{1}{2} N(N-1) \int \int d\mathbf{X}_1 d\mathbf{X}_2 \hat{O}^{(2)}(\mathbf{X}_1, \mathbf{X}_2) \rho(\mathbf{X}_1, \mathbf{X}_2, \ldots, \mathbf{X}_N, t)$$

$$= \mathrm{tr}\left(\hat{O}^{(2)} \rho\right) \tag{4.27b}$$

Here, we have used the invariance of integration under the exchange of $\mathbf{X}_i$ and $\mathbf{X}_j$. Clearly, we can apply the same procedure for s-body operators.

We have thus found the form of the equation of motion for ρ. Let us see now how to obtain from this the equation of motion for the reduced densities. We define

$$q_s(\mathbf{X}_1, \mathbf{X}_2, \ldots, \mathbf{X}_N, t) \equiv V^s \int \ldots \int d\mathbf{X}_{s+1} \ldots d\mathbf{X}_N \rho(\mathbf{X}_1, \mathbf{X}_2, \ldots, \mathbf{X}_N, t) \tag{4.28}$$

with the relation

$$q_N(\mathbf{X}_1, \mathbf{X}_2, \ldots, \mathbf{X}_N, t) \equiv V^N \rho(\mathbf{X}_1, \mathbf{X}_2, \ldots, \mathbf{X}_N, t).$$

Through

$$V^{-N} q_N(\mathbf{X}_1, \mathbf{X}_2, \ldots, \mathbf{X}_N, t) d\mathbf{X}_1 d\mathbf{X}_2 \ldots d\mathbf{X}_N$$

we have that q_N, corresponds to the probability of finding the system within the volume $d\mathbf{X}_1 d\mathbf{X}_2 \ldots d\mathbf{X}_N$, of the Γ-phase space. Hence,

$$\int \ldots \int d\mathbf{X}_1 d\mathbf{X}_2 \ldots d\mathbf{X}_N q_N(\mathbf{X}_1, \mathbf{X}_2, \ldots, \mathbf{X}_N, t) = V^N \tag{4.29}$$

gives the normalization for these functions. We will assume that the system is driven by a Hamiltonian of the form

$$\mathcal{H} = \sum_i \frac{\mathbf{p}_i^2}{2m} + \sum_<^N V_2^2(|\mathbf{q}_i - \mathbf{q}_j|) \tag{4.30}$$

where V_2 is a two-body operator, corresponding to a spherically symmetric interaction potential between the i-th and j-th particles. We could also add to the previous form an external (one-body) potential $V_1(\mathbf{q}_i)$, but, as this does not alter the results we will not include it here. The equation of motion for ρ is given by Eq. (4.21a), that is explicitly written as

$$\frac{\partial \rho}{\partial t} = -\sum_i \frac{\mathbf{p}_i}{m} \cdot \frac{\partial \rho}{\partial \mathbf{q}_i} + \sum_{<j} \mathbb{U}_{i,j} \rho \tag{4.31a}$$

where, calling $\mathcal{V}_{ij} = V_2(|\mathbf{q}_i - \mathbf{q}_j|)$, we have defined

$$\mathbb{U}_{i,j} = \frac{\partial \mathcal{V}_{ij}}{\partial \mathbf{q}_i} \frac{\partial}{\partial \mathbf{p}_i} + \frac{\partial \mathcal{V}_{ij}}{\partial \mathbf{q}_j} \frac{\partial}{\partial \mathbf{p}_j} \tag{4.31b}$$

Now, integrating Eq. (4.31a) over $\mathbf{X}_{s+1} \ldots \mathbf{X}_N$ and multiplying by V^s [in order to normalize the result, see Eq. (4.28)], we obtain

$$\frac{\partial q_s}{\partial t} + \hat{\mathcal{H}}^{(s)} q_s = V^s \int \ldots \int d\mathbf{X}_{s+1} \ldots d\mathbf{X}_N \tag{4.32}$$

$$\left(-\sum_{i=s+1}^{N} \frac{\mathbf{p}_i}{m} \frac{\partial}{\partial \mathbf{q}_i} + \sum_{i<s; s+1<j<N} \mathbb{U}_{i,j} + \sum_{s+1<k<l} \mathbb{U}_{k,l} \right) \rho(\mathbf{X}_1, \ldots, \mathbf{X}_N, t)$$

where $\hat{\mathcal{H}}^{(s)}$ is an operator like the one appearing in Eq. (4.21a), with a Hamiltonian like that of Eq. (4.30), except that the index for the particle number only runs to "s" instead of "N". From the normalization of q or ρ, we have

$$\int \ldots \int d\mathbf{X}_1 \ldots d\mathbf{X}_s q_s(\mathbf{X}_1, \ldots, \mathbf{X}_s, t) =$$

$$= V^s \int \ldots \int d\mathbf{X}_1 \ldots d\mathbf{X}_N \rho(\mathbf{X}_1, \ldots, \mathbf{X}_N, t) = V^s \tag{4.33}$$

As q_s (and ρ) decays to zero at the system boundary (at infinity), we have

$$\int \ldots \int d\mathbf{X}_{s+1} \ldots d\mathbf{X}_N \sum_{i}^{N} \frac{\mathbf{p}_i}{m} \frac{\partial}{\partial \mathbf{q}_i} \rho =$$

$$= \int d\mathbf{q}_{s+1} d\mathbf{p}_{s+1} \ldots d\mathbf{q}_N d\mathbf{p}_N \sum_{i}^{N} \frac{\mathbf{p}_i}{m} \frac{\partial}{\partial \mathbf{q}_i} \rho =$$

$$= \int d\mathbf{p}_{s+1} \ldots d\mathbf{p}_N \sum_{i}^{N} \frac{\mathbf{p}_i}{m} \rho(\mathbf{X}_1, \ldots, \mathbf{X}_s, t) \bigg|_{\mathbf{q}_i = -\infty}^{\mathbf{q}_i = +\infty} = 0 \tag{4.34a}$$

In a similar form, we can prove that

$$\int \ldots \int d\mathbf{X}_{s+1} \ldots d\mathbf{X}_N \sum_{s+1<k<l} \mathbb{U}_{k,l} \rho(\mathbf{X}_1, \ldots, \mathbf{X}_s, t) = 0 \tag{4.34b}$$

For the remaining term on the right of Eq. (4.32), we proceed as follows

$$
V^s \int \ldots \int d\mathbf{X}_{s+1} \ldots d\mathbf{X}_N \sum_{i<s; s+1<j<N} \mathbb{U}_{i,j} \rho(\mathbf{X}_1, \ldots, \mathbf{X}_s, t) =
$$

$$
= V^s (N-s) \sum_{i=1}^{s} \int d\mathbf{X}_{s+1} \mathbb{U}_{i,s+1} \int d\mathbf{X}_{s+2} \ldots d\mathbf{X}_N \rho(\mathbf{X}_1, \ldots, \mathbf{X}_s, t) =
$$

$$
= \frac{(N-s)}{V} \sum_{i=1}^{s} \int d\mathbf{X}_{s+1} \mathbb{U}_{i,s+1} q_{s+1}(\mathbf{X}_1, \ldots, \mathbf{X}_s, \mathbf{X}_{s+1}, t) \tag{4.35}
$$

Then, Eq. (4.32) reduces to

$$
\frac{\partial q_s}{\partial t} + \hat{\mathcal{H}}^{(s)} q_s = \frac{(N-s)}{V} \sum_{i=1}^{s} \int d\mathbf{X}_{s+1} \mathbb{U}_{i,s+1} q_{s+1}(\mathbf{X}_1, \ldots, \mathbf{X}_s, \mathbf{X}_{s+1}, t) \tag{4.36}
$$

We can now take the thermodynamic limit: $N \to \infty$, $V \to \infty$, in such a way that $1 = N/V = $ constant and finite. Hence, we have obtained a hierarchy of evolution equations for the reduced probability densities $_s$, known as the *BBGKY hierarchy*. The first few equations of the hierarchy read

$$
\frac{\partial}{\partial t} q_1(\mathbf{X}_1, t) + \frac{\mathbf{p}_1}{m} \frac{\partial q_1}{\partial \mathbf{q}_1} = \frac{1}{v} \int d\mathbf{X}_2 \mathbb{U}_{1,2} q_2(\mathbf{X}_1, \mathbf{X}_2, t) \tag{4.37a}
$$

$$
\frac{\partial}{\partial t} q_2(\mathbf{X}_1, \mathbf{X}_2, t) + \frac{\mathbf{p}_1}{m} \frac{\partial q_2}{\partial \mathbf{q}_1} + \frac{\mathbf{p}_2}{m} \frac{\partial q_2}{\partial \mathbf{q}_2} - \mathbb{U}_{1,2} q_2 =
$$

$$
= \frac{1}{v} \int d\mathbf{X}_3 \left(\mathbb{U}_{1,3} + \mathbb{U}_{2,3} \right) q_3(\mathbf{X}_1, \mathbf{X}_2, \mathbf{X}_3, t) \tag{4.37b}
$$

We have then obtained that within this hierarchy, the equation for q_s involves contributions that depend on q_{s+1}. This fact prevents, in general, the possibility of finding the solution of the complete hierarchy. An exception are those cases in which we find some way to cut the hierarchy. For instance, if we could express q_2 in terms of q_1, replacing it in equation Eq. (4.37a), we would obtain a closed kinetic equation for q_1.

4.4   Simple kinetic equations: Vlasov and Boltzmann

As we show in Sec. 4.9, the calculation of transport coefficients presupposes a knowledge of nonequilibrium single-particle and two-particle distribution functions. It is clear that such a problem cannot be solved in general but only for certain simplified models. One such case is the so called Boltzmann gas, corresponding to a dilute gas of neutral particles that rarely interact with each other. In order to derive Boltzmann's equation for the single-particle density, we start from the BBGKY hierarchy (instead of following the original intuitive approach). The first equation of the hierarchy, Eq. (4.37a), reads

$$\frac{\partial q_1(\mathbf{X}_1, t)}{\partial t} + \frac{\mathbf{p}_1}{m} \frac{\partial q_1}{\partial \mathbf{q}_1} = \left(\frac{\partial q_1}{\partial t}\right)_{\text{coll}} \tag{4.38a}$$

where the term

$$\left(\frac{\partial q_1}{\partial t}\right)_{\text{coll}} = \frac{1}{v} \int d\mathbf{X}_2 \mathbb{U}_{1,2} q_2(\mathbf{X}_1, \mathbf{X}_2, t) \tag{4.38b}$$

is usually referred to as the *collision integral*. In order to decouple the first equation from the rest of the hierarchy, our task is to find suitable approximations of this quantity in terms of single-particle distributions alone.

But before discussing the derivation of the Boltzmann equation, we will consider a simpler approximation, assuming that $q_2(\mathbf{X}_1, \mathbf{X}_2, t)$, the probability of finding two particles at $\mathbf{X}_1$ and $\mathbf{X}_2$ simultaneously, is simply the product of the probability of finding a particle at $\mathbf{X}_1$ times the probability of finding the other particle at $\mathbf{X}_2$. We then write

$$q_2(\mathbf{X}_1, \mathbf{X}_2, t) = q_1(\mathbf{X}_1, t) q_1(\mathbf{X}_2, t). \tag{4.38c}$$

The result for the collision integral is then

$$\left(\frac{\partial q_1}{\partial t}\right)_{\text{coll}} = \frac{1}{v} \int d\mathbf{X}_2 \mathbb{U}_{1,2} q_1(\mathbf{X}_1, t) q_1(\mathbf{X}_2, t)$$
$$= \frac{\partial}{\partial \mathbf{q}_1} \mathbb{Q}(\mathbf{q}_1, t) \frac{\partial}{\partial \mathbf{p}_1} q_1(\mathbf{X}_1, t) \tag{4.38d}$$

where

$$\mathbb{Q}(\mathbf{q}_1, t) = \frac{1}{v} \int d\mathbf{X}_2 V_2(|\mathbf{q}_i - \mathbf{q}_j|) q_1(\mathbf{X}_1, t) \tag{4.38e}$$

is known as the *mean-field potential*, and plays the role of an external potential in the *Vlasov kinetic equation*

$$\left(\frac{\partial}{\partial t} + \frac{\mathbf{p}_1}{m} \frac{\partial q_1}{\partial \mathbf{q}_1} - \frac{\partial}{\partial \mathbf{q}_1} \mathbb{Q}(\mathbf{q}_1, t) \frac{\partial}{\partial \mathbf{p}_1}\right) q_1(\mathbf{X}_1, t) = 0 \tag{4.38f}$$

Here, the meaning of the approximation in Eq. (4.38c) is clear. It consists in assuming that each particle moves independently, or uncorrelated, from the others, and that the effect of their mutual interactions is such that each particle experiences an average potential field produced by the others. This last point is reflected by the dependence of $\mathbb{Q}$ on q_1 itself, making Eq. (4.38f) a nonlinear equation, to be solved in a self-consistent way. A physical system for which the Vlasov equation offers a useful description is a dilute plasma. However, due to the fact that it is still a time reversible equation (i.e.: the change $t \to -t$, $\mathbf{p} \to -\mathbf{p}$ renders an identical equation), it could not describe an approach to equilibrium, and then could be valid only in the initial stage of the evolution, and during a period short compared to typical macroscopic evolution times.

We turn now to discussing how to derive the Boltzmann equation. We will not go into all the details of this derivation, but only sketch it. To start with, as indicated before, we restrict ourselves to studying a dilute gas of neutral particles interacting via short range forces. This assumption suggest several simplifications. The first, and most relevant, is that we can neglect the occurrence of triple collisions, as the probability of such events is extremely small. The implication of this approximation is that we can neglect the r.h.s. in Eq. (4.37b) and reduce it to an equation for q_2 alone:

$$\left(\frac{\partial}{\partial t} + \frac{\mathbf{p}_1}{m} \frac{\partial}{\partial \mathbf{q}_1} + \frac{\mathbf{p}_2}{m} \frac{\partial}{\partial \mathbf{q}_2} - \mathbb{U}_{1,2} \right) q_2(\mathbf{X}_1, \mathbf{X}_2, t) = 0 \qquad (4.39a)$$

truncating this way the BBGKY hierarchy. A further simplification is realizing that in Eq. (4.38b) we will only need to know $q_2(\mathbf{X}_1, \mathbf{X}_2, t)$ for those values of $\mathbf{r}_2$ that are within the range of the inter-particle interaction of the particle at $\mathbf{r}_1$. In the case of having a true external potential acting on the particles, we can assume that within that range it is constant (having no effect on the collision processes), and that its effect on the whole evolution is achieved through its inclusion in the l.h.s. of Eq. (4.38a) for q_1. Furthermore, as the partial time derivative in Eq. (4.39a) accounts for the explicit time dependence of q_2 due to the overall evolution of the gas over a period of the order of the time between collisions, and in the collision integral we follow q_2 over an even shorter period (of the order of the two-body collision time), we can argue that this partial time derivative can be dropped, simplifying still more this equation. After doing some integrations by parts, and some rearrangements, we arrive at

$$\left(\frac{\partial q_1}{\partial t} \right)_{\text{coll}} = \frac{1}{v} \int d\mathbf{R} d\mathbf{p}_2 \left(\frac{\mathbf{p}_1}{m} - \frac{\mathbf{p}_2}{m} \right) q_2(\mathbf{r}_1, \mathbf{p}_1, \mathbf{R}, \mathbf{p}_2, t) \qquad (4.39b)$$

($\mathbf{R} = \mathbf{r}_1 - \mathbf{r}_2$). Here, we must remember that the two particles included in q_2 only interact via short range forces (of typical radius r_0). Hence, we can find a separation distance $R > r_0$ such that the interaction for larger separations becomes negligible. Outside this range, both behave as free particles and we can write for $R = |\mathbf{p}_1 - \mathbf{p}_2| > r_0$

$$q_2(\mathbf{r}_1, \mathbf{p}_1, \mathbf{r}_2, \mathbf{p}_2, t) = q_1(\mathbf{r}_1, \mathbf{p}_1, t)q_1(\mathbf{r}_2, \mathbf{p}_2, t) + \Lambda(\mathbf{r}_1, \mathbf{p}_1, \mathbf{r}_2, \mathbf{p}_2, t) \quad (4.39c)$$

This factorization corresponds to the famous Boltzmann *Stosszahlansatz*, or assumption of *molecular chaos* (in fact when assuming $\Lambda = 0$). It is worth remarking that this assumption is only valid in the pre-collisional configuration, as the interaction processes introduce dynamical correlations among the particles in the post-collisional stage. The difference in the behavior between both (pre– and post-collisional) stages is the origin of irreversibility.

Coming back to the collision integral Eq. (4.39b), using Gauss's theorem we convert the volume integral over $d\mathbf{R}$ into a surface integral over a sphere of radius $R_0 > r_0$. Considering that q_2 essentially factorizes according to Eq. (4.39c), and keeping only the factorized part (that is neglecting the contribution from the correlated part: $\Lambda(\mathbf{r}_1, \mathbf{p}_1, \mathbf{r}_2, \mathbf{p}_2, t)$), we can simplify further the form of the kinetic equation.

At this point it is necessary to resort to some notions of scattering theory involving, among other properties, momentum and energy conservation. After some algebraic operations, and calling $\sigma(\vartheta, |\mathbf{p}_1 - \mathbf{p}_2|)$ the scattering cross section which, for a spherically symmetric interaction, is a function of the relative initial momentum $|\mathbf{p}_1 - \mathbf{p}_2|$ and the scattering angle ϑ (this scattering cross section being the only trace that remains of the scattering process), the final form of the Boltzmann equation reads

$$\left(\frac{\partial}{\partial t} + \frac{\mathbf{p}_1^f}{m} \frac{\partial}{\partial \mathbf{q}_1} \right) q_1(\mathbf{r}_1, \mathbf{p}_1^f, t) = \frac{1}{v} \int d\mathbf{p}_2 d\mathbf{s} |\mathbf{p}_1 - \mathbf{p}_2| \sigma(\vartheta, |\mathbf{p}_1 - \mathbf{p}_2|)$$

$$\left[q_1(\mathbf{r}_1, \mathbf{p}_1^f, t)q_1(\mathbf{r}_1, \mathbf{p}_1^f, t) - q_1(\mathbf{r}_1, \mathbf{p}_1^f, t)q_1(\mathbf{r}_1, \mathbf{p}_1^f, t) \right] \quad (4.39d)$$

where $\mathbf{s}$ is a unitary vector indicating the direction between both final momentum vectors.

Although the Boltzmann equation is not a Master Equation (since its origin and derivation are quite different), it does have points of coincidence with it in that its physical interpretation is also that of a *balance equation*. Here $q_1(\mathbf{r}, \mathbf{p}, t)d\mathbf{r}d\mathbf{p}$ is the probability of finding a gas particle in a volume $d\mathbf{r}d\mathbf{p}$, around the point $(\mathbf{r}, \mathbf{p})$ in the single particle phase space, and the different contributions that arise in the equation correspond to the different *gain* and *loss* terms for this probability (drift and collisional contributions).

4.5   Microscopic balance equations

When we consider systems of particles with short range interactions and long wave length inhomogeneities, it is possible to derive microscopic balance equations for the particle, momentum and energy densities. The interest in these quantities rests on the fact that, when the dynamics is driven by a Hamiltonian like the one shown in Eq. (4.30), all these quantities are conserved during the time evolution of the system. These equations have the form of *hydrodynamic equations*, such as the equations governing the dynamic of a fluid.

As we saw earlier, the expectation value of an n-body observable $O^{(N)}(\mathbf{X}_1, \dots \mathbf{X}_N, t)$ at time t, is given by

$$\langle O^{(N)}(t) \rangle = \text{Tr}[O^{(N)}(t)\rho(t)] \equiv \qquad (4.40a)$$

$$\equiv \int \dots \int d\mathbf{X}_1 \dots d\mathbf{X}_N O^{(N)}(\mathbf{X}_1, \dots, \mathbf{X}_N, t)\rho(\mathbf{X}_1, \dots, \mathbf{X}_N, t)$$

$$= \int \dots \int d\mathbf{X}_1 \dots d\mathbf{X}_N O^{(N)}(\mathbf{X}_1, \dots, \mathbf{X}_N)e^{-\hat{\mathcal{H}}t}\rho(\mathbf{X}_1, \dots, \mathbf{X}_N, 0)$$

Expanding the exponential and integrating by parts (due to the $\partial/\partial \mathbf{q}_j$ and $\partial/\partial \mathbf{p}_j$ terms included in $\hat{\mathcal{H}}$), and summing the series, we obtain

$$= \int \dots \int d\mathbf{X}_1 \dots d\mathbf{X}_N \left[e^{-\hat{\mathcal{H}}t}O^{(N)}(\mathbf{X}_1, \dots, \mathbf{X}_N) \right] \rho(\mathbf{X}_1, \dots, \mathbf{X}_N, 0)$$
$$(4.40b)$$

that indicates the difference between the representation in a Schrödinger-like picture as in Eq. (4.40a) and a Heisenberg-like picture as in Eq. (4.40b). For instance, the equation of motion for a function in Γ-space (phase space) such as $O^{(N)}(\mathbf{X}_1, \dots \mathbf{X}_N, t)$, is

$$\frac{\partial}{\partial t}O^{(N)}(\mathbf{X}^N, t) = \hat{\mathcal{H}}O^{(N)}(t)$$

$$= -\{O^{(N)}(\mathbf{X}^N, t), \mathcal{H}\} = \{\mathcal{H}, O^{(N)}(\mathbf{X}^N, t)\} \qquad (4.41)$$

at variance with Eq. (4.21a) for ρ. In order to find the microscopic balance equations, we first write some useful relations:

$$\dot{\mathbf{q}}_j = \frac{\mathbf{p}_j}{m} \qquad (4.42a)$$

$$\dot{\mathbf{q}}_j = -\sum_{i \neq j} \frac{\partial}{\partial \mathbf{q}_i} V_2(|\mathbf{q}_i - \mathbf{q}_j|) = \sum_{i \neq j} \mathbf{F}_{ji} \qquad (4.42b)$$

Here, it is clear that the force $\mathbf{F}_{ij}$ has the property

$$\mathbf{F}_{ji} \equiv -\frac{\partial}{\partial \mathbf{q}_j} V_2(|\mathbf{q}_i - \mathbf{q}_j|) = \frac{\partial}{\partial \mathbf{q}_i} V_2(|\mathbf{q}_i - \mathbf{q}_j|) = -\mathbf{F}_{ij} \qquad (4.42c)$$

4.5.1 *Balance equation for the particle density*

The relevant microscopic operator in this case, the particle density at position $\mathbf{q}$, at time t, is defined as

$$n(\mathbf{q}_1, \mathbf{q}_2, \ldots, \mathbf{q}_n, \mathbf{R}) \equiv \sum_i \delta(\mathbf{q}_i - \mathbf{R}) \qquad (4.43)$$

Using Eqs. (4.7), (4.40b) and (4.43), we obtain for the equation of motion of this density

$$\frac{\partial n}{\partial t} = \sum_i \dot{\mathbf{q}}_i \frac{\partial}{\partial \mathbf{q}_i} \delta(\mathbf{q}_i - \mathbf{R}) = -\nabla_{\mathbf{R}} \left[\sum_i \frac{\mathbf{p}_i}{m} \delta(\mathbf{q}_i - \mathbf{R}) \right] \qquad (4.44a)$$

We also have

$$\frac{\partial}{\partial t} n(\mathbf{q}_1, \mathbf{q}_2, \ldots, \mathbf{q}_n, \mathbf{R}) = -\nabla_{\mathbf{R}} \mathbb{J}^{(N)}(\mathbf{q}_1, \mathbf{p}_1, \mathbf{q}_2, \mathbf{p}_2, \ldots, \mathbf{q}_n, \mathbf{p}_n, \mathbf{R}) \qquad (4.44b)$$

where $\nabla_{\mathbf{R}}$ indicates the gradient with respect to $\mathbf{R}$, and the microscopic current $\mathbb{J}^{(N)}$ is given by

$$\mathbb{J}^{(N)}(\mathbf{q}_1, \mathbf{p}_1, \mathbf{q}_2, \mathbf{p}_2, \ldots, \mathbf{q}_n, \mathbf{p}_n, \mathbf{R}) = \sum_i \frac{\mathbf{p}_i}{m} \delta(\mathbf{q}_i - \mathbf{R}) \qquad (4.44c)$$

Eq. (4.44b) is a microscopic balance equation that expresses the conservation of the number of particles, and is invariant against time inversion.

4.5.2 *Balance equation for the momentum density*

The momentum density is given by the expression

$$m\mathbb{J}^{(N)}(\mathbf{q}_1, \mathbf{p}_1, \mathbf{q}_2, \mathbf{p}_2, \ldots, \mathbf{q}_n, \mathbf{p}_n, \mathbf{R}) \equiv \sum_i \mathbf{p}_i \delta(\mathbf{q}_i - \mathbf{R}) \qquad (4.45)$$

and satisfying the following equation of motion

$$\frac{\partial}{\partial t} m\mathbb{J}^{(N)} = \sum_i \left[\dot{\mathbf{p}}_i \delta(\mathbf{q}_i - \mathbf{R}) + \mathbf{p}_i \dot{\mathbf{q}}_i \frac{\partial}{\partial \mathbf{q}_i} \delta(\mathbf{q}_i - \mathbf{R}) \right] \qquad (4.46)$$

Let us analyze the contribution of each term on the r.h.s.

$$\sum_i \dot{\mathbf{p}}_i \delta(\mathbf{q}_i - \mathbf{R}) = \sum_i \sum_{l \neq i} \mathbf{F}_{il} \delta(\mathbf{q}_i - \mathbf{R}) =$$

$$= \frac{1}{2} \sum_i \sum_l \mathbf{F}_{il} \left[\delta(\mathbf{q}_i - \mathbf{R}) - \delta(\mathbf{q}_l - \mathbf{R}) \right] \tag{4.47}$$

where we have used that $\mathbf{F}_{il} = -\mathbf{F}_{li}$. We restrict ourselves to the case of short range interactions, as well as to long wave length inhomogeneities (corresponding to the hydrodynamic limit). Also we assume that the particles are enclosed in a cubic box of side L (with $V = L^3$). Now, we expand the densities in Fourier series

$$n(\mathbf{q}_1, \ldots, \mathbf{q}_n, \mathbf{R}) = \sum_i \delta(\mathbf{q}_i - \mathbf{R}) = V^{-1} \sum_{l,\mathbf{k}} e^{i\mathbf{k} \cdot (\mathbf{q}_l - \mathbf{R})}$$

$$= \sum_{\mathbf{k}} n_k(\mathbf{q}_1, \ldots, \mathbf{q}_n) e^{-i\mathbf{k} \cdot \mathbf{R}} \tag{4.48}$$

Hence, for the case of long wave length perturbations, we have that the small $k = |\mathbf{k}|$ components (i.e. $\mathbf{k} = (2\pi/L)[l, n, m]$, with l, n and m small) will give larger contributions, leading to

$$[\delta(\mathbf{q}_i - \mathbf{R}) - \delta(\mathbf{q}_l - \mathbf{R})] = V^{-1} \sum_{\mathbf{k}} e^{-i\mathbf{k} \cdot \mathbf{R}} \left[e^{i\mathbf{k} \cdot \mathbf{q}_i} - e^{i\mathbf{k} \cdot \mathbf{q}_l} \right]$$

$$\simeq V^{-1} \sum_{\mathbf{k}} e^{-i\mathbf{k} \cdot \mathbf{R}} e^{i\mathbf{k} \cdot \mathbf{q}_i} \left[i\mathbf{k} \cdot (\mathbf{q}_i - \mathbf{q}_l) + \mathcal{O}(\mathbf{k}^2) \right]$$

$$= -\nabla_{\mathbf{R}}(\mathbf{q}_i - \mathbf{q}_l) \delta(\mathbf{q}_i - \mathbf{R}) + \mathcal{O}(\mathbf{k}^2) \tag{4.49}$$

Hence we get

$$\sum_i \dot{\mathbf{p}}_i \delta(\mathbf{q}_i - \mathbf{R}) \simeq \frac{1}{2} \nabla_{\mathbf{R}} \left[\sum_i \sum_l (\mathbf{q}_i - \mathbf{q}_l) \mathbf{F}_{il} \delta(\mathbf{q}_i - \mathbf{R}) \right] \tag{4.50}$$

On the other hand, the second term on the r.h.s. of Eq. (4.46), gives

$$\sum_i \mathbf{p}_i \dot{\mathbf{q}}_i \frac{\partial}{\partial \mathbf{q}_i} \delta(\mathbf{q}_i - \mathbf{R}) = -\frac{1}{m} \nabla_{\mathbf{R}} \sum_i \mathbf{p}_i \mathbf{p}_i \delta(\mathbf{q}_i - \mathbf{R}). \tag{4.51}$$

Putting the previous results together, we obtain

$$\frac{\partial}{\partial t} m \mathbb{J}^{(N)}(\mathbf{q}_1, \mathbf{p}_1, \mathbf{q}_2, \mathbf{p}_2, \ldots, \mathbf{q}_n, \mathbf{p}_n, \mathbf{R}) =$$

$$= -\nabla_{\mathbf{R}} \mathbb{J}^{(P)}(\mathbf{q}_1, \mathbf{p}_1, \mathbf{q}_2, \mathbf{p}_2, \ldots, \mathbf{q}_n, \mathbf{p}_n, \mathbf{R}) \tag{4.52}$$

corresponding to the microscopic balance equation for the momentum density. The flux of momentum tensor $\mathbb{J}^{(P)}$ appears here, defined through

$$\mathbb{J}^{(P)}(\mathbf{q}_1, \mathbf{p}_1, \ldots, \mathbf{q}_n, \mathbf{p}_n, \mathbf{R}) =$$
$$\frac{1}{m} \sum_i \mathbf{p}_i \mathbf{p}_i \delta(\mathbf{q}_i - \mathbf{R}) + \frac{1}{2} \sum_{il} (\mathbf{q}_i - \mathbf{q}_l) \mathbf{F}_{il} \delta(\mathbf{q}_i - \mathbf{R}) \qquad (4.53)$$

According to what we said before, and to the derivation procedure, Eq. (4.52) is valid only for long wave length. For instance, to describe viscous fluids, where short wave length phenomena are relevant, it is necessary to introduce modifications into this derivation.

4.5.3 *Balance equation for the energy density*

The energy density is defined by

$$E(\mathbf{q}_1, \mathbf{p}_1, \mathbf{q}_2, \mathbf{p}_2, \ldots, \mathbf{q}_n, \mathbf{p}_n, \mathbf{R}) \equiv \sum_i E_i \delta(\mathbf{q}_i - \mathbf{R}) \qquad (4.54)$$

where E_i, the energy for the i–th particle, is given by

$$E_i = \frac{\mathbf{p}_i^2}{2m} + \frac{1}{2} \sum_{i \neq j} V_2(|\mathbf{q}_i - \mathbf{q}_j|) \qquad (4.55)$$

Proceeding as before and also considering the case of long wave length perturbations, we get for the kinetic equation of the energy density

$$\frac{\partial}{\partial t} E(\mathbf{q}_1, \mathbf{p}_1, \ldots, \mathbf{q}_n, \mathbf{p}_n, \mathbf{R}) = -\nabla_{\mathbf{R}} \mathbb{J}^{(E)}(\mathbf{q}_1, \mathbf{p}_1, \ldots, \mathbf{q}_n, \mathbf{p}_n, \mathbf{R}) \qquad (4.56)$$

where the energy flux (or energy current) is given by

$$\mathbb{J}^{(E)}(\mathbf{q}_1, \mathbf{p}_1, \mathbf{q}_2, \mathbf{p}_2, \ldots, \mathbf{q}_n, \mathbf{p}_n, \mathbf{R}) = \qquad (4.57)$$
$$= \sum_i E_i \frac{\mathbf{p}_i}{m} \delta(\mathbf{q}_i - \mathbf{R}) + \frac{1}{2} \sum_{i \neq j} \frac{\mathbf{p}_i + \mathbf{p}_l}{m} \mathbf{F}_{il}(\mathbf{q}_i - \mathbf{q}_l) \delta(\mathbf{q}_i - \mathbf{R})$$

These microscopic balance equations are the basis for the microscopic hydrodynamic equations.

4.5.4 *Balance equation for the entropy density*

In an entirely similar way to the previous cases, we can also derive a microscopic balance equation for the entropy density. To do this, we need to assume that the system is locally in equilibrium; this means that the fundamental relation

$$T ds = dE + p d\rho^{-1} \tag{4.58a}$$

is valid locally. The equation for the entropy density s, is then

$$\frac{\partial s}{\partial t} = -\nabla_{\mathbf{R}}(s\mathbb{U} + \mathbb{J}_s) + \sigma_s \tag{4.58b}$$

where $s\mathbb{U}$ is the convective entropy flux (that exists even without dissipation), while $\mathbb{J}_s$ is the entropy current due to dissipative effects, and σ_s is the entropy source (that also arises due to dissipative effects).

4.6   Density operator

For quantum systems, the phase space coordinates do not commute. Hence, if we want to make a statistical description of quantal system, as is taught in courses of statistical physics, it is necessary to resort to a more general kind of object called the *density operator* (or *density matrix*) $\hat{\rho}(t)$. It turns out to be a Hermitian operator, positive definite, that may be used in order to obtain mean values of physical observables. The knowledge of this operator makes it possible to calculate the expectation value of an observable $\hat{O}$ at time t, as:

$$\langle \hat{O}(t) \rangle = \mathrm{Tr}\left[\hat{O}\hat{\rho}(t)\right] = \mathrm{Tr}\left[\hat{O}(t)\hat{\rho}\right] \tag{4.59a}$$

with the normalization

$$\mathrm{Tr}\hat{\rho}(t) = 1. \tag{4.59b}$$

Due to the Hermitian character and the positivity of the density operator $\hat{\rho}(t)$, it may be diagonalized, and all its eigenvalues are real and positive. Let us call $\{|\aleph_i\rangle\}$ a set of eigenvectors of this operator, with eigenvalues $\{\lambda_i\}$ $(\lambda_i > 0)$. We can then write

$$\hat{\rho}(t) = \sum_j \lambda_j |\aleph_j(t)\rangle\langle\aleph_j(t)| \tag{4.60a}$$

Considering Eq. (4.59b),

$$\sum_j \lambda_j = 1 \tag{4.60b}$$

Hence, we write Eq. (4.59a) as

$$\langle \hat{\rho}(t) \rangle = \mathrm{Tr}\left[\hat{O}\hat{\rho}(t)\right] = \sum_j \lambda_j \langle\aleph_j(t)|\hat{O}|\aleph_j(t)\rangle \tag{4.60c}$$

where $\langle \aleph_j(t)|\hat{O}|\aleph_j(t)\rangle$ is the expectation value in the state $|\aleph_j(t)\rangle$ of the operator $\hat{O}$, and λ_j is the probability of its being in such a state. For another (arbitrary) orthogonal set $\{|\eta\rangle\}$, the probability $P_\eta(t)$ for the system to be in a state $|\eta\rangle$ is

$$P_\eta(t) = \langle\eta|\hat{\rho}(t)|\eta\rangle = \sum_j \lambda_j\langle\eta\aleph_j(t)\rangle\langle\aleph_j(t)\eta\rangle \qquad (4.61a)$$

and the expectation value of the operator reads

$$\langle\hat{O}(t)\rangle = \mathrm{Tr}\left[\hat{O}\hat{\rho}(t)\right] = \sum_{\eta,\eta'}\langle\eta|\hat{O}|\eta'\rangle\langle\eta'|\hat{\rho}(t)|\eta\rangle \qquad (4.61b)$$

where $\langle\eta'|\hat{\rho}(t)|\eta\rangle$ is an element of the *density matrix*.

In order to describe the dynamics of the system, we remember that the Schrödinger equation with Hamiltonian must hold:

$$i\hbar\frac{\partial}{\partial t}|\aleph_j(t)\rangle = \hat{\mathcal{H}}|\aleph_j(t)\rangle \qquad (4.62a)$$

which gives

$$i\hbar\frac{\partial}{\partial t}\hat{\rho}(t) = \hat{\mathcal{H}}\hat{\rho}(t) - \hat{\rho}(t))\hat{\mathcal{H}} = \hbar\hat{\mathcal{L}}\hat{\rho}(t) \qquad (4.62b)$$

where

$$\hat{\mathcal{L}} = \hbar^{-1}[\hat{\mathcal{H}},\cdot]. \qquad (4.62c)$$

Equation (62b) is the quantal version of Liouville's equation. If we know the value of $\hat{\rho}$ at time $t = 0$, we can evaluate it at time $t > 0$, i.e.

$$\hat{\rho}(t) = \exp\{-i\hat{\mathcal{L}}t\}\hat{\rho}(0) = e^{-i\hat{\mathcal{H}}t/\hbar}\hat{\rho}(0)e^{i\hat{\mathcal{H}}t/\hbar} \qquad (4.62d)$$

and similarly for the operator $\hat{O}$,

$$\hat{O}(t) = e^{i\hat{\mathcal{H}}t/\hbar}\hat{O}(0)e^{-i\hat{\mathcal{H}}t/\hbar} \qquad (4.62e)$$

which obeys the (*von Neumann*) equation

$$-i\frac{\partial}{\partial t}\hat{O}(t) = \hbar^{-1}[\hat{\mathcal{H}},\hat{O}(t)] = \hat{\mathcal{L}}\hat{O}(t). \qquad (4.62f)$$

Note the sign change in relation to Eq. (4.62c).

If we choose a basis $\{|\epsilon_k\rangle\}$ that diagonalizes $\hat{\mathcal{H}}$, with eigenvalues $\{\epsilon_k\}$, corresponding to the *energy representation*, we have

$$\sum_k |\epsilon_k\rangle\langle\epsilon_k| = 1 \qquad (4.63a)$$

and

$$\hat{\rho}(t) = \sum_{k,k'} \langle \epsilon_k | \hat{\rho}(0) | \epsilon'_k \rangle e^{-i(\epsilon_k - \epsilon'_k)t/\hbar} | \epsilon_k \rangle \langle \epsilon'_k |. \tag{4.63b}$$

A stationary state will occur when all non-diagonal terms of $\hat{\rho}(0)$ are zero. Hence, if there are no degenerate levels, the stationary state will be diagonal in the *energy representation*, meaning that $\hat{\rho}$ is a function of $\hat{\mathcal{H}}$.

4.7 Reduced density operator

As in the classical case, it is often necessary to evaluate the expectation values of one- and two-body operators. Hence, as for the classical case, it is useful to introduce the concept of the *reduced one-* (or *two-*) *body density matrix*. Let us consider a one–body observable described by an operator $\hat{O}^{(1)}$, then

$$\langle \hat{O}^{(1)}(t) \rangle = \mathrm{Tr}\left[\hat{O}^{(1)} \hat{\rho}(t) \right]$$

which, using the notation of second quantization, can be written as

$$\langle \hat{O}^{(1)}(t) \rangle = \sum_{n_\alpha} \sum_{k,k'} \langle \mathbf{k}^{(1)} \mathbf{k}' \rangle \langle \{n_\alpha\}^{(1)}(t) \rangle^\dagger_{k,k'}(t) \{n_\alpha\} \rangle$$

$$= \sum_{k,k'} \langle \mathbf{k}^{(1)} \mathbf{k}' \rangle \langle \mathbf{k}' \mathbf{k} \rangle \tag{4.64a}$$

where $|\{n_\alpha\}\rangle$ indicates the set of states corresponding to different occupation (or quantum) numbers $\{n_\alpha\}$, and $\hat{a}^\dagger_{\mathbf{k}}$ ($\hat{a}_{\mathbf{k}}$) is the creation (annihilation) operator as usual. Here we have introduced the *one-body reduced density matrix*

$$\langle \mathbf{k}' | \hat{\rho}^{(1)}(t) | \mathbf{k} \rangle = \mathrm{Tr}\,(\hat{a}_{\mathbf{k}}\dagger \hat{a}'_{\mathbf{k}} \hat{\rho}(t))$$

$$= \sum_{\{n_\alpha\}} \langle \{n_\alpha\} | \hat{a}^\dagger_{\mathbf{k}} \hat{a}'_{\mathbf{k}} \hat{\rho}(t) | \{n_\alpha\} \rangle \tag{4.64b}$$

In the position representation we have

$$\langle \hat{O}^{(1)}(t) \rangle = \int \int d\mathbf{r} d\mathbf{r}' \langle \mathbf{r}_1 | \hat{O}^{(1)}(t) | \mathbf{r}'_1 \rangle \langle \mathbf{r}'_1 | \hat{\rho}^{(1)}(t) | \mathbf{r}_1 \rangle \tag{4.65a}$$

where

$$\langle \mathbf{r}'_1 | \hat{\rho}^{(1)}(t) | \mathbf{r}_1 \rangle = \mathrm{Tr}\left(\hat{\Psi}(\mathbf{r}_1)\dagger \hat{\Psi}(\mathbf{r}'_1) \hat{\rho}(t) \right) = \langle n(\mathbf{r}, t) \rangle$$

$$= \sum_{\{n_\alpha\}} \langle \{n_\alpha\} | \hat{\Psi}(\mathbf{r}_1)\dagger \hat{\Psi}(\mathbf{r}'_1) \hat{\rho}(t) | \{n_\alpha\} \rangle \tag{4.65b}$$

It is also possible to write out all these expressions explicitly indicating the symmetric and antisymmetric contributions (from boson and fermion terms respectively).

For two-body operators we can obtain similar results, for instance

$$\langle \mathbf{k}_1', \mathbf{k}_2' | \hat{\rho}^{(2)}(t) | \mathbf{k}_1, \mathbf{k}_2 \rangle = \text{Tr} \left(\hat{a}_{\mathbf{k}_1}{}^\dagger \hat{a}_{\mathbf{k}_2}{}^\dagger \hat{a}_{\mathbf{k}_1'} \hat{a}_{\mathbf{k}_2'} \hat{\rho}(t) \right)$$

$$= \sum_{\{n_\alpha\}} \langle \{n_\alpha\} | \hat{a}_{\mathbf{k}_1}{}^\dagger \hat{a}_{\mathbf{k}_2}{}^\dagger \hat{a}_{\mathbf{k}_1'} \hat{a}_{\mathbf{k}_2'} \hat{\rho}(t) | \{n_\alpha\} \rangle \quad (4.66)$$

Let us look at the properties of these operators. First, we have that

$$\langle \mathbf{k} | \hat{\rho}^{(1)}(t) | \mathbf{k} \rangle = \text{tr} \left(\hat{a}_{\mathbf{k}}{}^\dagger \hat{a}_{\mathbf{k}} \hat{\rho}(t) \right) = \langle n(\mathbf{k}, t) \rangle \quad (4.67a)$$

$$\langle \mathbf{r} | \hat{\rho}^{(1)}(t) | \mathbf{r} \rangle = \text{tr} \left(\hat{\Psi}(\mathbf{r})^\dagger \hat{\Psi}(\mathbf{r}) \hat{\rho}(t) \right) = \langle n(\mathbf{r}, t) \rangle \quad (4.67b)$$

where $\mathbf{k}$ and $\mathbf{r}$ are variables associated to conjugate spaces. We also have

$$\sum_{\mathbf{k}} \langle \mathbf{k} | \hat{\rho}^{(1)}(t) | \mathbf{k} \rangle = \int d\mathbf{r} \langle \mathbf{r} | \hat{\rho}^{(1)}(t) | \mathbf{r} \rangle = N \quad (4.68a)$$

where N is the total number of particles in the system. In a similar way

$$\sum_{\mathbf{k}, \mathbf{k}'} \langle \mathbf{k}', \mathbf{k} | \hat{\rho}^{(2)}(t) | \mathbf{k}', \mathbf{k} \rangle = \int\int d\mathbf{r} d\mathbf{r}' \langle \mathbf{r}, \mathbf{r}' | \hat{\rho}^{(2)}(t) | \mathbf{r}, \mathbf{r}' \rangle = N(N-1) \quad (4.68b)$$

According to these results we have that the eigenvalues of $\rho^{(1)}$ are $\lambda_i < N$, and those of $\rho^{(2)}$ are $\lambda_j < N(N-1)$. Calling $\{|\aleph_j^{(1)}\rangle\}$ the set of eigenvectors of $\rho^{(1)}$, we can write

$$\hat{\rho}^{(1)}(t) = \sum_j \lambda_j |\aleph_j^{(1)}\rangle \langle \aleph_j^{(1)}| \quad (4.69a)$$

Then we have that $\langle \aleph_j^{(1)} | \hat{\rho}^{(1)}(t) | \aleph_j^{(1)} \rangle = \lambda_j$ corresponds to the occupation number of state $|\aleph_j^{(1)}\rangle$. Taking the trace in the $\mathbf{r}$-representation (assuming that the wave functions $\langle \mathbf{r} | \aleph_j^{(1)} \rangle = \aleph_j^{(1)}(\mathbf{r})$ and $\langle \aleph_j^{(1)} | \mathbf{r} \rangle = \aleph_j^{(1)}(\mathbf{r})^*$, are orthogonal) we find

$$\int d\mathbf{r} \langle \mathbf{r} | \hat{\rho}^{(1)}(t) | \mathbf{r} \rangle = \sum_j \lambda_j \int d\mathbf{r} \langle \mathbf{r} | \aleph_j^{(1)} \rangle \langle \aleph_j^{(1)} | \mathbf{r} \rangle$$

$$= \sum_j \lambda_j \int d\mathbf{r} \aleph_j^{(1)}(\mathbf{r}) \aleph_j^{(1)}(\mathbf{r})^* = \sum_j \lambda_j = N \quad (4.69b)$$

where, due to normalization, $\aleph_j^{(1)}(\mathbf{r}) \simeq V^{-1/2}$. Let us now take the nondiagonal elements

$$\langle \mathbf{r}' | \hat{\rho}^{(1)}(t) | \mathbf{r} \rangle = \sum_j \lambda_j \langle \mathbf{r}' | \aleph_j^{(1)} \rangle \langle \aleph_j^{(1)} | \mathbf{r} \rangle = \sum_j \lambda_j \aleph_j^{(1)}(\mathbf{r}') \aleph_j^{(1)}(\mathbf{r})^* \qquad (4.69c)$$

In the thermodynamic limit $N \to \infty$, $V \to \infty$, the density N/V must remain constant. How would do this affect the non-diagonal matrix element for $\mathbf{R} = |\mathbf{r} - \mathbf{r}'| \to \infty$?. If the λ_j remain finite

$$\lim_{R \to \infty} \langle \mathbf{r}' | \hat{\rho}^{(1)}(t) | \mathbf{r} \rangle = 0 \qquad (4.69d)$$

because $\aleph_j^{(1)}(\mathbf{r}') \aleph_j^{(1)}(\mathbf{r})^* \simeq V^{-1}$. But, if one of the eigenvalues, let us say λ_0, is such that $\lambda_0 \simeq \alpha N$ (with α finite), then

$$\lim_{R \to \infty} \langle \mathbf{r}' | \hat{\rho}^{(1)}(t) | \mathbf{r} \rangle = \alpha N V^{-1} f(\mathbf{r}, \mathbf{r}') \qquad (4.69e)$$

with $f(\mathbf{r}, \mathbf{r}')$ a certain function of both coordinates. Such a system has *non-diagonal long range order* in $\rho^{(1)}$. For instance, a system of fermions cannot have this kind of order (due to the Pauli principle), but it is possible in the case of systems of bosons (i.e. in superfuid Helium). However, for the case of fermions, such kind of order can arise in $\rho^{(2)}$, as happens in superconductivity.

As a final point it is worth remarking that within this quantum formalism we can also proceed through a BBGKY-like hierarchy and obtain microscopic *hydrodynamic-like* balance equations, as we have done in the classical formalism.

4.8 $\mathcal{H}$–theorem

In order to show that Boltzmann's equation describes the correct behavior of decay to equilibrium, we will present Boltzmann's $\mathcal{H}$-theorem. We start defining the function $\mathcal{H}(t)$ as

$$\mathcal{H}(t) = \int \int d\mathbf{r}_1 d\mathbf{p}_1 \, q_1(\mathbf{r}_1, \mathbf{p}_1, t) \ln[q_1(\mathbf{r}_1, \mathbf{p}_1, t)] \qquad (4.70)$$

What we are going to show is that if $q_1(\mathbf{r}, \mathbf{p}, t)$ satisfies the Boltzmann equation, then $\mathcal{H}(t)$ always decreases with time due to the effect of collisions. Let us take the time derivative of $\mathcal{H}(t)$ as

$$\frac{\partial}{\partial t} \mathcal{H}(t) = \int \int d\mathbf{r}_1 d\mathbf{p}_1 \left[\frac{\partial}{\partial t} q_1(\mathbf{r}_1, \mathbf{p}_1, t) \right] \{ \ln[q_1(\mathbf{r}_1, \mathbf{p}_1, t)] + 1 \} \qquad (4.71)$$

Now, as we have assumed that $q_1(\mathbf{r}_1, \mathbf{p}_1, t)$ satisfies the Boltzmann equation, we replace Eq. (4.39d) into Eq. (4.71), that yields

$$
\frac{\partial}{\partial t}\mathcal{H}(t) = -\int\int d\mathbf{r}_1 d\mathbf{p}_1 \left[\dot{\mathbf{r}}_1 \frac{\partial}{\partial \mathbf{r}_1} q_1(\mathbf{r}_1, \mathbf{p}_1, t)\right] \{\ln[q_1(\mathbf{r}_1, \mathbf{p}_1, t)] + 1\}
$$

$$
+\int\int\int d\mathbf{r}_1 d\mathbf{p}_1 d\mathbf{p}_2 \int ds|\mathbf{p}_1 - \mathbf{p}_2|\sigma(\vartheta, |\mathbf{p}_1 - \mathbf{p}_2|)\{\ln[q_1(\mathbf{r}_1, \mathbf{p}_1, t)] + 1\}
$$

$$
[q_1(\mathbf{r}_1, \mathbf{p}_1, t)q_1(\mathbf{r}_1, \mathbf{p}_2', t) - q_1(\mathbf{r}_1, \mathbf{p}_1', t)q_1(\mathbf{r}_1, \mathbf{p}_2, t)] \tag{4.72}
$$

Now, we can change the first term on the r.h.s. into a surface integral, which, assuming that for $\mathbf{p}$ and $\mathbf{r}$ large enough we have $q_1(\mathbf{r}_1, \mathbf{p}_1, t) \to 0$, gives no contribution, and we have

$$
\frac{\partial}{\partial t}\mathcal{H}(t) = \int\int\int d\mathbf{r}_1 d\mathbf{p}_1 d\mathbf{p}_2 \int ds|\mathbf{p}_1 - \mathbf{p}_2|\sigma(\vartheta, |\mathbf{p}_1 - \mathbf{p}_2|) \tag{4.73}
$$

$$
[q_1(\mathbf{r}_1, \mathbf{p}_1, t)q_1(\mathbf{r}_1, \mathbf{p}_2', t) - q_1(\mathbf{r}_1, \mathbf{p}_1', t)q_1(\mathbf{r}_1, \mathbf{p}_2, t)] \{\ln[q_1(\mathbf{r}_1, \mathbf{p}_1, t)] + 1\}
$$

The last equation can be rewritten by exchanging $\mathbf{p}_1$ and $\mathbf{p}_2$, and obtain

$$
\frac{\partial}{\partial t}\mathcal{H}(t) = \int\int\int d\mathbf{r}_1 d\mathbf{p}_1 d\mathbf{p}_2 \int ds|\mathbf{p}_1 - \mathbf{p}_2|\sigma(\vartheta, |\mathbf{p}_1 - \mathbf{p}_2|) \tag{4.74}
$$

$$
[q_1(\mathbf{r}_1, \mathbf{p}_1, t)q_1(\mathbf{r}_1, \mathbf{p}_2', t) - q_1(\mathbf{r}_1, \mathbf{p}_1', t)q_1(\mathbf{r}_1, \mathbf{p}_2, t)] \{\ln[q_1(\mathbf{r}_2, \mathbf{p}_2, t)] + 1\}
$$

If we now add both previous equations and divide by two, we obtain

$$
\frac{\partial}{\partial t}\mathcal{H}(t) = \frac{1}{2}\int\int\int d\mathbf{r}_1 d\mathbf{p}_1 d\mathbf{p}_2 \int ds|\mathbf{p}_1 - \mathbf{p}_2|\sigma(\vartheta, |\mathbf{p}_1 - \mathbf{p}_2|)
$$

$$
[q_1(\mathbf{r}_1, \mathbf{p}_1', t)q_1(\mathbf{r}_1, \mathbf{p}_2', t) - q_1(\mathbf{r}_1, \mathbf{p}_1, t)q_1(\mathbf{r}_1, \mathbf{p}_2, t)]
$$

$$
\{\ln[q_1(\mathbf{r}_1, \mathbf{p}_1, t)] + \ln[q_1(\mathbf{r}_1, \mathbf{p}_1, t)] + 2\} \tag{4.75}
$$

As the final step we exchange the dummy variables $\mathbf{p}_1 \to \mathbf{p}_1'$ and $\mathbf{p}_2 \to \mathbf{p}_2'$ in Eq. (4.75), add the result to the last equation, and divide by two. Remembering that $d\mathbf{p}_1 d\mathbf{p}_2 = d\mathbf{p}_1' d\mathbf{p}_2'$, we find

$$
\frac{\partial}{\partial t}\mathcal{H}(t) = \frac{1}{2}\int\int\int d\mathbf{r}_1 d\mathbf{p}_1 d\mathbf{p}_2 \int ds|\mathbf{p}_1 - \mathbf{p}_2|\sigma(\vartheta, |\mathbf{p}_1 - \mathbf{p}_2|)
$$

$$
[q_1(\mathbf{r}_1, \mathbf{p}_1', t)q_1(\mathbf{r}_1, \mathbf{p}_2', t) - q_1(\mathbf{r}_1, \mathbf{p}_1, t)q_1(\mathbf{r}_1, \mathbf{p}_2, t)]
$$

$$
\ln\left[\frac{q_1(\mathbf{r}_1, \mathbf{p}_1', t)q_1(\mathbf{r}_1, \mathbf{p}_2', t)}{q_1(\mathbf{r}_1, \mathbf{p}_1, t)q_1(\mathbf{r}_1, \mathbf{p}_2, t)}\right] \le 0 \tag{4.76}
$$

The last inequality comes from the fact that a function of the form $(y - x)\ln(x/y)$ is always negative or zero. Hence, we have found that the time derivative of $\mathcal{H}(t)$ will be zero only if (in a short hand notation) $q_1(1')q_1(2') = q_1(1)q_1(2)$ in *every* collision. This is a *detailed balance condition*, corresponding to the equilibrium condition for the gas. The previous result indicates that Boltzmann's equation has the correct behavior of decay to equilibrium. Another form we can adopt to write it, is

$$\ln q_1(\mathbf{r}_1, \mathbf{p}_1, t) + \ln q_1(\mathbf{r}_2, \mathbf{p}_2, t) = \ln q_1(\mathbf{r}_1, \mathbf{p}_1') + \ln q_1(\mathbf{r}_2, \mathbf{p}_2', t).$$

This last form allow us to finally obtain the form of the equilibrium Maxwell distribution.

Since $\mathcal{H}(t)$ always decreases with time, the negative of this function will always increase with time. Boltzmann has identified this quantity as proportional to the *nonequilibrium entropy* $S(t)$

$$S(t) = -k_B \mathcal{H}(t) = -k \int \int d\mathbf{r}_1 d\mathbf{p}_1 q_1(\mathbf{r}_1, \mathbf{p}_1, t) \ln[q_1(\mathbf{r}_1, \mathbf{p}_1, t)] \quad (4.77)$$

which differs from Gibbs' entropy in that the latter depends on the full distribution function, while the present one only depends on the reduced distribution $q_1(\mathbf{r}_1, \mathbf{p}_1, t)$.

4.9   Transport coefficients

In this section we want to present a few simple examples of how to calculate transport coefficients when the one particle distribution function $q_1(\mathbf{r}, \mathbf{p}, t)$ is known (in what follows, and in order to alleviate notation we drop the subscript 1 for the function q), at least approximately. In order to simplify the evaluation we will consider the simplest approximation for the collision term, that is the *relaxation time approximation*, that consists in adopting

$$\left(\frac{\partial q_1}{\partial t}\right)_{\text{coll}} = -\tau_r^{-1}\left[q(\mathbf{r}, \mathbf{p}, t) - q^0(\mathbf{r}, \mathbf{p})\right] \quad (4.78)$$

where τ_r is the typical system's relaxation time (it is not excluded that in a general situation τ_r may be a function of $\mathbf{r}$ and $\mathbf{p}$), and $q^0(\mathbf{r}, \mathbf{p})$ is the equilibrium distribution function.

We start studying the electrical conductivity of a gas of charged particles subject to an external electric field $\mathcal{E}$ and a temperature gradient $\partial T/\partial x$, both applied in the x direction. The idea is to solve in an approximately form the Boltzmann equation for $q(\mathbf{r}, \mathbf{p}, t)$ and afterwards to find the flux of electric charge and energy. We assume steady state conditions, that is $\partial q(\mathbf{r}, \mathbf{p}, t)/\partial t = 0$. In this case the Boltzmann equation becomes

$$e\mathcal{E}\frac{\partial q}{\partial \mathbf{p}} + \frac{\mathbf{p}}{m}\frac{\partial q}{\partial \mathbf{q}} = \frac{e\mathcal{E}}{m}\frac{\partial q}{\partial v} + v\frac{\partial q}{\partial x} = -\tau_r^{-1}(q - q^0) \qquad (4.79)$$

where v is the x–component of the velocity, e is the charge of the particle (for instance electrons), and m their mass. Equation (4.79) can be arranged as

$$q(\mathbf{r}, \mathbf{p}, t) = q^0(\mathbf{r}, \mathbf{p}) - \tau_r\left(\frac{e\mathcal{E}}{m}\frac{\partial q}{\partial v} + v\frac{\partial q}{\partial x}\right). \qquad (4.80)$$

We will assume that the field is weak and the temperature gradient small, so that changes in the distribution function q will be small and second order terms in these quantities in an expansion of q may be neglected. In other words we are assuming that $(q - q^0)/q^0 \ll 1$. Hence, we can approximate the previous equation as

$$q(\mathbf{r}, \mathbf{p}, t) = q^0(\mathbf{r}, \mathbf{p}) - \tau_r\left(\frac{e\mathcal{E}}{m}\frac{\partial q^0}{\partial v} + v\frac{\partial q^0}{\partial x}\right). \qquad (4.81)$$

However, by an iterative procedure we can get higher order effects.

Now, at equilibrium, q^0 is a function of the energy E, the temperature T, and the chemical potential μ. Also, the energy is a function of the velocity. We then have

$$\frac{\partial q^0}{\partial x} = \frac{\partial q^0}{\partial \mu}\frac{\partial \mu}{\partial x} + \frac{\partial q^0}{\partial T}\frac{\partial T}{\partial x} \quad \text{and} \quad \frac{\partial q^0}{\partial v} = \frac{\partial q^0}{\partial E}\frac{\partial E}{\partial v} = mv\frac{\partial q^0}{\partial E}. \qquad (4.82)$$

Usually, the electrical conductivity is defined under the conditions that there are neither temperature nor chemical potential gradients. This implies $\partial q^0/\partial x = 0$, reducing Eq. (4.81) to

$$q(\mathbf{r}, \mathbf{p}, t) = q^0(\mathbf{r}, \mathbf{p}) - \tau_r\frac{e\mathcal{E}}{m}\frac{\partial q^0}{\partial v}. \qquad (4.83)$$

On the other hand, the electric current density is given by

$$\mathbb{J} = \int dv\, e v q(\mathbf{r}, \mathbf{p}, t) = -\tau_r e^2 \mathcal{E} \int dv\, v^2 \frac{\partial q^0}{\partial E}, \qquad (4.84)$$

because $\mathbf{v}q^0 d\mathbf{v} = 0$, due to the spherical symmetry of q^0. Here we assumed that τ_r is constant (independent of $\mathbf{r}$ and $\mathbf{p}$), but this is not relevant (for instance for a Fermi gas, only the value of τ_r at $E = \mu_F$ is what matters). Now we will evaluate Eq. (4.84) for two cases: Maxwellian and Fermi–Dirac equilibrium distributions.

(a) Maxwell distribution is given by

$$q^0(v) = N \left(\frac{m}{2\pi kT}\right)^{3/2} \exp\left(-mv^2/2kT\right) \qquad (4.85)$$

with $v^2 = v_x^2 + v_y^2 + v_z^2$, and N the particle density. For this case, Eq. (4.79), according to the second of Eqs. (4.82) has the form

$$\frac{\partial q^0}{\partial E} = -(kT)^{-1}q^0 \qquad (4.86)$$

so that Eq. (4.84) results in

$$\mathbf{J} = \tau_r e^2 \mathcal{E}(kT)^{-1} \int d\mathbf{v}\, v_x^2 + q^0(v). \qquad (4.87)$$

Replacing the form of the Maxwell distribution, Eq. (4.85), this yields

$$\mathbf{J} = \frac{Ne^2\tau_r}{m}\mathcal{E} \qquad (4.88)$$

From here we have for the electrical conductivity

$$\sigma = \frac{Ne^2\tau_r}{m} \qquad (4.89)$$

(b) The Fermi–Dirac distribution, in an adequate normalized form, is given by

$$q^0(E) = 2 \left(\frac{m}{2\pi\hbar}\right)^3 \left[\exp\left(\frac{E - \mu_F}{kT}\right) + 1\right]^{-1}, \qquad (4.90)$$

where we have included the factor 2 due to the electron spin. The electric current density in Eq. (4.84) will be given by

$$\mathbf{J} = -2\tau_r e^2 \mathcal{E} \left(\frac{m}{2\pi\hbar}\right)^3 \int d\mathbf{v}\, v_x^2 \frac{\partial}{\partial E} \left[\exp\left(\frac{E - \mu_F}{kT}\right) + 1\right]^{-1}. \qquad (4.91)$$

Now, we have the relation

$$v_x^2 d\mathbf{v} = \frac{4\pi}{3}v^4 dv = \frac{4\pi}{3}\left(\frac{2E}{m}\right)^{3/2}\frac{1}{m}dE = \frac{8\pi}{3}\sqrt{2}m^{-5/2}E^{3/2}dE \qquad (4.92)$$

And also, for a highly degenerate electron gas ($kT \ll \mu_F$), we can approximate

$$\frac{\partial}{\partial E}\left[\exp\left(\frac{E - \mu_F}{kT}\right) + 1\right]^{-1} \approx -\delta(E - \mu_F). \qquad (4.93)$$

According to the previous results, Eq. (4.91) reduces to

$$\mathbf{J} = \tau_r(\mu_F)e^2\mathcal{E}\left(\frac{m}{2\pi\hbar}\right)^3 \mu_F^{3/2}\frac{16\pi}{3}\sqrt{2}m^{-5/2} \qquad (4.94)$$

But, as for the Fermi energy of the system at equilibrium we have that

$$\mu_F = \left(\frac{2\pi\hbar}{8m\pi^2} \right) (3\pi^2 N)^{2/3} \tag{4.95}$$

the electrical conductivity (in the degenerate limit) has the form

$$\sigma = \frac{\mathbb{J}}{\mathcal{E}} = \frac{Ne^2 \tau_r(\mu_F)}{m} \tag{4.96}$$

which is identical with the result Eq. (4.89) for the Maxwellian distribution, but where we have explicitly written that the relaxation time is evaluated on the Fermi surface.

Now we turn to study the thermal conductivity of this system. In chapter 5 we will see that we can write the relations

$$\mathbb{J} = L_{11}\mathcal{E} + L_{12}\frac{\partial T}{\partial x}$$

$$\mathbb{Q} = L_{21}\mathcal{E} + L_{22}\frac{\partial T}{\partial x} \tag{4.97}$$

where L_{ij} are the Onsager coefficients and $\mathbb{Q}$ the heat current. So far, we have determined the coefficient L_{11}, corresponding to the electric conductivity. Now, we want to determine L_{12}, which is related with the thermal conductivity.

Bearing in mind that $\mathcal{E} = -\frac{\partial \phi(\mathbf{r})}{\partial x}$ (since we have assumed that the electric field is applied on the x direction) and assuming that the relaxation time is independent of $\mathbf{r}$ and $\mathbf{p}$, for the case of the Maxwell distribution [Eq. (4.85)], the first of Eqs. (4.82) will have the form

$$\frac{\partial q^0}{\partial x} = \left(\frac{E}{kT} - \frac{3}{2} \right) q^0(v) \frac{1}{kT} \frac{\partial T}{\partial x} \tag{4.98}$$

Hence, Eq. (4.81) takes the form

$$q(\mathbf{r}, \mathbf{p}, t) = q^0(v) - \tau_r e \mathcal{E} v \frac{\partial q^0}{\partial E} - \tau_r v \left(E - \frac{3}{2}kT \right) q^0(v) \left(\frac{1}{kT} \right)^2 \frac{\partial T}{\partial x}. \tag{4.99}$$

From this we get

$$L_{12} = -\tau_r e(kT)^{-2} \int dv v_x^2 \left(E - \frac{3}{2}kT \right) q^0(v). \tag{4.100}$$

For the Maxwellian distribution we have

$$\int dv v_x^2 E q^0(v) = \frac{5}{2} N \frac{(kT)^2}{2m}, \tag{4.101}$$

yielding for L_{12}

$$L_{12} = -\frac{\tau_r e N}{m} \qquad (4.102)$$

For the thermal current density, or heat current we then obtain

$$\mathbb{Q} = \int d\mathbf{v} v_x E q^0() = \frac{5}{2} e \tau_r N \frac{(kT)}{2m} \mathcal{E} - \frac{5}{m} \tau_r N k T \frac{\partial T}{\partial x} \qquad (4.103)$$

Now, the thermal conductivity K is not simply given by L_{22}. Rather, as the thermal conductivity is usually measured not with $\mathcal{E} = 0$ but for $\mathbb{J} = 0$, it means

$$\mathcal{E} = -\frac{L_{12}}{L_{11}} \frac{\partial T}{\partial x} \qquad (4.104)$$

where we have assumed the validity of Onsager relations (see chapter 5). This finally yields

$$K = \frac{5}{2m} \tau_r N k T. \qquad (4.105)$$

We can also obtain other useful expressions for transport coefficients in several cases, for instance magneto-resistance, viscosity, Hall effect, etc. However, we stop our discussion here and leave those other cases as exercises for the reader.

Chapter 5

Linear nonequilibrium thermodynamics and Onsager relations

. . . .
not because the calculation of the apostle is wrong,
but because we have not learnt the art
on which such calculus is grounded.

. . . .
Umberto Eco

5.1 Introduction

It is usually assumed that irreversible thermodynamics (that is the set of thermodynamical methods developed to deal with irreversible processes) is adequate for the analysis of complex physicochemical processes away from thermodynamic equilibrium, although restricted to a small class of phenomena, essentially described by linear transport theory. Outside this range, a kinetic description based on a scheme such as the Navier-Stokes equations and the like ones should be the most convenient. However, a description based on linear phenomenological laws is worth doing as it gives an intuitive discrimination between forces and fluxes which is based in concepts of cause and effect. Also, through this kind of analysis we may obtain some general trends regarding the evolution of nonlinear systems far away form equilibrium such as those to be discussed in Chapters 7 and 9.

Summarizing, such a thermodynamical point of view is advantageous for the following reasons. It offers the most natural choice of variables and parameters. It imposes some physical constraints on the variables and on the structure of rate laws. It gives information on the role of the *departure from thermodynamic equilibrium*, a general parameter that arises in several physical problems. Finally, and most noteworthy, it makes the link with fluctuation theory more natural as, by definition, the notion of fluctuation is absent in the usual kinetic or phenomenological descriptions such as those discussed before.

In Chapter 4 we have seen transport or kinetic approaches as well as hydrodynamical relations describing long wave length and low frequency phenomena adequate to describe different systems: dilute gases, liquids, solids, etc. In the case of complicated systems different transport phenomena (for heat, charge, etc) turn out to be coupled. Within the thermodynamical approach described above, Onsager was the first to show that the irreversibility of the microscopic dynamical laws lead to relations among those transport coefficients describing coupled phenomena. In this chapter we shall describe and discuss the Onsager relations and approach. But first introduce in an elementary way Onsager's ideas about regression to equilibrium, that make it clear the role of fluctuations. Next, we give a more or less general presentation of the Onsager relations, and some examples of application in simple systems. Finally we discuss the *minimum entropy production theorem*, which shows that, for linear systems, steady states out of equilibrium play a role similar to that of usual equilibrium states in equilibrium thermodynamics. All these ideas form the basis on which the linear response theory (to be discussed in Chapter 6) is grounded.

5.2   Onsager's regression-to-equilibrium hypothesis

In order to discuss Onsager's ideas about regression to equilibrium, we consider a system slightly out of equilibrium, implying that the deviations and the perturbations that move the system out from equilibrium are linearly related. Let us consider, as an example, an electrolytic aqueous solution, that in equilibrium has zero net charge flux, and an average current $\langle j(t) \rangle = 0$. At a given time $t = t_1$ an electric field $\mathcal{E}$ is turned on, and the ions start to drift. At $t = t_2$, the field is turned off. Figure 5.1 shows the qualitative behavior of the observed current $j(t)$ as a function of time. The behavior of the system after the disturbance from equilibrium indicates that $\langle j(t) \rangle \neq 0$, and is linear whenever $j(t)$ is proportional to the field $\mathcal{E}$, i.e.

$$j(t, \lambda \mathcal{E}) = \lambda j(t, \mathcal{E}). \tag{5.1}$$

Another less general way to identify linear behavior focuses on the *thermodynamical forces* or *affinities*, instead of on the external fields. For instance, the existence of a chemical potential gradient is associated with a mass flux. For very small gradients (corresponding to *small departures*

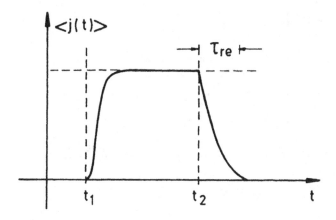

Fig. 5.1

from equilibrium), the mass current or flux becomes proportional to such gradient, i.e.

$$j(t) \propto \nabla \left(\frac{\mu}{T} \right), \tag{5.2}$$

which corresponds to *Fick's law*. This proportionality no longer holds for large gradients. However, even in the linear regime it is an approximation because $j(t)$ will have a *delay* relative to the gradient (see the example on diffusion to be discussed later). Since such a delay is of the order of τ_{rel} (a typical *relaxation time*), it is negligible for macroscopic properties.

Let us introduce now the *regression hypothesis*:

> The relaxation of perturbative macroscopic departures from equilibrium is ruled by the same laws as the regression of spontaneous microscopic fluctuations in an equilibrium system.

The temporal behavior of a state variable $A(t)$ is qualitatively depicted in Fig. 5.2. This regression hypothesis is a (very important) consequence of a fundamental theorem in statistical physics: *the fluctuation-dissipation theorem*, to be discussed in Chapter 6, within the context of the *linear response theory*.

In order to describe qualitatively the meaning of this hypothesis we need to speak about the correlation of *spontaneous fluctuations*, requiring the language of *temporal correlation functions*, which we have met in Chapter 1 and which will be discussed in detail in Chapter 6. Here, we shall only consider them in the simplest way. Let us consider a certain *state variable*

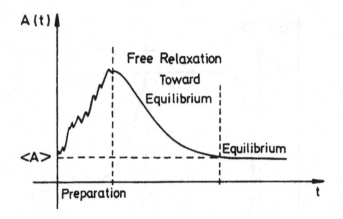

Fig. 5.2

$A(t)$, and call $\delta A(t) = A(t) - \langle A(t) \rangle$ its departure from equilibrium, whose temporal evolution is ruled by microscopic laws. In general, the behavior of $\delta A(t)$ will be of the type indicated in Fig. 5.3.

Except when A is a constant of motion, $\delta A(t)$ will appear *random*, as in the figure, even in case of an equilibrium state. Because in equilibrium $\langle \delta A(t) \rangle = 0$, an uninteresting result, it is possible to extract *non-random* information by considering the correlations of fluctuations at different times

$$C(t) = \langle \delta A(0) \delta A(t) \rangle = \langle A(0) A(t) \rangle - \langle A \rangle^2$$

$$= \int d\mathbf{r}_1 d\mathbf{p}_1 d\mathbf{r}_2 d\mathbf{p}_2 \ldots d\mathbf{r}_n d\mathbf{p}_n f(\mathbf{r}_1, \mathbf{p}_1, \ldots, \mathbf{r}_n, \mathbf{p}_n)$$

$$= \delta A(0, \mathbf{r}_1, \mathbf{p}_1, \ldots, \mathbf{r}_n, \mathbf{p}_n) \delta A(t, \mathbf{r}_1, \mathbf{p}_1, \ldots, \mathbf{r}_n, \mathbf{p}_n) \qquad (5.3)$$

In equilibrium, this process will be stationary, which means that, calling $t = t' - t''$,

$$C(t) = \langle \delta A(t') \delta A(t'') \rangle$$

$$= \langle \delta A(0) \delta A(t) \rangle = \langle \delta A(-t) \delta A(0) \rangle$$

$$= \langle \delta A(0) \delta A(-t) \rangle = C(-t) \qquad (5.4)$$

if $A(0)$ and $A(t)$ commute. A more detailed discussion on correlation functions will be given in Chapter 6; here we only need to use a couple of useful properties. The following relations are fulfilled:

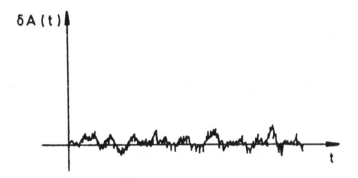

Fig. 5.3

(a) at short times

$$C(0) = \langle \delta A(0) \delta A(0) \rangle = \langle \delta A(0)^2 \rangle \tag{5.5a}$$

(b) at long times, $\delta A(t)$ shall be uncorrelated with $\delta A(0)$ and

$$\lim_{t \to \infty} C(t) = \langle \delta A(0) \rangle \langle \delta A(t) \rangle \to 0. \tag{5.5b}$$

This reduction in the magnitude of the fluctuations for long times is the *spontaneous regression of fluctuations*, mentioned in connection to the Onsager hypothesis. The averages indicated in the correlation functions may be obtained through the *ergodic hypothesis*, as $(t = t' - t'')$

$$\langle \delta A(0) \delta A(t) \rangle = \lim_{t \to \infty} \frac{1}{t} \int_0^t ds \delta A(t' + s) \delta A(t'' + s) \tag{5.5c}$$

The physical meaning of Onsager's regression hypothesis can be stated as follows: the time behavior of the correlation between $A(t)$ and $A(0)$ in a system in equilibrium, is the same as the time behavior of the average of $A(t)$ when a *natural* fluctuation occurs at $t = 0$. This corresponds to a given (*specific*) initial distribution in the phase space, but *out* of equilibrium. More specifically, in a system near equilibrium it is not possible to distinguish between spontaneous fluctuations and externally produced departures from equilibrium. It is because of this impossibility that the relaxation of $\langle \delta A(0) \delta A(t) \rangle$ and the decay to equilibrium of $\langle A(t) \rangle$ (a departure from equilibrium produced ad hoc) are coincident.

We consider now a couple of simple examples in order to clarify the meaning and implications of the above indicated hypothesis.

5.2.1   *Chemical kinetics*

We consider the simple reversible reaction

$$A \leftrightarrow B \tag{5.6}$$

where k_1 and k_2 are the reaction rates for the direct and inverse reactions. The usual kinetic equations are

$$\dot{n}_A(t) = -k_1 n_A(t) + k_2 n_B(t)$$
$$\dot{n}_B(t) = k_1 n_A(t) - k_2 n_B(t) \tag{5.7}$$

where $n_A(t)$ and $n_B(t)$ indicate the concentrations of A and B reactants respectively. They satisfy the condition $\dot{n}_A(t) + \dot{n}_B(t) = 0$. The equilibrium concentrations must obey the *detailed balance condition*:

$$0 = -k_1 \langle n_A(t) \rangle + k_2 \langle n_B(t) \rangle \tag{5.8}$$

or equivalently

$$K_{ef} \equiv \frac{\langle n_B \rangle}{\langle n_A \rangle} = \frac{k_1}{k_2}$$

The solution of Eq. (5.7) leads us to

$$\Delta n_A(t) = n_A(t) - \langle n_A \rangle$$
$$= \Delta n_A(0) e^{-t/\tau_c} \tag{5.9}$$

where

$$\tau_c^{-1} = k_1 + k_2 \tag{5.10}$$

is the relaxation time of $n_A(t)$ from an initial nonequilibrium state. The population $n_A(t)$ can be described as a nonequilibrium average of a dynamical variable N_A: $\bar{N}_A(t) \propto n_A(t)$, the *Onsager's regression hypothesis* tells us that

$$\frac{\Delta n_A(t)}{\Delta n_A(0)} = \frac{\langle \delta N_A(0) \delta N_A(t) \rangle}{\langle \delta N_A(0)^2 \rangle} \tag{5.11a}$$

from where we obtain

$$\frac{t}{\tau_c} = -\ln \left[\frac{\langle \delta N_A(0) \delta N_A(t) \rangle}{\langle \delta A(0)^2 \rangle} \right] \tag{5.11b}$$

The last equation expresses the relation between a phenomenological constant τ_c [involving the reaction rates k_1 and k_2 as indicated in Eq. (5.10)] on the l.h.s., and the microscopic dynamics in terms of the correlation that appears on the r.h.s. We see then that the *regression hypothesis* provides us with a procedure to evaluate macroscopic reaction rates from microscopic laws.

5.2.2 Self-diffusion

We consider now a very dilute solute (that is in a very low concentration) in a fluid solvent. We call $n(\mathbf{r}, t)$ the solute density out of equilibrium (corresponding to the dynamical variable $\rho(\mathbf{r}, t)$, the instantaneous density). As we have seen in Chapter 4, conservation of the particle number is expressed through the continuity equation

$$\frac{\partial}{\partial t} n(\mathbf{r}, t) = -\nabla \mathbf{j}(\mathbf{r}, t) \tag{5.12}$$

$\mathbf{j}(\mathbf{r}, t)$ being the average current of the out of equilibrium solute. The macroscopic thermodynamical driving term for mass flux is a gradient in the chemical potential. Equivalently, for a dilute solution, it is a gradient in the solute concentration. From a phenomenological point of view this is described by Fick's law

$$\mathbf{j}(\mathbf{r}, t) = -\mathcal{D} \nabla n(\mathbf{r}, t) \tag{5.13}$$

where $\mathcal{D}$ is the (transport) *self-diffusion* coefficient. According to the Onsager hypothesis, the (space-time) correlation function

$$C(\mathbf{r}, t) = \langle \delta \rho(\mathbf{r}, t) \delta \rho(\mathbf{0}, 0) \rangle$$

shall obey the same equation as $n(\mathbf{r}, t)$, that is

$$\frac{\partial}{\partial t} n(\mathbf{r}, t) = \mathcal{D} \nabla^2 n(\mathbf{r}, t) \tag{5.14a}$$

and

$$\frac{\partial}{\partial t} C(\mathbf{r}, t) = \mathcal{D} \nabla^2 C(\mathbf{r}, t) \tag{5.14b}$$

On the other hand, as $\langle \delta \rho(\mathbf{r}, t) \delta \rho(\mathbf{0}, 0) \rangle$ is proportional to $\mathcal{P}(\mathbf{r}, t)$, the conditional probability distribution of finding a particle in $(\mathbf{r}, t)$ if it was at $(\mathbf{r} = \mathbf{0}, t = 0)$, we also have

$$\frac{\partial}{\partial t} \mathcal{P}(\mathbf{r}, t) = \mathcal{D} \nabla^2 \mathcal{P}(\mathbf{r}, t) \tag{5.14c}$$

However, these equations cannot be correct at all times (in fact, Fick's law fails at short times). Let us consider the following quantity

$$\Delta R^2(t) = \langle |\mathbf{r}(t) - \mathbf{r}(0)|^2 \rangle = \int d\mathbf{r} \, r^2 \mathcal{P}(\mathbf{r}, t) \tag{5.15a}$$

whose time behavior is expressed by the equation

$$\frac{d}{dt}\Delta R^2(t) = \int d\mathbf{r} r^2 \frac{\partial}{\partial t}\mathcal{P}(\mathbf{r}, t) = \int d\mathbf{r} r^2 \mathcal{D}\nabla^2 \mathcal{P}(\mathbf{r}, t)$$

$$= 6\mathcal{D}\int d\mathbf{r}\mathcal{P}(\mathbf{r}, t) = 6\mathcal{D} \qquad (5.15b)$$

Then

$$\Delta R^2(t) = 6\mathcal{D}t \qquad (5.15c)$$

This is *Einstein's relation* (which we have met in Chapter 2), when addressing to the Brownian motion problem, and which will meet again in Chapter 6 in relation with linear response theory), that holds only after an initial transient has elapsed. The behavior of $\Delta R^2(t)$ depends on the motion being diffusive or inertial:

$$diffusion \Rightarrow \Delta R^2(t) \propto t \qquad (5.16a)$$

$$inertial \Rightarrow \Delta R^2(t) \propto t^2 \qquad (5.16b)$$

This kind of behavior is schematically shown in Fig. 5.4.

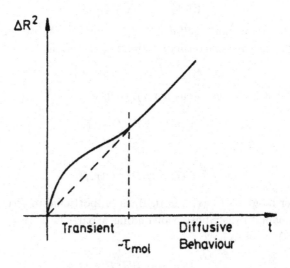

Fig. 5.4

This result indicates that in the inertial regime, the particle wanders further away than in the diffusive case (remember that in the diffusive case there is an underlying *random-walk*). On the other hand, we have

$$\mathbf{r}(t) - \mathbf{r}(0) = \int_0^t dt' \mathbf{v}(t')$$ (5.17)

where we have used $d\mathbf{r}(t)/dt = \mathbf{v}(t)$. We then have

$$\Delta R^2(t) = \int_0^t dt' \int_0^t dt'' \langle \mathbf{v}(t')\mathbf{v}(t'') \rangle$$ (5.18)

Its temporal evolution is given by

$$\frac{d}{dt}\Delta R^2(t) = 2\langle \mathbf{v}(t) \cdot [\mathbf{r}_1(t) - \mathbf{r}_1(0)] \rangle = 2\langle \mathbf{v}(0) \cdot [\mathbf{r}_1(0) - \mathbf{r}_1(-t)] \rangle$$

$$= 2 \int_{-t}^0 dt' \langle \mathbf{v}(0)\mathbf{v}(t') \rangle$$ (5.19a)

On the other hand, from Eq. (5.15b), $\frac{d}{dt}\Delta R^2(t) = 6\mathcal{D}$. Hence, at long times we find

$$\mathcal{D} = \frac{1}{3} \int_0^\infty dt \langle \mathbf{v}(0)\mathbf{v}(t) \rangle$$ (5.19b)

Relations of this type, connecting a transport coefficient (such as the diffusion coefficient $\mathcal{D}$) with an integral of a correlation function, are known as *Green–Kubo formulas*, which we will meet again within the context of linear response theory. Here we shall also define

$$\tau_{\text{rel}} = \int_0^\infty dt \frac{\langle \mathbf{v}(t)\mathbf{v}(0) \rangle}{\langle \mathbf{v}^2 \rangle} = \frac{m\mathcal{D}}{kT} \simeq \tau_{\text{mol}}$$ (5.20)

which is the time required to reach the diffusive regime.

5.3 Onsager's relations

In this section we are going to give a more general presentation of Onsager's ideas and their origin, namely, *microscopic reversibility*. The first point is to show that *Hamiltonian dynamical microscopic reversibility* (or invariance under *temporal inversion*) implies that the microscopic time dependent correlation functions fulfills

$$\langle \delta A_i \delta A_j(t) \rangle = \langle \delta A_i(t) \delta A_j \rangle$$ (5.21)

where $\delta A_i = A_i - \langle A_i \rangle$ are the fluctuations of the set of state variables $\{A_i\}$ around equilibrium. In order to obtain the above result Eq. (5.21) we start by noting that

$$\langle\{\delta A\}\{\delta A(t)\}\rangle = \int\int d\{\delta A\}d\{\delta A'\}$$

$$= \{\delta A\}\{\delta A'\}\mathcal{P}(\{\delta A'\})\mathcal{P}(\{\delta A\}, t|\{\delta A'\}) \qquad (5.22)$$

where $\{\delta A\} = (\delta A_1, \delta A_2, \ldots, \delta A_N)$ and $\mathcal{P}(\{\delta A\}, t|\{\delta A'\})$ is the conditional probability of finding the set of values $\{\delta A\}$ at time t, provided we had $\{\delta A'\}$ at time $t = 0$, and [see Sec. 5.6, Eqs. (5.74) and (5.75)]

$$\mathcal{P}(\{\delta A\}) = ((2\pi k_B)^n \det \mathbf{G})^{1/2} e^{-\{\delta A\}^T \circ \mathbf{G} \circ \{\delta A\}/2k_B} \qquad (5.23a)$$

The entropy variation due to the fluctuations is [see Eq. (5.73)]

$$\Delta S = -\frac{1}{2}\{\delta A\}^T \mathbf{G}\{\delta A\} \qquad (5.23b)$$

It will be useful to introduce the *generalized forces and fluxes* defined respectively through

$$\mathbb{X} = \mathbf{G}\{\delta A\} = -\frac{\partial \Delta S}{\partial \{\delta A\}} \qquad (5.24a)$$

$$\mathbb{J} = \frac{d}{dt}\{\delta A\} \qquad (5.24b)$$

and the entropy change will then obey the equation

$$\frac{d}{dt}\Delta S = -\mathbb{J}\mathbb{X} \qquad (5.24c)$$

Remember that $\{\delta A\}$ are a set of macroscopic variables and that for each choice of this set there are a large number of possible microscopic states, which could be indicated by

$$\mathcal{P}(\{\delta A'\})\mathcal{P}(\{\delta A\}, t|\{\delta A'\}) =$$

$$= \frac{1}{\Omega_{\Delta E}(E)}\int\int dq_1 dp_1 \ldots dq_N dp_N \int\int dq'_1 dp'_1 \ldots dq'_N dp'_N$$

$$P(q_1, p_1, \ldots, q_N, p_N, t|q'_1, p'_1, \ldots, q'_N, p'_N) \qquad (5.25)$$

where we have used that $\rho(q_1, p_1, \ldots, q_N, p_N) = \Omega_{\Delta E}(E)^{-1}$ for a closed and isolated system ($\Omega_{\Delta E}(E)$ being the *energy shell* volume). Because a classical system is completely deterministic, we have

$$P(q_1, p_1, \ldots, q_N, p_N, t|q, p, \ldots, q, p) =$$

$$= \delta[q_1 - q - \Delta q_1(q_1, p_1, \ldots, q_N, p_N, t)] \ldots$$

$$\ldots \delta[p_N - p - \Delta p_N(q_1, p_1, \ldots, q_N, p_N, t)] \qquad (5.26a)$$

where $\Delta q_i(q_1, p_1, \ldots, q_N, p_N, t)$ and $\Delta p_i(q_1, p_1, \ldots, q_N, p_N, t)$ are univocally determined by Hamilton equations. Because the latter are invariant under time inversion, it follows that

$$P(q_1, p_1, \ldots, q_N, p_N, t | q\,, p\,, ..q\,, p\,) =$$
$$= P(q\,, -p\,, \ldots, q\,, -p\,, t | q_1, -p_1, \ldots, q_N, -p_N) \qquad (5.26b)$$

Replacing this in Eq. (5.25) we find

$$\mathcal{P}(\{\delta A\})\mathcal{P}(\{\delta A\}, t | \{\delta A'\}) = \mathcal{P}(\{\delta A'\})\mathcal{P}(\{\delta A'\} | \{\delta A\}, -t) \qquad (5.26c)$$

Using this result in Eq. (5.22), Eq. (5.21) follows immediately.

Let us see now how these results, together with the hypothesis of fluctuations regression, led to expressions for the transport coefficients, known as *Onsager's relations*. Following Onsager's derivation, the average of fluctuations obeys an equation having the form

$$\frac{d}{dt}\langle\{\delta A\}_0\{\delta A(t)\}\rangle = \frac{d}{dt}\langle\{\delta A(t)\}\rangle_0 = -\mathbb{M}\circ\langle\{\delta A(t)\}\rangle_0 \qquad (5.27a)$$

whose solution is

$$\langle\{\delta A(t)\}\rangle_0 = e^{-\mathbb{M}\circ t}\{\delta A\}_0 \qquad (5.27b)$$

where $\{\delta A\}_0$ indicates the set of values at $t = 0$, and $\langle\{\delta A(t)\}\rangle_0$ the average value of $\{\delta A(t)\}$, provided we have the set of values $\{\delta A\}_0$ at $t = 0$. It is clear that the time derivative in Eq. (5.27a) must be taken with caution, in the sense of considering, say through a finite difference scheme, that the time increment Δt shall fulfill the relation

$$\tau_{\text{coll}} \ll \Delta t \ll \tau_{\text{rel}}$$

with τ_{coll} the average time between collisions (if we have in mind a fluid system), τ_{rel} the relaxation time.

In order to impose conditions over $\mathbb{M}$, we do a short time expansion of the Eq. (5.27b)

$$\langle\{\delta A(t)\}\rangle_0 = \{\delta A\}_0 - t\mathbb{M}\circ\{\delta A\}_0 \qquad (5.28a)$$

which gives

$$\langle\{\delta A\}_0\mathbb{M}\circ\{\delta A\}_0\rangle = \langle\mathbb{M}\circ\{\delta A\}_0\{\delta A\}_0\rangle \qquad (5.28b)$$

when substituted into Eq. (5.21) or (5.22). Here we have used the relation between transposed matrices

$$\mathbb{M}\circ\{\delta A\} = \{\delta A\}^T\circ\mathbb{M}^T$$

Using the relation indicated in Sec. 5.6 [Eqs. (5.75), (5.76)] for the variance of the fluctuations, we have

$$\mathbb{G}^{-1} \circ \mathbb{M}^T = \mathbb{M} \circ \mathbb{G}^{-1} \qquad (5.28c)$$

We define now the matrix

$$\mathbb{L} = \mathbb{M} \circ \mathbb{G}^{-1} \qquad (5.29a)$$

that, according to Eq. (5.28c), has the properties

$$\mathbb{L} = \mathbb{L}^T \quad \text{or} \quad L_{ij} = L_{ji} \qquad (5.29b)$$

that are the *Onsager relations* we were looking for. According to Eq. (5.24a), Eq. (5.27a) will be written as

$$\mathbb{J} = \frac{d}{dt}\langle\{\delta A(t)\}\rangle_0 = -\mathbb{L}\langle\mathbb{X}(t)\rangle_0 \qquad (5.30)$$

So far, in our discussion we have said nothing about the presence of magnetic or rotational fields in the system. However, whenever either or both are present, we shall separate the variables A_i into two sets, those that are invariant under time inversion, A_g, and those that are not, A_u (for instance a velocity). Keeping this in mind and working as with the Hamilton equations, we find that *Onsager's relations* adopt the form

$$\mathbb{L}^{gg}(-\mathbb{B})^T = \mathbb{L}^{gg}(\mathbb{B})$$

$$\mathbb{L}^{gu}(-\mathbb{B})^T = -\mathbb{L}^{ug}(\mathbb{B})$$

$$\mathbb{L}^{ug}(-\mathbb{B})^T = -\mathbb{L}^{gu}(\mathbb{B})$$

$$\mathbb{L}^{uu}(-\mathbb{B})^T = \mathbb{L}^{uu}(\mathbb{B}) \qquad (5.31)$$

where $\mathbb{B}$ is a magnetic field and g and u indicate the *matrix-blocks* of even or odd variables respectively. Similar results are obtained if we consider a system rotating with an angular velocity Ω, instead of being subject to a magnetic field.

We now proceed to consider some illustrative examples of application of these results in practical situations. After that we will discuss the important theorem of *minimum production of entropy*, for systems in a steady state out of equilibrium.

5.4   Examples of Onsager's relations

Let us consider the system depicted in Fig. 5.5, composed of two containers or boxes (called I and II respectively) containing the same ideal gas,

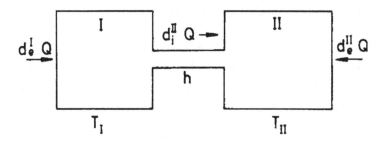

Fig. 5.5

connected through a pipe (or hole) indicated by h. The total mass and the energy of the particles is constant, though it could be transferred from one box to the other (each compartment being in contact with its respective *thermal bath*). Without taking into account the mass transfer, the increase of entropy dS in both boxes will be

$$dS = d_e S + d_i S \tag{5.32a}$$

$$d_e S = \frac{d_e S^I}{T_I} + \frac{d_e S^{II}}{T_{II}} \tag{5.32b}$$

$$d_i S = d_i Q^I \left(\frac{1}{T_{II}} - \frac{1}{T_I} \right) \tag{5.32c}$$

The first contribution in Eq. (5.32a), $d_e S$, corresponds to the change of entropy due to the interaction of volumes I and II, with their respective thermal baths (entropy flux coming from outside the system). The second one, $d_i S$, is the entropy increase inside the system (entropy production). Thermodynamics tells us that

$$d_i S > 0 \tag{5.33a}$$

because heat transfer inside a (closed) system, always increases the entropy. The rate of entropy change is

$$\frac{dS}{dt} = \frac{d_e S}{dt} + \frac{d_i S}{dt} \tag{5.33b}$$

Then, in a stationary state (heat flowing from the *hot* to the *cold* box)

$$\frac{dS}{dt} = 0 \qquad \left[\begin{array}{l} \frac{d_i S}{dt} > 0 \\ \frac{d_e S}{dt} > 0 \end{array} \right. \tag{5.34}$$

The heat flux (or current) $\mathbb{J}$ is defined as

$$\mathbb{J} = \frac{d_i Q^{II}}{dt} \tag{5.35a}$$

and, correspondingly, the generalized force $\mathbb{X}$ is

$$\mathbb{X} = \frac{1}{T_{II}} - \frac{1}{T_I} \tag{5.35b}$$

Near equilibrium we have seen that it is possible to take

$$\mathbb{J} = \mathbb{L}\mathbb{X} \tag{5.35c}$$

leading to

$$\frac{d_i S}{dt} = \mathbb{L}\mathbb{X}^2 \tag{5.35d}$$

This relation, together with Eq. (5.35a), leads to $\mathbb{L} > 0$, because a positive production of entropy is characteristic of a true irreversible process.

We assume now that there is a mass flux between boxes I and II. Due to the conservation of the total mass we have $-dM_I = dM_{II}$, and this transfer comes jointly with an increase of entropy. Then we have for the entropy change rate:

$$\frac{d_i S}{dt} = \frac{d_i Q^{II}}{dt}\left(\frac{1}{T_{II}} - \frac{1}{T_I}\right) - \frac{dM_{II}}{dt}\left(\frac{\mu_{II}}{T_{II}} - \frac{\mu_I}{T_I}\right) \tag{5.36a}$$

where the thermodynamic forces are given by

$$\mathbb{X}_u = \left(\frac{1}{T_{II}} - \frac{1}{T_I}\right) = \Delta\left(\frac{\partial S}{\partial u}\right)_M$$

$$\mathbb{X}_M = \left(\frac{\mu_{II}}{T_{II}} - \frac{\mu_I}{T_I}\right) = \Delta\left(\frac{\partial S}{\partial M}\right)_u \tag{5.36b}$$

where μ_i indicate the appropriate *chemical potentials*. The fluxes or currents are defined by

$$\mathbb{J}_u = \frac{d_i^{II} Q}{dt} \qquad \mathbb{J}_M = \frac{dM_{II}}{dt} \tag{5.36c}$$

The linear relations between thermodynamical fluxes and forces give

$$\mathbb{J}_M = \mathbb{L}_{MM}\mathbb{X}_M + \mathbb{L}_{Mu}\mathbb{X}_u$$

$$\mathbb{J}_u = \mathbb{L}_{uM}\mathbb{X}_M + \mathbb{L}_{uu}\mathbb{X}_u \tag{5.37a}$$

This assumption gives the following form for the entropy production

$$\frac{d_i S}{dt} = \mathbb{L}_{MM}\mathbb{X}_M^2 + \mathbb{L}_{Mu}\mathbb{X}_u\mathbb{X}_M + \mathbb{L}_{uM}\mathbb{X}_M\mathbb{X}_u + \mathbb{L}_{uu}\mathbb{X}_u^2 \tag{5.37b}$$

where, since $d_i S/dt$ must be positive definite, we have

$$\mathbb{L}_{MM} > 0 \qquad \mathbb{L}_{uu} > 0$$

$$\mathbb{L}_{MM}\mathbb{L}_{uu} - \mathbb{L}_{Mu}\mathbb{L}_{uM} > 0 \tag{5.37c}$$

From the *Onsager relations*, given in Eqs. (5.29b) and (5.31), we have

$$\mathbb{L}_{Mu} = \mathbb{L}_{uM} \tag{5.38}$$

According to the examples we want to analyze, it will be convenient to rewrite the previous equations with the temperature T and pressure p as independent variables. Hence

$$\mathbb{X}_u = \left(\frac{1}{T_{II}} - \frac{1}{T_I}\right) = \Delta\left(\frac{1}{T}\right) = -\frac{\Delta T}{T^2}$$

$$\mathbb{X}_M = \left(\frac{\mu_{II}}{T_{II}} - \frac{\mu_I}{T_I}\right) = -\Delta\left(\frac{\mu}{T}\right) = -\frac{v\Delta p}{T} + h\frac{\Delta T}{T^2} \tag{5.39a}$$

where we have used the *Gibbs-Duhem equation*

$$d\mu = -sdT + vdp \tag{5.39b}$$

while the enthalpy per unit mass is

$$h = u + vp = sT + \mu \tag{5.39c}$$

and s, v and u indicate the entropy, volume and internal energy per unit mass. From Eqs. (5.37a) and Onsager relations, we have

$$\mathbb{J}_u = -\mathbb{L}_{uM}\frac{v}{T}\Delta p + \frac{\mathbb{L}_{uM}h - \mathbb{L}_{uu}}{T^2}\Delta T \tag{5.40}$$

$$\mathbb{J}_M = -\mathbb{L}_{MM}\frac{v}{T}\Delta p + \frac{\mathbb{L}_{MM}h - \mathbb{L}_{Mu}}{T^2}\Delta T$$

in terms of T and p as independent variables. We shall now see how to use the Onsager relations to connect two, in principle completely different, effects. We start discussing the mechano-caloric effect.

5.4.1   *Mechano-caloric effect*

We assume that **both boxes are at the same temperature**, $\Delta T = 0$, but at different pressures, $\Delta p \neq 0$. Hence, the energy flux is induced by the mass flux, with a transfer of energy *, that is obtained from

$$\mathbb{J}_u = -\mathbb{L}_{uM}\frac{v}{T}\Delta p$$

$$\mathbb{J}_M = -\mathbb{L}_{MM}\frac{v}{T}\Delta p \tag{5.41a}$$

giving

$$\mathbb{J}_u = \frac{\mathbb{L}_{uM}}{\mathbb{L}_{MM}}\mathbb{J}_M \Rightarrow \frac{\mathbb{L}_{uM}}{\mathbb{L}_{MM}} = u^* \tag{5.41b}$$

We have then found that the pressure gradient causes a matter flow ($\mathbb{J}_M$), which carries heat. The quantity u^* is the proportionality constant determining the amount of energy transferred per unit of mass by the particles that move through the hole. This quantity may be calculated microscopically. Two possible limiting cases are:

(a) *Knudsen Gas*, related to the effusion problem discussed in Chapter 3; the particle mean free path is longer than the hole size;
(b) *Boyle Gas*, when the size of the hole is much larger than the particle mean free path;

Let us take the first case, a *Knudsen Gas*. Here every particle reaching the hole goes freely through it. The transferred energy u^* is

$$u^* = \frac{\int_0^\infty dv_x \int \int_{-\infty}^{+\infty} dv_y dv_z \frac{m}{2} v^2 n v_x \mathcal{F}(\mathbf{v})}{\int_0^\infty dv_x \int \int_{-\infty}^{+\infty} dv_y dv_z n v_x \mathcal{F}(\mathbf{v})} = 2k_B T \qquad (5.42a)$$

and within the framework discussed in Chapter 3, this is an obvious result. Here, n is the molecular density and $\mathcal{F}(\mathbf{v})$ is the Maxwell distribution. Hence, we have

$$u^* = \frac{2RT}{m} \qquad (5.42b)$$

with R the gas constant ($R = N_A k_B$, N_A being Avogadro's number).

In the second case, a *Boyle gas*, the particles are not able to go through the hole freely, and a certain amount of work will be done. The amount of work is given by vp, a quantity that will be added to the energy transfer u,

$$u^* = u + vp = h \qquad (5.42c)$$

where h is the enthalpy per unit mass.

5.4.2　*Thermo-molecular pressure effect*

We now consider the relation between temperature and pressure in the case when there is no net flux of mass (i.e. $\mathbb{J}_M = 0$), but there is a net flux of energy ($\mathbb{J}_u \neq 0$). In such a case

$$\frac{\Delta p}{\Delta T} = \frac{\mathbb{L}_{MM} h - \mathbb{L}_{Mu}}{\mathbb{L}_{MM} vT} = \frac{h - \mathbb{L}_{Mu}/\mathbb{L}_{MM}}{vT} = \frac{h - u^*}{vT} = -\frac{q^*}{vT} \qquad (5.43)$$

that has the form of a Clausius–Clapeyron equation (*though it is not!*), with $q^* = u^* - h$. Here we see that, in a solution, the pressure gradient has to be transformed into a concentration gradient. The possibility of maintaining the diffusion-free state by an adequate concentration gradient is called the *Ludwig–Soret effect*. The opposite phenomenon, that the diffusion of two substances into each other can produce a temperature gradient, the *Dufour effect*, is not discussed here. All these effects are particularly relevant for semipermeable membranes which allow the solvent, but not the dissolved substance, to pass. When the membrane is rigid, what will appear is not a temperature gradient but a pressure gradient, called *osmotic pressure*.

Now we look the two cases discussed previously:

(a) *Knudsen Gas*: Since the enthalpy per unit mass for an ideal gas is

$$h = \frac{5}{2} \frac{RT}{m},$$

it follows from Eq. (5.42b) that

$$\frac{\Delta p}{\Delta T} = -\frac{1}{vT} \left(2 - \frac{5}{2} \right) \frac{RT}{m} = \frac{1}{2} \frac{RT}{vTm} = \frac{1}{2} \frac{pv}{vT} = \frac{p}{2T} \qquad (5.44a)$$

and after integration (taking care of the fact that Δp and ΔT are finite) we get

$$\frac{p_1}{T_1^{1/2}} = \frac{p_2}{T_2^{1/2}} \qquad (5.44b)$$

Hence, for a Knudsen gas it is possible to have a difference of temperature and pressure between both boxes, even without a mass flux. This phenomenon is called *thermo-osmosis*, and is analogous to the *source effect* in a superfluid.

(b) *Boyle Gas*: Taking into account Eqs. (5.42b), (5.42c), the Onsager relations (5.38) and the previous result, we find

$$\frac{\Delta p}{\Delta T} = 0 \qquad (5.44c)$$

This indicates that for a Boyle gas, if there is no mass flux, there cannot exist a pressure difference between the boxes, even if there is a temperature difference.

As a final example we consider a simple thermocouple consisting of two wires of different metals with a capacitance C, with one of them interrupted as indicated in Fig. 5.6. It is known from experiments that keeping the points denoted by 1 and 2 at different temperatures T and $T + \Delta T$, produces not only a heat current $\mathbf{j}_q$, but also an electrical current $\mathbf{j}_e$ in the wires, as well as a potential difference $\Delta \Psi$ across the capacitance. In order to analyze the coupled thermal and electrical effects, within the linear theory, we shall choose the thermodynamical forces and currents with some freedom, but always requiring that their product must correspond to a term in the entropy production.

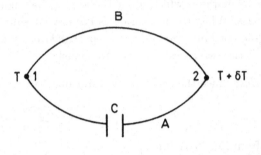

Fig. 5.6

For the heat conduction term we choose as the thermodynamical force

$$\mathbb{X}_q = \nabla \frac{1}{T} \tag{5.45}$$

Now, since

$$S_q = -\frac{1}{T^2} \mathbf{j}_q \cdot \nabla T = \mathbb{J}_q \mathbb{X}_q \tag{5.46a}$$

our heat flux coincides with the heat current

$$\mathbb{J}_q = \mathbf{j}_q \tag{5.46b}$$

which, through the use of Fourier's law, becomes

$$\mathbf{j}_q = -\lambda \nabla T = \lambda T^2 \nabla \frac{1}{T} = \lambda T^2 \mathbb{X}_q. \tag{5.46c}$$

Now, for the thermodynamical flux corresponding to electrical conduction we have as the electric current

$$\mathbf{j}_e = \langle e \sum_i \dot{\mathbf{r}}_i(t) \rangle \tag{5.47a}$$

where $\mathbf{r}_i(t)$ is the position of the i–th particle at time t (we assume that there are N electrons). Clearly, this flux is the time derivative of an extensive thermodynamical variable of the system, namely the electric dipole moment. This equation can be rewritten as

$$\mathbf{j}_e = -\frac{e}{m} \rho_e \mathbf{v}_e \tag{5.47b}$$

where ρ_e is the electron number density and $\mathbf{v}_e$ their drift velocity relative to the ionic (fixed) background. The entropy production associated with this current appears as

$$S_e = -\left(\frac{e}{T}\mathcal{E} + \nabla\frac{\mu}{T}\right)\mathbb{J}_e \tag{5.48a}$$

from which it is possible to identify the associated thermodynamical force as

$$\mathbb{X}_e = -\frac{e}{T}\mathcal{E} - \nabla\frac{\mu}{T}. \tag{5.48b}$$

The linear equations relating fluxes and forces, equivalent to Eq. (5.37a), become

$$\mathbf{j}_e = -\mathbb{L}_{ee}\left(\frac{e}{T}\mathcal{E} + \nabla\frac{\mu}{T}\right) \circ \mathbf{j}_e + \mathbb{L}_{eq}\nabla\frac{1}{T}$$

$$\mathbf{j}_q = -\mathbb{L}_{qe}\left(\frac{e}{T}\mathcal{E} + \nabla\frac{\mu}{T}\right) \circ \mathbf{j}_e + \mathbb{L}_{qq}\nabla\frac{1}{T}. \tag{5.49}$$

The isothermal electrical conductivity is $\sigma = -(e\mathbb{L}_{ee}/T)$ and the thermal conductivity can be obtained taking $\mathbf{j}_e = 0$,

$$\lambda = \{\mathbb{L}_{qq}\mathbb{L}_{ee} - \mathbb{L}_{qe}\mathbb{L}_{eq}\}\mathbb{L}_{ee}^{-1} \tag{5.50}$$

For our model of metal wires, we could treat the system as one-dimensional, and simplify the expressions eliminating vector notation.

Now we are in position of obtaining some useful relations. We first analyze the stationary state with $\Delta T =$const and $j_e = 0$, implying that

$$\mathbb{L}_{ee}\mathbb{X}_e + \mathbb{L}_{eq}\mathbb{X}_q = 0 \tag{5.51a}$$

giving

$$\frac{\Delta\Psi}{\Delta T} = -\frac{\mathbb{L}_{eq}}{\mathbb{L}_{ee}}T \tag{5.51b}$$

This corresponds to the *Seebeck effect*, giving the potential difference that results in association with a given temperature difference in a thermocouple, when there is no electric current flow. Next, if we impose a fixed potential difference $\Delta\Psi =$const across the capacitance and keep $\Delta T = 0$, we may find the dependence of the generated heat current upon the applied electric current, that is

$$\frac{j_q}{j_e} = \frac{\mathbb{L}_{eq}}{\mathbb{L}_{ee}} = \Pi \tag{5.52}$$

corresponding to the *Peltier effect*. At this point by using the Onsager relations, we can relate the Peltier and Seebeck effects, and the related parameters:

$$\frac{\Delta\Psi}{\Delta T} = -\frac{\Pi}{T} \tag{5.53}$$

which is known as the *second Thompson relation*.

5.5   Minimum production of entropy

In irreversible phenomena, there is an important class of processes that play a role analogous to that of equilibrium states in reversible thermodynamics. These are the steady state processes which are subject to external constraints and are characterized by time-independent forces and fluxes. Just as isolated systems in equilibrium are characterized by a maximum of entropy, Prigogine has proved a theorem stating that stationary nonequilibrium states are characterized by a minimum of the entropy production.

The *minimum entropy production theorem* states that

> "*In a steady state, a system where an irreversible process takes place is in such a state that the rate of entropy production has the minimum value, consistent with the external constraints which prevent the system from reaching equilibrium. When there are no constrains the system evolves towards a state in which the rate of entropy production is zero, i.e. to the equilibrium state.*"

However, this principle gives only the steady state solution under the restrictive condition that the temperature be high in comparison with the typical energy level difference. In order to prove the theorem, rather restrictive assumptions are made about the system:

- that it is described by linear phenomenological laws with constant coefficients satisfying the Onsager relations,
- that it is subject to time independent boundary conditions.

We present now a rather general proof of the theorem and afterwards, a simple example.

In a system with n components having densities ρ_i, where there are mass fluxes and reactions, it is possible to rewrite the mass balance equation as a set of n coupled equations

$$\frac{\partial \rho_i}{\partial t} = \nabla \sum_j L_{ij} \nabla \frac{\mu_i}{T} + \sum_{kl} \nu_{ik} \ell_{kl} \frac{a_l}{T} \qquad (5.54a)$$

where ν_{ij} are the *stoichiometric reaction factors*, a_l the chemical *affinities*, and L_{ij} and ℓ_{ij} the phenomenological transport coefficients that fulfill the Onsager relations

$$L_{ij} = L_{ji} \qquad \ell_{kl} = \ell_{lk} \qquad (5.54b)$$

The total entropy production is given by [see Eq. (5.35d)]

$$S = \int dV s = \frac{1}{T^2} \int dV \left[\sum_{ij} L_{ij} \nabla \mu_i \nabla \mu_j + \sum_{ij} \ell_{ij} a_i a_j \right] \qquad (5.55)$$

Let us study how the system evolves toward a stationary state (not necessarily an equilibrium one). In order to do this, we need to evaluate $\frac{dS}{dt}$, and thus the coefficients L_{ij} and ℓ_{ij}. These must take their equilibrium values if we assume them constant. Recalling the relation that holds between the affinities a_l and the chemical potentials μ_k

$$a_l = - \sum_k \mu_k \nu_{kl}$$

we have

$$\frac{dS}{dt} = \frac{2}{T^2} \int dV \left[\sum_{ij} L_{ij} \nabla \mu_i \nabla \frac{\partial \mu_j}{\partial t} - \sum_{ijk} \ell_{ij} a_i \nu_{jk} \frac{\partial \mu_k}{\partial t} \right] \qquad (5.56)$$

But, according to local thermodynamics $\mu_i = \mu_i(\{\rho_j\})$. Then

$$\frac{\partial \mu_j}{\partial t} = \sum_i \left(\frac{\partial \mu_j}{\partial \rho_i} \right)_\rho \frac{\partial \rho_i}{\partial t}, \qquad (5.57)$$

leading to the expression

$$\frac{dS}{dt} = \frac{2}{T^2} \int dV \qquad (5.58)$$

$$\left[\sum_{ijk} L_{ij} \nabla \mu_i \nabla \left(\frac{\partial \mu_i}{\partial \rho_k} \right)_\rho \frac{\partial \rho_k}{\partial t} - \sum_{ijkl} \ell_{ij} a_i \nu_{jk} \left(\frac{\partial \mu_k}{\partial \rho_l} \right)_\rho \frac{\partial \rho_l}{\partial t} \right]$$

where the subindex ρ indicates that this quantity remains constant. The first term, including a divergence, may be transformed into a surface integral through a partial integration, yielding

$$\left(\frac{dS}{dt}\right)_{\text{surf}} = \frac{2}{T^2} \int_\Sigma d\sigma \mathbf{n} \circ \sum_{ijk} \left(\frac{\partial \mu_i}{\partial \rho_k}\right)_\rho \frac{\partial \rho_k}{\partial t} L_{ij} \nabla \mu_i. \tag{5.59}$$

However, imposing time independent boundary conditions (in order to let the system reach a stationary state), this term vanishes. Using Eq. (5.54a), the other contribution leads us to

$$\frac{dS}{dt} = -\frac{2}{T^2} \int dV \sum_{kl} \left(\frac{\partial \mu_k}{\partial \rho_l}\right)_\rho \frac{\partial \rho_l}{\partial t} \frac{\partial \rho_k}{\partial t}. \tag{5.60}$$

At this point it is necessary to resort to some properties of state functions at equilibrium. In an open thermodynamical system in equilibrium it is known that the generalized thermodynamical potential $\theta = \theta(T, V, \{\mu_i\})$ is a minimum, i.e.:

$$(\delta\theta)_{eq} = 0 \quad ; \quad (\delta^2\theta)_{eq} \geq 0 \tag{5.61}$$

where $\delta\theta$ indicates the variation of the potential θ. Introducing the potential density ϕ_v through

$$\theta = \int dV \phi_v \tag{5.62a}$$

we have, for an isothermal system without convection, that the variation in ϕ_v is

$$\delta\phi_v = \sum_i \rho_i \delta\mu_i \tag{5.62b}$$

and

$$\left(\delta^2\phi_v\right)_{eq} = \sum_i \delta\rho_i \delta\mu_i = \sum_{ij} \left(\frac{\partial \mu_i}{\partial \rho_j}\right)_{eq} \delta\rho_i \delta\mu_i \geq 0 \tag{5.62c}$$

which follows from the inequality in Eq. (5.61). Such conditions are usually called *thermodynamical stability conditions*. Going back to the last equation, the variation of the variables, $\delta\rho_i$, are arbitrary and may be represented by a variation of ρ_i arising from their time evolution. This point leads us to consider that the quadratic forms in Eqs. (5.60) and (5.62c) are identical, and then

$$\int dV \sum_{ij} \left(\frac{\partial \mu_i}{\partial \rho_j}\right)_{eq} \delta\rho_i \delta\mu_i \geq 0 \tag{5.63}$$

and correspondingly $dS/dt \leq 0$, indicating that for a non-equilibrium steady state (of *a linear system*) the entropy production becomes a minimum, compatible with the constraints applied to the system and provided the equilibrium state itself is stable. We note that if the system is removed by some external disturbance from the steady state, it will return to it in virtue of the minimum entropy theorem. We then say that the steady state is *asymptotically stable*. As a corollary, it is interesting to point out that sustained or even damped oscillations of the state variables are not possible near thermodynamical equilibrium.

As an example we analyze now a system composed of N particles each one having two states with energies $\epsilon_1 = 0$ and $\epsilon_2 = \epsilon > 0$ respectively. The system is in contact with a thermal bath at temperature T, and subject to monochromatic radiation with frequency $\omega = \epsilon/\hbar$. The irreversible process that takes place is the conversion of the energy of this monochromatic radiation into thermal energy of the bath. We call p_1 and p_2 the probabilities of finding a given particle in the lower or upper states respectively. A particle of the system can make a transition between these two states exchanging an energy ϵ with the thermal bath. Also, since the system is irradiated with monochromatic radiation whose quanta have energy ϵ, a particle can make a transition by exchanging this amount of energy with the radiation field. Keeping this in mind, the equation for the time variation of p_1 has the form

$$\frac{dp_1}{dt} = \left[\mathcal{W}_t e^{\epsilon/kT} + \mathcal{W}_r \right] p_2 - \left[\mathcal{W}_t + \mathcal{W}_r \right] p_1 \tag{5.64}$$

where $\mathcal{W}_t$ is the transition probability per unit time for a transition from the lower to the upper state due to the coupling *with the heat bath*, and $\mathcal{W}_r$ is the (symmetric) transition probability per unit time due to the coupling *with the radiation field*. Here we explicitly use the fact that, for transitions due to exchange of energy with the thermal bath, downward transitions are more probable than upward ones by a Boltzmann–like factor. In thermal equilibrium, with $\mathcal{W}_r = 0$, we know that $p_1/p_2 = \exp\{\epsilon/kT\}$. Clearly, we have $dp_2/dt = -dp_1/dt$, because $p_1 + p_2 = 1$. Setting $dp_1/dt = 0$, we get the steady state of the system with $\mathcal{W}_r \neq 0$ and a value of p_1 given by

$$p_1^s = \left[\mathcal{W}_t e^{\epsilon/kT} + \mathcal{W}_r \right] \left[\mathcal{W}_t \left(e^{\epsilon/kT} + 1 \right) + 2\mathcal{W}_r \right]^{-1}$$

$$= \left[e^{\epsilon/kT} + \beta \right] \left[e^{\epsilon/kT} + 1 + \beta \right]^{-1} \tag{5.65}$$

where $\beta = \mathcal{W}_r/\mathcal{W}_t$. The rate of entropy production will be given by the sum of two terms: the entropy production in the system and the entropy production in the heat bath. As usual, we obtain for the entropy production within the system

$$\frac{dS_i}{dt} = -N\frac{d}{dt}\left(p_1 \ln p_1 + p_2 \ln p_2\right)$$

$$= -N \ln\left(\frac{p_1}{p_2}\right)\left[\mathcal{W}_t e^{\epsilon/kT} + \mathcal{W}_r\right]p_2 - \left[\mathcal{W}_t + \mathcal{W}_r\right]p_1 \qquad (5.66a)$$

while for the entropy production of the heat bath we obtain

$$\frac{dS_t}{dt} = N\left(\frac{\epsilon}{kT}\right)\left[\mathcal{W}_t e^{\epsilon/kT}p_2 - \mathcal{W}_t p_1\right]. \qquad (5.66b)$$

The last result arises because the heat bath gains or losses an amount of entropy of the order of $\epsilon/k_B T$ for each downward or upward transition respectively. The total production of entropy will be

$$\frac{dS}{dt} = N\mathcal{W}_t\left[\left(e^{\epsilon/kT}p_2 - p_1\right)\ln\left(e^{\epsilon/kT}\frac{p_2}{p_1}\right) + \beta(p_2 - p_1)\ln\left(\frac{p_2}{p_1}\right)\right]$$
$$(5.66c)$$

The state of minimum production of entropy is obtained by minimizing Eq. (5.66c) subject to the constraint $p_1 + p_2 = 1$. This results in

$$\left(e^{\epsilon/kT} + 1\right)\ln\left(e^{\epsilon/kT}\frac{p_2}{p_1}\right) + 2\beta\ln\left(\frac{p_2}{p_1}\right)$$
$$+ \left[\left(e^{\epsilon/kT} + \beta\right)p_2 - (1 + \beta)p_1\right]\left(p_2^{-1} + p_1^{-1}\right) = 0 \qquad (5.67)$$

A careful analysis of this equation shows that it has a simplified form when both $e^{\epsilon/kT}p_2/p_1$ and p_2/p_1 are of the order of unity (differing from one by a very small, second order quantity). This lead us to the previously found steady-state values of the occupation probabilities, corresponding to Eq. (5.56), which is the limit where the principle of minimum production of entropy holds. In this case, the rate of entropy production is of second order in $1 - e^{\epsilon/kT}p_2/p_1$ and $1 - p_2/p_1$, which will be small quantities if $e^{\epsilon/kT}$ is near one. This will occur at high enough temperature, $k_B T \gg \epsilon$. We emphasize again that, as indicated by previous results, the steady state we were talking about turns out to be very near the equilibrium one as it should be.

From the point of view of a more general framework to analyze stability (to be introduced in Chapter 7), as the entropy production S has a negative time derivative indicating a monotonous approach toward an equilibrium state (in the same spirit as the discussion on the Boltzmann's $\mathcal{H}$–theorem of Chapter 4), this functional has the property of being a *Lyapunov functional*, an idea that we are going to briefly discuss in Chapter 7.

5.6 Some concepts of fluctuations around equilibrium

In this section we shall review some concepts about (Einstein's) theory for the probability distribution of fluctuations around equilibrium. Let us consider a closed and isolated system with an energy value within the interval $(E, E + dE)$. We call $\Gamma(E)$ the number of microscopic states in such interval. The entropy of the system is

$$S(E) = k_B \ln \Gamma(E) \tag{5.68}$$

Let us assume that, besides the energy, the state of the system is fixed through the knowledge of the values of other n *state variables* (independent, experimentally measurable variables, such as energy, mass, magnetization and charge densities, etc), that we denote by $A_1, A_2, \ldots, A_n$. Then, $\Gamma(E, A_1, A_2, \ldots, A_n)$ will indicate the number of microstates of the system for a given set of values of the parameters, and the probability of the system being in a given macroscopic state is

$$\mathcal{P}(E, A_1, A_2, \ldots, A_n) = \Gamma(E, A_1, A_2, \ldots, A_n)\Gamma(E)^{-1} \tag{5.69}$$

Correspondingly, the entropy is

$$S(E, A_1, A_2, \ldots, A_n) = k_B \ln \Gamma(E, A_1, A_2, \ldots, A_n) \tag{5.70}$$

Both expressions will be related through

$$\mathcal{P}(E, A_1, A_2, \ldots, A_n) = \Gamma(E)^{-1} e^{S(E, A_1, \ldots, A_n)/k_B} \tag{5.71}$$

As is well known, the entropy becomes a maximum when the system is in an equilibrium state given by the values $A_1^{(0)}, A_2^{(0)}, \ldots, A_n^{(0)}$. Each fluctuation near this state will produce a decrease in the entropy. Let us call δA_i the fluctuations defined as

$$\delta A_i = A_i - A_i^{(0)}. \tag{5.72}$$

We can expand the entropy around the equilibrium state according to

$$S(E, A_1, A_2, \ldots, A_n) = S(E, A_1^{(0)}, A_2^{(0)}, \ldots, A_n^{(0)}) + \sum_j \left(\frac{\partial S}{\partial A_j} \right)_{\{A^{(0)}\}} \delta A_j$$

$$+ \sum_{jl} \left(\frac{\partial^2 S}{\partial A_j \partial A_l} \right)_{\{A^{(0)}\}} \delta A_j \delta A_l + \ldots \tag{5.73}$$

It is clear that the first order terms, due to the equilibrium condition (maximum of the entropy), must be identically zero:

$$\left(\frac{\partial S}{\partial A_j}\right)_{\{A^{(0)}\}} \equiv 0.$$

Since the entropy must decrease when the fluctuations increase,

$$g_{jl} = -\left(\frac{\partial^2 S}{\partial A_j \partial A_l}\right)_{\{A^{(0)}\}}$$

will be the elements of a positive definite symmetric matrix. Replacing this result in Eq. (5.71) and calling $\delta\mathbf{A} = \{\delta A_1, \delta A_2, \ldots, \delta A_n\}$, we have (up to second order in the fluctuations) that the probability density is

$$\mathcal{P}(\delta\mathbf{A}) = C \exp -\frac{\delta\mathbf{A}^T \mathbb{G}\, \delta\mathbf{A}}{2k_B} \tag{5.74}$$

where $(\mathbb{G})_{jl} = g_{jl}$ and $\delta\mathbf{A}^T$ is the transpose of $\delta\mathbf{A}$. The constant C is obtained from the normalization condition $\int d\delta\mathbf{A}\mathcal{P}(\delta\mathbf{A}) = 1$, rendering

$$C = [(2\pi k_B)^n \det \mathbb{G}]^{1/2} \tag{5.75}$$

Now, since the distribution $\mathcal{P}(\delta\mathbf{A})$ results to be Gaussian, it is clear that

$$\langle \delta A_j \delta A_l\rangle = \int d\delta\mathbf{A}\delta A_j \delta A_{lj}\mathcal{P}(\delta\mathbf{A}) = k_B(\mathbb{G}^{-1})_{jl} \tag{5.76}$$

and all the higher moments may be written in terms of cumulants.

5.7   Fluctuation theorems

Although this subject lies somewhat outside the scope of this book, we felt it was not an option to plainly ignore it, because of its timeliness and importance (it seeks a formulation of thermodynamics' 2nd law suiting 21th century's lore). So our treatment will be by necessity brief and somewhat superficial.

The *fluctuation theorem* is related to the relative probability that the entropy of a system, which is currently away from thermodynamic equilibrium (that is far from its maximum entropy value), will increase or decrease over a given amount of time. Clearly its origin can be traced back to statistical mechanics. While the 2nd law of thermodynamics predicts that the entropy of an isolated system should increase until it reaches equilibrium, it became apparent that the 2nd law is only a statistical one, suggesting that there should always be some nonzero (however small) probability that the entropy of an isolated system might spontaneously decrease. The fluctuation theorem quantifies this probability.

In a simple form we can say that the fluctuation theorem (FT) is related to the probability distribution of the time-averaged irreversible entropy production, indicated by Ω_t. The theorem states that, for systems away from equilibrium over a finite time t, the ratio between the probability that Ω can take on a value A and the probability of taking the opposite value, $-A$, will be exponential in At. In other words, it means that for a finite non-equilibrium system and in a finite time, the FT gives an expression for the probability that the entropy will flow in a direction opposite to the one dictated by the 2nd law of thermodynamics. It is expressed as

$$\frac{P(\Omega_t = A)}{P(\Omega_t = -A)} = e^{At}, \tag{5.77}$$

meaning that, as the time or the system size increases (since Ω is extensive), the probability of observing an entropy production opposite to that dictated by the 2nd law of thermodynamics, decreases exponentially. The FT is one of the few expressions in non-equilibrium statistical mechanics that is valid far from equilibrium.

The FT was first proposed and tested using computer simulations and afterwards a mathematical proof was given [Evans and Searles (2002)]. Much mathematical and computational work has been done to show that the FT applies to a variety of statistical ensembles, and was experimentally verified by pulling with a laser a plastic bead through a solution: fluctuations in the velocity were recorded showing a behavior opposite to what the 2nd law of thermodynamics dictates for macroscopic systems [Wang *et al.* (2002)].

It is worth remarking that the FT does not state that the 2nd law of thermodynamics is wrong or invalid. The 2nd law of thermodynamics is a statement about macroscopic systems, but the FT is more general. It can be applied to both microscopic and macroscopic systems and when applied to macroscopic systems, it is equivalent to the 2nd Law of thermodynamics.

5.7.0.1 *Second-law inequality*

A simple consequence of the fluctuation theorem given above is that if we carry out an arbitrarily large ensemble of experiments starting at some initial time $t = 0$, and perform an ensemble average of time averages of the entropy production then an exact consequence of the FT is that the ensemble average cannot be negative for any value of the averaging time t

$$\langle \Omega_t \rangle \geq 0, \tag{5.78}$$

and for all t. This inequality is called the *2nd Law Inequality*, and can be proved for systems with time dependent fields of arbitrary magnitude and arbitrary time dependence[Evans and Searles (2002)]. It is important to understand that the 2nd law inequality does not imply that the ensemble averaged entropy production is non-negative at all times. This is false, as consideration of the entropy production in a viscoelastic fluid subject to a sinusoidal time dependent shear rate could show. In such an example the ensemble average of the entropy production's time integral is however non negative - as expected from the 2nd law inequality.

5.7.1   *Nonequilibrium partition identity*

A remarkably simple and elegant consequence of the FT is the so-called *nonequilibrium partition identity*

$$\langle e^{-\Omega_t \, t} \rangle = 1, \tag{5.79}$$

and for all t. Thus, in spite of the 2nd law inequality which might lead to expect that the average would decay exponentially with time, the exponential probability ratio given by the FT exactly cancels the negative exponential in the average above leading to an average which is unity for all time! There are many important implications from the FT. One is that small machines (such as nanomachines or even mitochondria in a cell) will spend part of their time actually running in "reverse". By "reverse", it is meant that they function so as to run in a way opposite to that for which they were presumably designed [Evans and Searles (2002)].

5.7.1.1   *Dissipation function*

Strictly speaking, FT refers to a quantity known as the *dissipation function*. In thermostatic nonequilibrium states close to equilibrium, the long time average of the dissipation function is equal to the average entropy production. However the FT refers to fluctuations rather than averages. The dissipation function is defined as

$$\Upsilon_t(\Gamma) = \int_0^t ds \Upsilon_t(\Gamma; s) \equiv \ln \left[\frac{f(\Gamma, 0)}{f(\Gamma(t), 0)} \right] + \frac{\Delta\Omega(\Gamma, t)}{kT}, \tag{5.80}$$

where k is Boltzmann's constant, $f(\Gamma, 0)$ is the initial $(t = 0)$ distribution of molecular states Γ, and $\Gamma(t)$ is the molecular state arrived at after time t, under the exact time reversible equations of motion. $f(\Gamma(t), 0)$ is the initial distribution of those time evolved states.

Let us remark that in order for the FT to be valid we require that $f(\Gamma(t), 0) \neq 0$ ($\forall \Gamma(0)$). This is known as the condition of ergodic consistency, and is widely satisfied in common statistical ensembles [Evans and Searles (2002)].

5.7.2 *The fluctuation theorem and Loschmidt's paradox*

The 2nd law of thermodynamics, which predicts that the entropy of an isolated system out of equilibrium should tend to increase rather than decrease or stay constant, stands in apparent contradiction with the time-reversible equations of motion for classical and quantum systems. The time reversal symmetry of the equations of motion show that if one films a given time dependent physical process, then playing the movie of that process backwards does not violate the laws of mechanics. It is often argued that for every forward trajectory in which entropy increases, there exists a time reversed anti-trajectory where entropy decreases, thus if one picks an initial state randomly from the system's phase space and evolves it forward according to the laws governing the system, decreasing entropy should be just as likely as increasing entropy. It might seem that this is incompatible with the second law of thermodynamics which predicts that entropy tends to increase. The problem of deriving irreversible thermodynamics from time-symmetric fundamental laws is referred to as Loschmidt's paradox.

The proof of the Fluctuation Theorem and in particular the 2nd Law Inequality shows that, given a non-equilibrium starting state, the probability of seeing its entropy increase is greater than the probability of seeing its entropy decrease[Evans and Searles (2002)]. However, one could also use the same laws of mechanics to extrapolate backwards from a later state to an earlier state, and in this case the same reasoning used in the proof of the FT would lead us to predict the entropy was likely to have been greater at earlier times than at later times. This second prediction would be frequently violated in the real world,since it is often true that a given nonequilibrium system was at an even lower entropy in the past (although the prediction would be correct if the nonequilibriumstate were the result of a random fluctuation in entropy in an isolated system that had previously been at equilibrium - in this case, if you happen to observe the system in a lower-entropy state, it is most likely that you are seeing the minimum of the random dip in entropy, in which case entropy would be higher on either side of this minimum).

So, it seems that the problem of deriving time-asymmetric thermodynamic laws from time-symmetric laws cannot be solved by appealing to statistical derivations which show entropy is likely to increase when you start from a nonequilibrium state and project it forwards. Many modern physicists believe the resolution to this puzzle lies in the low-entropy state of the universe shortly after the big bang, although such an explanation for this initial low entropy is still under debate.

5.7.3 *Jarzynski's equality*

The Jarzynski equality (JE) is an equation that relates free energy differences between two equilibrium states and non-equilibrium processes [Jarzynski (1997)]. In thermodynamics, the free energy difference $\triangle F = F_B - F_A$ between two states A and B is connected to the work W done on the system through the inequality

$$\triangle F \leq W, \tag{5.81}$$

the equality happening only in the case of a quasi-static process, that is when one "moves" the system from A to B infinitely slowly.

In contrast to the thermodynamic statement above, the JE remains valid no matter how fast the process happens. The equality itself can be straightforwardly derived from the Crooks FT [Crooks (1998)]. The JE states

$$e^{\frac{\triangle F}{kT}} = \overline{e^{-\frac{W}{kT}}}. \tag{5.82}$$

Here k is the Boltzmann constant and T is the temperature of the system in the equilibrium state A or, equivalently, the temperature of the heat reservoir with which the system was thermalize before the process took place. The over-line indicates an average over all possible realizations of an external process that takes the system from the equilibrium state A to a new, generally nonequilibrium state under the same external conditions as that of the equilibrium state B. In the limit of an infinitely slow process, the work W performed on the system in each realization is numerically the same, so the average becomes irrelevant and the Jarzynski equality reduces to the thermodynamic equality $\triangle F = W$ (see above). In general, however, W depends upon the specific initial microstate of the system, though its average can still be related to $\triangle F$ through an application of Jensen's inequality in the JE,

$$\triangle F \leq \overline{W}, \tag{5.83}$$

in accordance with the second law of thermodynamics.

Since its original derivation, the Jarzynski equality has been verified in a variety of contexts, ranging from experiments with biomolecules to numerical simulations. Many other theoretical derivations have also appeared, lending further confidence to its generality.

5.7.4 *Crooks' fluctuation theorem*

The Crooks equation (CE)[Crooks (1998)] relates the work done on a system during a non-equilibrium transformation to the free energy difference between the final and the initial state of the transformation. During the non equilibrium transformation the system is at constant volume and in contact with a heat reservoir.

If we define a generic reaction coordinate of the system as a function of the Cartesian coordinates of the constituent particles (for instance the distance between two particles), we can characterize every point along the reaction coordinate path by a parameter λ, such that $\lambda = 0$ and $\lambda = 1$ correspond to two ensembles of microstates for which the reaction coordinate is constrained to different values. A dynamical process where λ is externally driven from zero to one, according to an arbitrary time scheduling, is referred as a forward transformation, while the time reversal path is indicated as a backward transformation. Given these definitions, the CE sets a relation between the quantities

- $P(A \rightarrow B)$ i.e. the joint probability of taking a microstate from the canonical ensemble A corresponding to $\lambda = 0$ and of performing a forward transformation to the microstate B corresponding to $\lambda = 1$;
- $P(A \leftarrow B)$, i.e. the joint probability of taking the microstate B from the canonical ensemble corresponding to $\lambda = 1$ and performing the backward transformation to the microstate A corresponding to $\lambda = 0$;
- $W_{A,B}$, the work done on the system during the forward transformation (from A to B);
- $\triangle F = F(B) - F(AB)$, the Helmholtz free energy difference between the state A and B, represented by the canonical distribution of microstates having $\lambda = 0$ and $\lambda = 1$, respectively).

The CE equation reads

$$\frac{P(A \to B)}{P(A \leftarrow B)} = e^{\beta[W_{A,B} - \triangle F]}, \tag{5.84}$$

where $\beta = (kT)^{-1}$, k is the Boltzmann constant and T the temperature of the reservoir. In the previous equation the difference $W_d = W_{A,B} - \triangle F$, corresponds to the work dissipated in the forward transformation. The probabilities $P(A \to B)$ and $P(A \leftarrow B)$ become identical when the transformation is performed at infinitely slow speed, that is for equilibrium transformations. In such case $W_{A,B} = \triangle F$ and $W_d = 0$.

Using the time reversal relation $W_{A,B} = -W_{B,A}$, and grouping together all the trajectories yielding the same work (in the forward and backward transformation), we can write the above equation in terms of the work distribution functions

$$P_{A \to B)}(W) = P_{A \leftarrow B}(-W) e^{\beta[W_{A,B} - \triangle F]}. \tag{5.85}$$

Note that for the backward transformation, the work distribution function must be evaluated by taking the work with the opposite sign. The two work distributions for the forward and backward processes cross at $W_{A,B} = \triangle F$. This fact has been experimentally verified using optical tweezers for the process of unfolding and refolding of a small RNA hairpin and an RNA three-helix junction.

The CE implies the Jarzynski equality.

5.7.5   *Final comments*

The fluctuation theorem is of fundamental importance to nonequilibrium statistical mechanics. The FT (together with the Axiom of Causality) gives a generalization of the second law of thermodynamics which includes as a special case, the conventional second law. When combined with the central limit theorem, the FT also implies the famous Green-Kubo relations for linear transport coefficients, close to equilibrium. The FT is however, more general than the Green-Kubo Relations because unlike them, the FT applies to fluctuations far from equilibrium. In spite of this fact, scientists have not yet been able to derive the equations for nonlinear response theory from the FT. The FT does not imply or require that the distribution of time averaged dissipation be Gaussian. There are many examples known where the distribution of time averaged dissipation is non-Gaussian and yet the FT (of course) still correctly describes the probability ratios. Lastly the theoretical constructs used to prove the FT can be applied to nonequilibrium transitions between two different equilibrium states. When this is done the so-called Jarzynski equality or nonequilibrium work relation, can be derived. This equality shows how equilibrium free energy differences can be computed or measured (in the laboratory), from nonequilibrium path integrals. Previously quasi-static (equilibrium) paths were required. The reason why the fluctuation theorem is so fundamental is that its proof requires so little. It requires: knowledge of the mathematical form of the initial distribution of molecular states, that all time evolved final states at time t, must be present with nonzero probability in the distribution of initial states ($t = 0$) - the so-called condition of ergodic consistency and, an assumption of time reversal symmetry. In regard to the latter "assumption", all the equations of motion for either classical or quantum dynamics are in fact time reversible [Bustamante *et al.* (2005); Boksenbojm *et al.* (2010)]

Several extensions have been recently accomplished of the FT. Among them, we outline the ones in [Van den Broeck (2010); Esposito and Van den Broeck (2010)].

Chapter 6

Linear response theory, fluctuation–dissipation theorem

with each breeze
the butterfly changes its place
over the willow
Bashô

6.1 Introduction

The subjects to be discussed in this chapter involve, fundamentally, the properties of time dependent correlation functions. With these powerful tools one can study the dynamics of microscopic processes in gases, liquids, solids, plasmas, close to or far from equilibrium. Such functions are, on one hand, a gauge of the intrinsic microscopic fluctuations and, depending on which microscopic dynamical variables we choose to correlate, they provide a test of the relative importance of single particle motions and of collective modes (due to the coherent motion of a large number of particles). On the other hand, they offer an adequate bridge between the microscopic and macroscopic descriptions in many body systems, making it possible to obtain microscopic exact expressions for the phenomenological transport coefficients. They are also important from the experimental point of view, because the frequency spectra obtained experimentally either through inelastic scattering (i.e.: light, neutrons), absorption processes (i.e.: infrared radiation), or some other relaxation processes (i.e.: dielectric relaxation), may be written in terms of simple correlation functions. Furthermore, for a given correlation function there exist different experimental procedures to measure spectra related to it, implying that different techniques could test identical microscopic processes (i.e.: local density fluctuations) at different time or spatial scales.

We shall start reviewing definitions and deriving some properties of these functions. Next, we shall see how it is possible to obtain some useful results within the framework of the *linear response theory*, and we shall present the famous *fluctuation-dissipation theorem*. Finally we shall discuss some examples.

6.2   Correlation functions: Definitions and properties

We consider an isolated system composed of N particles, each one with ν *degrees of freedom*. We shall call a dynamical variable of the system (of scalar, vectorial or tensorial character) to any function of the instantaneous values of some of, or eventually all, the νN coordinates $\{\mathbf{q}_i\}$ and νN momenta $\{\mathbf{p}_i\}$

$$A(t) = A(\{\mathbf{q}_i\}, \{\mathbf{p}_i\}, t) \tag{6.1}$$

When considering a quantum problem, A must be an operator (usually Hermitian) that could also be a function of spin variables. Depending on A being an *even* or *odd* function of the momenta it has a given *signature*, $\epsilon_A = \pm 1$ respectively, under time inversion.

The time evolution of A is given by

$$\frac{\partial}{\partial t} A(t) = i\mathcal{L}A(t) \tag{6.2a}$$

where $\mathcal{L}$ is the Liouville operator, given by

$$\mathcal{L} = i\{\mathcal{H}, \cdot\} = i\sum \frac{\partial \mathcal{H}}{\partial \mathbf{q}_i} \frac{\partial}{\partial \mathbf{p}_i} - \frac{\partial \mathcal{H}}{\partial \mathbf{p}_i} \frac{\partial}{\partial \mathbf{q}_i} \qquad \text{CLASSICAL} \tag{6.2b}$$

$$= \frac{1}{\hbar}[\mathcal{H}, \cdot] \qquad \text{QUANTUM} \tag{6.2c}$$

where

- $\mathcal{H}$ is hamiltonian of the isolated system,
- $\{\cdot, \cdot\}$ denotes the classical *Poisson bracket*, and
- $[\cdot, \cdot]$ the *quantum mechanical commutator*.

The formal solution of Eq. (6.2a) is

$$A(t) = e^{i\mathcal{L}t} A(0) \tag{6.3a}$$

where $A(0)$ is the initial value (at $t = 0$) of the dynamical variable. The quantum result reads

$$A(t) = e^{i\mathcal{H}t/\hbar} A(0) e^{-i\mathcal{H}t/\hbar} \tag{6.3b}$$

Let us consider a *local* dynamical variable. Its general form is

$$A(\mathbf{r}, t) = \sum_i a_i(t)\delta(\mathbf{r} - \mathbf{r}_i(t)) \tag{6.4}$$

where the sum extends over all particles, a_i is some physical quantity associated with the dynamical variable (i.e.: mass, momentum, ...) corresponding to the i–th particle, located at $\mathbf{r}_i(t)$. In a quantum case, Eq. (6.4) must be adequately symmetrized.

A local dynamical variable is said to be *conserved* if it obeys a *continuity equation*

$$\frac{\partial}{\partial t}A(\mathbf{r}, t) + \nabla \mathbf{J}_A(\mathbf{r}, t) = 0 \tag{6.5}$$

where $\mathbf{J}_A(\mathbf{r}, t)$ is the *current* associated with the *density* $A(\mathbf{r}, t)$. Equation (6.5) is a way to express the condition $A^{\text{tot}} = \sum_i a_i(t) = \text{const}$. The continuity Eq. (6.5) could be written in a simple form using the Fourier decomposition of $A(\mathbf{r}, t)$. The Fourier spatial components of $A(\mathbf{r}, t)$ are defined through

$$A(\mathbf{k}, t) = \int d\mathbf{r} A(\mathbf{r}, t)e^{-i\mathbf{k}\cdot\mathbf{r}} = \sum_i a_i(t)e^{-i\mathbf{k}\cdot\mathbf{r}_i(t)} \tag{6.6a}$$

Taking this into account, Eq. (6.5) takes the form

$$\frac{\partial}{\partial t}A(\mathbf{k}, t) + i\mathbf{k} \cdot \mathbf{J}_A(\mathbf{k}, t) = 0 \tag{6.6b}$$

A classical example is the *number density* $(a_i = 1)$

$$\rho(\mathbf{r}, \mathbf{t}) = \sum_i \delta(\mathbf{r} - \mathbf{r}_i(t)) \tag{6.7a}$$

and it is easy to see that the associated current is given by

$$\mathbf{J}_\rho(\mathbf{r}, t) = \sum_i \frac{\mathbf{p}_i(t)}{m}\delta[\mathbf{r} - \mathbf{r}_i(t)] \tag{6.7b}$$

In classical equilibrium statistical mechanics, the time-correlation function of two dynamical variables A and B is defined as an ensemble average (see Chapter 3), with a general expression

$$C_{AB}(t_1, t_2) = \langle A(t_1)B(t_2)\rangle = \int d^N\mathbf{q}(t_1)\int d^N\mathbf{p}(t_1)\int d^N\mathbf{q}(t_2)\int d^N\mathbf{p}(t_2)$$

$$A(\{\mathbf{q}_i(t_1)\}, \{\mathbf{p}_i(t_1)\})B(\{\mathbf{q}_i(t_2)\}, \{\mathbf{p}_i(t_2)\})f_0(\{\mathbf{q}_i(t_1)\}, \{\mathbf{p}_i(t_1)\})$$

$$P(\{\mathbf{q}_i(t_2)\}, \{\mathbf{p}_i(t_2)\}|(\{\mathbf{q}_i(t_1)\}, \{\mathbf{p}_i(t_1)\}) \tag{6.8a}$$

or as a time average

$$C_{AB}(t_1, t_2) = \langle A(t_1)B(t_2) \rangle = i\frac{1}{\tau} \int_0^\tau ds A(t_1 + s)B(t_2 + s) \qquad (6.8b)$$

In equilibrium, both averages give the same result, provided the system is *ergodic*, a fact that we shall assume from now on.

If the dynamical variables are not only time-dependent but also depend on position the correlation function becomes nonlocal both in time and space:

$$C_{AB}(\mathbf{r}_1, t_1; \mathbf{r}_2, t_2) = \langle A(\mathbf{r}_1, t_1)B(\mathbf{r}_2, t_2) \rangle \qquad (6.9a)$$

In such a case, it is also useful to know the Fourier components, that now became complex quantities, and may be defined according to

$$C_{AB}(\mathbf{k}_1, t_1; \mathbf{k}_2, t_2) = \langle A(\mathbf{k}_1, t_1)B^*(\mathbf{k}_2, t_2) \rangle = \langle A(\mathbf{k}_1, t_1)B(-\mathbf{k}_2, t_2) \rangle \quad (6.9b)$$

as the dynamical variables A and B are hermitian. If the system is homogeneous in time and space we find

SPATIAL HOMOGENEITY

$$C_{AB}(\mathbf{k}_1, t_1; \mathbf{k}_2, t_2) =$$

$$= \int d\mathbf{r}_1 \int d\mathbf{r}_2 e^{-i\mathbf{k}_1 \cdot (\mathbf{r}_1 - \mathbf{r}_2) + i(\mathbf{k}_2 - \mathbf{k}_1) \cdot \mathbf{r}_2} \langle A(\mathbf{r}_1, t_1)B(\mathbf{r}_2, t_2) \rangle$$

$$= \int d\mathbf{r}_1 \int d\mathbf{r}_2 e^{-i\mathbf{k}_1 \cdot (\mathbf{r}_1 - \mathbf{r}_2) + i(\mathbf{k}_2 - \mathbf{k}_1) \cdot \mathbf{r}_2} \langle A(\mathbf{r}_1 - \mathbf{r}_2, t_1)B(\mathbf{0}, t_2) \rangle$$

TEMPORAL HOMOGENEITY

$$= \int d\mathbf{r}_1 \int d\mathbf{r}_2 e^{-i\mathbf{k}_1 \cdot (\mathbf{r}_1 - \mathbf{r}_2) + i(\mathbf{k}_2 - \mathbf{k}_1) \cdot \mathbf{r}_2} \langle A(\mathbf{r}_1 - \mathbf{r}_2, t_2 - t_1)B(\mathbf{0}, 0) \rangle$$

$$= \int d\mathbf{r} e^{-i\mathbf{k}_1 \circ \mathbf{r}} \langle A(\mathbf{r}, t_2 - t_1)B(\mathbf{0}, 0) \rangle \int d\mathbf{r}_2 e^{i(\mathbf{k}_2 - \mathbf{k}_1) \cdot \mathbf{r}_2}$$

$$= C_{AB}(\mathbf{k}_1, t_2 - t_1)\delta_{\mathbf{k}_2, \mathbf{k}_1} \qquad (6.9c)$$

For quantal systems, it is usual to define the so called *one sided correlation function* (in opposition to the symmetrized one to be defined immediately) as

$$C_{AB}(t_1, t_2) = \langle A(t_1)B(t_2) \rangle = \text{Tr}[\rho A(t_1)B(t_2)] \qquad (6.10a)$$

where ρ $(= \mathcal{Z}^{-1} \exp[-\beta\mathcal{H}])$ is the equilibrium density matrix. When expanded in terms of the basis of eigenstates $|\nu\rangle$ of the Hamiltonian $\mathcal{H}$, ρ is diagonal, and therefore

$$\langle A(t_1)B(t_2)\rangle = \sum_\nu \rho_\nu \langle \nu|A(t_1)B(t_2)|\nu\rangle$$

$$= \sum_\nu \rho_\nu \langle \nu|A(t_1)|\eta\rangle\langle\eta|B(t_2)|\nu\rangle e^{i\omega_{\nu\eta}(t_2-t_1)} \quad (6.10b)$$

where we have used Eq. (6.3b), and $\hbar\omega_{\nu\eta} = (E_\nu - E_\eta)$, E_ν being the energy eigenvalue corresponding to the state $|\nu\rangle$. If there is no degeneracy with respect to the total angular momentum, the eigenstates $|\nu\rangle$ are real, and taking into account the *signatures* of A and B under time inversion, Eq. (6.10b) yields

$$C_{AB}^*(t_1, t_2) = \epsilon_A \epsilon_B C_{AB}(t_1, t_2) \quad (6.11)$$

As it is usual in quantum systems, if $A(t_1)$ and $B(t_2)$ do not commute, it is not only convenient but necessary, to build up the symmetrized version of Eqs. (6.10)

$$C_{AB}^{(s)}(t_1, t_2) = \frac{1}{2}\langle\{A(t_1), B(t_2)\}\rangle \quad (6.12a)$$

where $\{,\}$ indicates the anticommutator. When A and B are both Hermitian we obtain, from Eq. (6.10b)

$$C_{AB}^{(s)}(t_1, t_2) = C_{AB}^{(s)*}(t_1, t_2) = \epsilon_A \epsilon_B C_{AB}^{(s)}(t_2, t_1) \quad (6.12b)$$

Before looking to some properties, we present another useful definition, introduced by Kubo within the context of the linear response theory: the so called *canonical* version of the time correlation function

$$C_{AB}^{(c)}(t_1, t_2) = \beta^{-1}\int_0^\beta d\lambda\langle\{A(t_1 - i\hbar\lambda), B(t_2)\}\rangle \quad (6.13)$$

where we have, in relation to Eq. (6.3b)

$$A(t_1 - i\hbar\lambda) = e^{\lambda\mathcal{H}}A(t_1)e^{-\lambda\mathcal{H}}$$

In the classical limit ($\hbar \to 0$), all definitions [Eqs. (6.10), (6.12) and (6.13)] reduce to the same (classical) correlation function.

There are several useful properties of the time correlation functions that, due to its very general validity, are worth to be discussed here. We are going now to present them in a general way, as well as a short discussion on the associated sum rules.

6.2.1 *Properties*

(i) *Stationarity*: Due to the fact that averages in equilibrium are stationary, that means independent of the initial time to compute the average, the correlation functions must be invariant under time translations, and will depend only on $T = t' - t''$:

$$C_{AB}(t', t'') = \langle A(t')B(t'') \rangle = \langle A(t' - t'')B(0) \rangle = C_{AB}(t' - t'') \quad (6.14)$$

This property is a consequence of Eq. (6.10b), as we can write

$$\frac{d}{ds}\langle A(t + s)B(s) \rangle = \langle \dot{A}(t + s)B(s) \rangle + \langle A(t + s)\dot{B}(s) \rangle = 0$$

Hence

$$\langle \dot{A}(t + s)B(s) \rangle = -\langle A(t + s)\dot{B}(s) \rangle \quad (6.15a)$$

In a similar manner, from $\frac{d^2}{ds^2}\langle A(t + s)B(s) \rangle = 0$ we obtain

$$\langle \ddot{A}(t + s)B(s) \rangle = -\langle \dot{A}(t + s)\dot{B}(s) \rangle \quad (6.15b)$$

On the other hand, from Eqs. (6.12b) and (6.14) we have

$$C_{AB}^{(s)}(t) = \epsilon_A \epsilon_B C_{AB}^{(s)}(-t) \quad (6.16)$$

As a result of the indicated relations, and due to the time stationarity, in particular we have that: the *symmetrized* (classical) *self-correlation function* is a **real** as well as an **even** function of time, while the *one sided* correlation function [Eq. (6.10a)] has one part **even** and **real**, and other part **imaginary** and **odd**.

(ii) From our knowledge of thermodynamics and equilibrium statistical mechanics, it is clear that asymptotically,

$$\lim_{\tau \to \infty} C_{AB}(\tau) = \langle AB \rangle \quad (6.17)$$

where $\langle AB \rangle$ corresponds to the *static* (*stationary*) correlation function. From this result and the *Schwartz inequality* it also follows immediately that

$$|C_{AB}^{(s)}(t)| \leq \left(\langle AA^\dagger \rangle \langle BB^\dagger \rangle \right)^{1/2} \quad (6.18a)$$

This inequality, sets a bound for the values of $C_{AB}^{(s)}(t)$; while, for the self-correlation, intuition suggests the following result

$$|C_{AA}^{(s)}(t)| \leq C_{AA}(0) = \langle AA^\dagger \rangle \qquad (6.18b)$$

(iii) Related with the previous result it also follows that both dynamic variables become asymptotically uncorrelated

$$\lim_{\tau \to \infty} C_{AB}(\tau) = \langle A \rangle \langle B \rangle \qquad (6.19a)$$

It is usually convenient to define the variables in such a way that the invariant (average) part is excluded and only the correlation of the fluctuating parts is considered:

$$C_{AB}(\tau) = \langle [A(\tau) - \langle A \rangle][B - \langle B \rangle] \rangle \qquad (6.19b)$$

This definition yields $\lim_{\tau \to \infty} C_{AB}(\tau) = 0$, which has a direct physical interpretation: asymptotically, there is a complete loss of correlation among the fluctuating parts.

(iv) There are some experiments whose outcome, instead of being given in terms of the time variable, are given in the frequency domain. For this reason it is useful to define, though Eq. (6.19b), the Fourier transform $C_{AB}(\omega)$. This quantity is called the *spectral function* (or *power spectrum*). The relation between the correlation and the spectral functions is usually known as the *Wiener-Kintchine theorem*. It is also convenient to define its Laplace transform $C_{AB}(z)$ (ω being real, and z complex). Both definitions are given by:

$$C_{AB}(\omega) = [2\pi]^{-1/2} \int_{-\infty}^{+\infty} dt C_{AB}(t) e^{i\omega t} \qquad (6.20a)$$

$$C_{AB}(z) = \int_{-\infty}^{\infty} dt C_{AB}(t) e^{zt} \qquad (6.20b)$$

Because $C_{AB}(t)$ is bounded [see Eq. (6.18)], it is clear that $C_{AB}(z)$ will be analytic in the upper part of the complex plane. Both transforms may be related through the *Hilbert transform*:

$$C_{AB}(z) = -i \int_{-\infty}^{\infty} d\omega C_{AB}(\omega)[\omega - z]^{-1} \qquad (6.20c)$$

From Eq. (6.11) it results that the spectrum of a self-correlation function is *always real*, and satisfies

$$C_{AA}(\omega) = \lim_{\epsilon \to 0} \pi^{-1} \Re\{C_{AA}(z = \omega + i\epsilon)\} \qquad (6.21a)$$

or more explicitly

$$C_{AA}(\omega) = \sum_{n,m} \rho_n |\langle m|A|n\rangle|^2 \delta(\omega - \omega_{nm}) \qquad (6.21b)$$

Since $\rho_n \rho_m^{-1} = \exp\{-\beta(E_n - E_m)\}$, if A is Hermitian, we have

$$C_{AA}(-\omega) = e^{-\beta\omega} C_{AA}(\omega) \qquad (6.21c)$$

And since $C_{AA}(t)$ is an even and real function of t, $C_{AA}(\omega)$ is an even and real function of ω. In order to fix ideas it is useful to consider a simple, but extensively used, example: the case of an exponential correlation function. The time correlation and spectral functions for this case are given, respectively, by

$$C_{AB}(\tau) \propto e^{-\tau/\tau_c} \qquad ; \qquad C_{AB}(\omega) \propto \frac{\tau_c}{1 + (\omega\tau_c)^2}.$$

The form of these functions is depicted in Fig. 6.1.

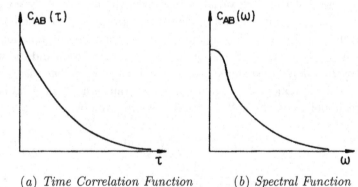

(a) *Time Correlation Function*   (b) *Spectral Function*

Fig. 6.1

(v) The spectral function of a self-correlation function is non-negative (this is left as an exercise).

(vi) In the case of a homogeneous liquid, the spatial translational invariance implies that the time and spatial correlation functions only depend on the relative position $\mathbf{r} = \mathbf{r}' - \mathbf{r}''$

$$C_{AB}(\mathbf{r}', \mathbf{r}'', t', t'') = C_{AB}(\mathbf{r}' - \mathbf{r}'', t' - t'') \qquad (6.22a)$$

The same translational invariance also implies that, for the Fourier transform, Eq. (6.9c) holds with

$$C_{AB}(\mathbf{k}, t' - t'') = \int d\mathbf{k} e^{-i\mathbf{k}\cdot\mathbf{r}} C_{AB}(\mathbf{r}, t' - t'') \qquad (6.22b)$$

(vii) If the Hamiltonian (or Liouville operator) has even parity (a typical situation), the correlation function of two dynamical variables having opposite spatial parities is zero at all times. In particular, the local number density and its associate current are uncorrelated at all times. Similarly, if the dynamical variables have a certain *signature* under reflection of all coordinates and momenta with respect to a given plane, the correlation function of two variables of opposite signatures is identically zero.

(viii) The self-correlation function of a *conserved variable* has the important property that in the *long wave length* limit ($k \to 0$),

$$\lim_{k \to 0} C_{AA}(\mathbf{k}, t) = \int d\mathbf{r} \langle A(\mathbf{r}, t) A(\mathbf{0}, 0) \rangle = \langle \left(\int d\mathbf{r} A(\mathbf{r}, t) \right) A(\mathbf{0}, 0) \rangle$$

$$= \langle A^{TOT} A(\mathbf{0}, 0) \rangle \qquad (6.23a)$$

Since A^{TOT} is constant in time, it follows that

$$\lim_{k \to 0} C_{AA}(\mathbf{k}, \omega) \cong \text{const.}\delta(\omega) \qquad (6.23b)$$

This result indicates that the *correlation time* in the *long wave length limit* for the self-correlation function of a conserved variable is *infinite* (i.e., for the case of an exponentially time-correlated function: $\tau_c \to \infty$), and its spectrum reduces to a δ–function. For decreasing wave length, it is expected that the correlation time will also decrease, becoming of the order of magnitude of the typical microscopic times, when the wave length becomes of the order of the intermolecular distances.

In quantum mechanics, when the sum of all the radiative–transition probabilities from a given level weighed by some power of each transition energy can be given in a closed form, generally as the expectation value of some operator, a sum rule is said to exist. Clearly such sum rule sets a bound to the size of matrix elements and thus provides a useful benchmark for model calculations. In the present context it is clear that a thermodynamical average shall be included as well. We will briefly show some relations existing between those sum rules and particular values of correlation functions.

6.2.2   *Sum rules*

For the case of nonsingular intermolecular potentials, the correlation function $C_{AB}(t)$, given for instance by Eqs. (6.8) and (6.9), can be expanded in *Taylor series* around $t = 0$

$$C_{AB}(t) = \sum_{n=0}^{\infty} \frac{t^n}{n!} C_{AB}^{(n)}(0) \qquad (6.24a)$$

which, using the stationarity condition and Eq. (6.2a), gives

$$C_{AB}^{(n)}(0) = \langle A^{(n)}(0)B(0)\rangle = (-1)^n \langle A(0)B^{(n)}(0)\rangle$$
$$= (-1)^n \langle A(0)[i\mathcal{L}]^n B(0)\rangle \qquad (6.24b)$$

leading to

$$C_{AB}(t) = \sum_{n=0}^{\infty} (-1)^n \frac{t^n}{n!} \langle A(0)[i\mathcal{L}]^n B(0)\rangle \qquad (6.24c)$$

On the other hand, from Eqs. (6.15), the symmetrized (classical) correlation functions have either only *even* or only *odd* powers of t. In particular, for the self-correlation we get only even powers of t

$$C_{AB}^{(s)}(t) = \sum_{n=0}^{\infty} \frac{t^{2n}}{2n!} C_{AB}^{(s)(2n)}(0) \qquad (6.25a)$$

Here it is possible to write

$$\langle A^{(2n)}(0)A(0)\rangle = (-1)^n \langle A^{(n)}(0)A^{(n)}(0)\rangle$$
$$= (-1)^n \langle [i\mathcal{L}]^n A(0)[i\mathcal{L}]^n A(0)\rangle = (-1)^n \langle ([i\mathcal{L}]^n A(0))^2\rangle \qquad (6.25b)$$

After differentiating $2n$ times with respect to t, and using the Fourier transform Eq. (6.20a), we obtain

$$\langle \omega^{2n}\rangle = \int_{-\infty}^{+\infty} d\omega\, \omega^{2n} C_{AA}^{(s)}(\omega) = (-1)^n C_{AA}^{(s)}(t = 0) \qquad (6.26)$$

Hence, the spectral function frequency moments are directly linked to the derivatives of the self-correlation function at $t = 0$. These last quantities correspond to static correlation functions, which may be generally written as integrals over the equilibrium distribution functions. The latter can be calculated for low n values. For the case of nonlocal correlation functions, those moments become functions of **k**.

The continuity equation for conserved dynamical variables leads to simple expressions for the second order moments within the sum rules, usually called *f–sum rules*

$$\langle \omega^2 \rangle = -\check{C}_{AA}(\mathbf{k}, 0) = \langle \dot{A}(\mathbf{k}, 0)\dot{A}(\mathbf{k}, 0) \rangle$$

$$= \langle [-i\mathbf{k} \cdot \mathbf{J}_A(\mathbf{k}, 0)][i\mathbf{k} \cdot \mathbf{J}_A(-\mathbf{k}, 0)] \rangle \tag{6.27a}$$

For the particular case of an isotropic fluid (with k taken along x), it reduces to

$$\langle \omega^2 \rangle = k^2 \langle |\mathbf{J}_A(k, 0)|^2 \rangle \tag{6.27b}$$

In the particular case of A being the local number density, the previous average, for N identical particles of mass m, yields

$$\langle \omega^2 \rangle = \frac{NkT}{m}k^2 = N\omega_0^2 \tag{6.28}$$

with the characteristic frequency $\omega_0 \to 0$ for $k \to 0$, coinciding with Eq. (6.23b). We also note that the spectral representation in Eq. (6.20c) gives directly an expansion in powers of z^{-1}, which is intimately related to the expansion in Eq. (6.24a). We shall meet this kind of expansion again when discussing the response functions [see Eq. (6.61a)].

A classical example that illustrates all the discussion above, is the velocity correlation function in the Langevin theory of Brownian motion. With this in mind, we suggest to turn back to Sec. 2.2, in Chapter 2, and to analyze that case as a useful exercise.

6.3 Linear response theory

Here we shall analyze the behavior of a system perturbed by an external test field, *weakly coupled* to the system. The main result will be that the *response* of the system to a weak perturbation may be entirely described in terms of time correlation functions of the system in equilibrium (i.e., without external field). Since under the influence of the external perturbation, and in correspondence to the energy dissipated during the process, the system usually heats, a result baptized *the fluctuation–dissipation theorem* will follow from the relation between response and correlation functions.

We write the Hamiltonian of the system under the action of an external field as

$$\mathcal{H} = \mathcal{H}_0 + \mathcal{H}' \tag{6.29a}$$

where $\mathcal{H}_0$ denotes the unperturbed system, and $\mathcal{H}'$ is the perturbation (usually time-dependent). It reads

$$\mathcal{H}'(t) = -\int d\mathbf{r} A(\mathbf{r}) \mathcal{F}(\mathbf{r}, t) \tag{6.29b}$$

where $\mathcal{F}(\mathbf{r}, t)$ is the time dependent external field (scalar, vector or tensor) and $A(\mathbf{r})$ is the dynamical variable coupled to the field (the minus sign being a convention). The kind of *product* between $A(\mathbf{r})$ and $\mathcal{F}(\mathbf{r}, \mathbf{t})$ depends on their tensorial character.

In order to introduce the formalism we start discussing a very useful example, and afterwards we introduce the response functions.

6.3.1   *Inelastic scattering cross section*

We take first a system interacting with an incident *monochromatic* beam of particles, represented by a plane wave. The particles could be neutrons interacting with the system nuclei, or electrons interacting with the system charges. The case of a beam of photons is more complicated and will not be treated here. In the case of neutrons, what is tested is $\rho_n(\mathbf{r})$, $\rho_n(\mathbf{r})$ being the microscopic nuclear density; while for electrons it is $\rho_z(\mathbf{r})$, the microscopic charge density. We shall show that the *inelastic scattering cross section* is directly related to the spectral function of the self-correlation of the densities. We assume that particles within the beam do not interact among them, allowing us to consider only one incident particle at $\mathbf{r}$. Then, the perturbation will have the general form given in Eq. (6.29b), with $A(\mathbf{r}) = \rho_n(\mathbf{r})$, or $\rho_z(\mathbf{r})$, and $\mathcal{F}(\mathbf{r}, t) = -\mathcal{V}(\mathbf{r} - \mathbf{r}')$, where $\mathcal{V}$ is the interaction potential between the particle and the nucleus or charge of the system:

$$\mathcal{H}'(t) = \int d\mathbf{r} \rho(\mathbf{r}) \mathcal{V}(\mathbf{r} - \mathbf{r}') \tag{6.29c}$$

For simplicity, we shall assume that in the case of neutron scattering, all nuclei are identical and without spin. In this way, all of them interact in the same way with the incident neutron. A similar assumption could be done for electron scattering. We shall represent this interaction through the *Fermi potential*, which has the simple form

$$\mathcal{V}(\mathbf{r}) = \frac{2\pi a \hbar^2}{m} \delta(\mathbf{r}) \tag{6.30}$$

indicating that the *scattering length* a is the same for all nuclei and then the scattering will be purely *coherent*. If a could vary from one nucleus to

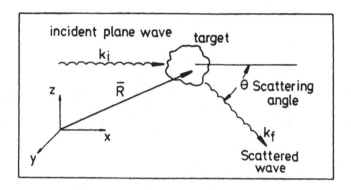

Fig. 6.2

another (i.e. for different isotopes), the scattering would become partially *incoherent*, but we shall not consider such a case here. For electrons the situation is simpler because the potential $\mathcal{V}(\mathbf{r})$ corresponds to the *Coulomb* case, $\mathcal{V}(\mathbf{r}) = -e/r$, e being the electron charge.

We call $\mathbf{k}_i$ and $\mathbf{k}_f$ the incident and final wave vectors of the test particle respectively (Fig. 6.2). Within the *first Born approximation*, the initial and final states of the beam particle are described by plane waves:

$$|\mathbf{k}_i\rangle \propto e^{i\mathbf{k}_i \cdot \mathbf{r}}; |\mathbf{k}_f\rangle \propto e^{i\mathbf{k}_f \cdot \mathbf{r}}. \tag{6.31}$$

We call θ the scattering angle, and $\mathbf{k} = \mathbf{k}_i - \mathbf{k}_f$ the scattering wave number vector. We denote the initial and final states of the target system by $|i\rangle$ and $|f\rangle$ respectively, E_i and E_f being their corresponding $\mathcal{H}_0$–energy eigenvalues (that is, the energies of the target system before and after the scattering process). Energy conservation implies:

$$\hbar\omega_{fi} = E_f - E_i = \frac{\hbar^2}{2m}(\mathbf{k}_i^2 - \mathbf{k}_f^2) \tag{6.32}$$

where $\hbar\omega_{fi}$ corresponds to the energy *gain* ($E_f < E_i$) or energy *loss* ($E_f > E_i$) of the (mass m) scattered particle. According to the *Fermi Golden Rule*, the transition probability per unit time from the initial state $|\mathbf{k}_i\rangle|i\rangle$ to the final state $|\mathbf{k}_f\rangle|f\rangle$, is given by

$$\mathcal{W}_{i\to f} = \frac{2\pi}{\hbar^2}|\langle f|\langle \mathbf{k}_f|\mathcal{H}'|\mathbf{k}_i\rangle|i\rangle|^2\delta(\omega - \omega_{fi}) \tag{6.33}$$

where the δ–function holds for energy conservation. Using the properties of the convolution product in the Fourier transform, this becomes

$$\mathcal{W}_{i\to f} = \frac{2\pi}{\hbar^2}|\langle f|\rho_k^\dagger|i\rangle|^2|\mathcal{V}(\mathbf{k})|^2\delta(\omega - \omega_{fi}) \tag{6.34a}$$

where

$$\mathcal{V}(\mathbf{k}) = \int d\mathbf{r} V(\mathbf{r}) e^{i\mathbf{k}\cdot\mathbf{r}} = \begin{cases} \frac{2\pi a \hbar^2}{m} & (neutrons) \\ \frac{-4\pi e}{k^2} & (electrons) \end{cases} \qquad (6.34\text{b})$$

and $\rho_k^\dagger = \rho_{-k}$ is the $\mathbf{k}$–th Fourier component of ρ_n or ρ_z, respectively.

In order to obtain the inelastic differential scattering cross section for the angle $d\Omega$ and an energy range $\hbar d\omega$, we must include in Eq. (6.34a)

- a sum over all final target states,
- the differential element factor

$$(2\pi)^{-3} d\mathbf{k}_f = (2\pi)^{-3} k_f^2 dk_f d\Omega = (2\pi)^{-3} \frac{m}{\hbar^2} k_f \hbar d\omega d\Omega,$$

- a factor equal to the inverse of the incident flux $k_i \hbar/m$, and finally
- the thermal average performed on the initial target states.

All this leads to:

$$\frac{d^2\sigma}{d\omega d\Omega} = \frac{k_f}{k_i} \left(\frac{m}{2\pi\hbar^2} \right)^2 |\mathcal{V}(\mathbf{k})|^2 \sum_{i,f} \rho_i |\langle f|\rho_k^\dagger|i\rangle|^2 \delta(\omega - \omega_{fi})$$

$$= \frac{k_f}{k_i} \left(\frac{m}{2\pi\hbar^2} \right)^2 |\mathcal{V}(\mathbf{k})|^2 \rho_i \int_{-\infty}^{+\infty} dt e^{i(\omega-\omega_{if})t} |\langle f|\rho_k^\dagger|i\rangle|^2 \qquad (6.35)$$

In the last equation we have used the Fourier representation of the δ–function. But, using Eqs. (6.3b) and (6.32), and the fact that $|i\rangle$ and $|f\rangle$ are eigenstates of $\mathcal{H}_0$, after a little of algebra we obtain

$$e^{-i\omega_{if}t} |\langle f|\rho_k^\dagger|i\rangle|^2 = \langle i|\rho_k(t)|f\rangle\langle f|\rho_k^\dagger(0)|i\rangle$$

Introducing all this into Eq. (6.35), and using the completeness relation of the $\mathcal{H}_0$ eigenstates, we arrive at

$$\frac{d^2\sigma}{d\omega d\Omega} = \frac{k_f}{k_i} \left(\frac{m}{2\pi\hbar^2} \right)^2 |\mathcal{V}(\mathbf{k})|^2 N S(\mathbf{k},\omega) \qquad (6.36)$$

where N is the number of particles of the scattering system (nuclei or charges in the target), and $S(\mathbf{k},\omega)$ is the spectral function corresponding to the density self-correlation function of nuclei or charges:

$$S(\mathbf{k},\omega) = \frac{1}{2\pi} \int_{-\infty}^{+\infty} dt e^{i\omega t} \frac{1}{N} \langle \rho_k(t)\rho_k^\dagger(0)\rangle \qquad (6.37\text{a})$$

In the neutron scattering literature, $S(\mathbf{k},\omega)$ is called *dynamical structure factor*, and the self-correlation function of the Fourier components of the density ρ_n is named *intermediate scattering function*. It reads

$$\mathcal{F}(\mathbf{k}, t) = \frac{1}{N}\langle \rho_k(t)\rho_k^\dagger(0)\rangle = \frac{1}{N}\langle \rho_k(t)\rho_{-k}(0)\rangle \qquad (6.37b)$$

For the simple case of a monoatomic fluid (rare gas, liquid metal), $\rho_n(\mathbf{r})$ coincides with the local number density $\rho(\mathbf{r})$ defined in Eq. (6.7a), whose Fourier components are

$$\rho_k = \sum_i e^{i\mathbf{k}\cdot\mathbf{r}_i} \qquad (6.38)$$

All the correlation functions indicated up to here are of utmost importance in the study of density fluctuations in liquids. The result given in Eq. (6.36) is the keystone of this formalism, since it points out the relation among inelastic scattering cross sections and time dependent correlation functions. Note that, according to the general properties of the self-correlation functions, $S(\mathbf{k}, \omega)$ and hence the differential scattering cross section, are real and positive functions, as expected. Finally, note also that, according to Eq. (6.26), the integrated intensity of the beam scattered in a given direction (that is, the differential scattering cross section integrated over all energies for a fixed Ω), turns out to be proportional to the value of the intermediate scattering function at $t = 0$, called the *static structure factor*:

$$\frac{d\sigma}{d\Omega} \propto \int_{-\infty}^{+\infty} d\omega\, S(\mathbf{k}, \omega) = \mathcal{F}(\mathbf{k}, t = 0) \equiv S(\mathbf{k}) \qquad (6.39a)$$

This function is familiar from discussions of crystal structures in solids, X–ray scattering, etc.

The qualitative connection between $S(\mathbf{k}, \omega)$ and $S(\mathbf{k})$ is schematically shown in Fig. 6.3.

We finally mention that the static structure function $S(\mathbf{k})$, which results fundamental for the calculation of equilibrium properties of a fluid, is related to the (equilibrium) *radial distribution function* $g(\mathbf{r})$. The latter function is defined by the (equilibrium) two-particle distribution function discussed in Chapter 4:

$$f_2(\mathbf{r}_1, \mathbf{r}_2, t) = \left(\frac{N}{V}\right)^2 g(|\mathbf{r}_1 - \mathbf{r}_2|) \qquad (6.39b)$$

and connected with $S(\mathbf{k})$ through

$$S(\mathbf{k}) = 1 + \int d\mathbf{r}\, e^{i\mathbf{k}\cdot\mathbf{r}} g(|\mathbf{r}|) \qquad (6.39c)$$

Typical forms of the *radial distribution function* $g(\mathbf{r})$ for solids, liquids and gases, are shown in Fig. 6.4.

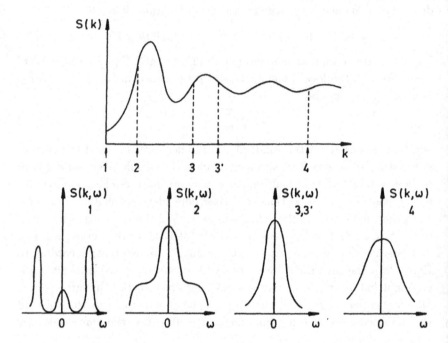

Fig. 6.3   Typical form of the Static Structure Factor as function of the wave vector k. The form of the Dynamical Structure Factor for different values of $\mathbf{k}$, as function of ω are also shown.

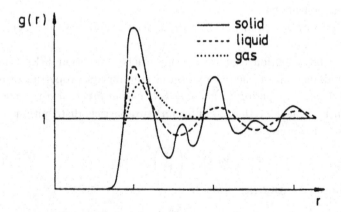

Fig. 6.4   Typical form of the Radial Distribution Function for Solids, Liquids and Gases.

6.3.2   Response functions

Here we want to generalize the results of the previous subsection by studying the response of a system to a weak external time dependent field. We
assume the perturbation has the general form of Eqs. (6.29), but we can
always consider that the external field is a superposition of monochromatic
plane waves. As we are only interested in the *linear* response of the system,
we may consider just a single plane wave of wave number $\mathbf{k}$ and frequency
ω (the response to an arbitrary—but weak!—perturbation will be a superposition of those corresponding to each of the Fourier components),

$$\mathcal{F}(\mathbf{r}, t) = \mathcal{F}(\mathbf{k})e^{i(\mathbf{k}\cdot\mathbf{r} - \omega t)} \tag{6.40a}$$

Replacing this in Eq. (6.29a) yields

$$\mathcal{H}' = -A^\dagger(\mathbf{k})\mathcal{F}(\mathbf{k})e^{-i\omega t} + \text{c.c.} \tag{6.40b}$$

In order to simplify the notation, we temporally disregard the dependence
on $\mathbf{k}$, to be introduced only at the end, that is, we take a spatially homogeneous external field.

We also assume that in the remote past ($t \to -\infty$):

- the perturbation was identically zero,
- the unperturbed system was in thermodynamical equilibrium.

We then write $\mathcal{H}'$ as

$$\mathcal{H}' = -A\mathcal{F}(t) = -A[\mathcal{F}_0 e^{-i\omega t}e^{\eta t} + \text{c.c.}] \tag{6.40c}$$

where $\eta > 0$. This guarantees that $\mathcal{F}(t \to -\infty) = 0$. At the end of the
calculation we must take the limit $\eta \to 0^+$.

Let us recall that the applied field $\mathcal{F}(t)$ (coupled to a dynamical variable
A) will produce changes in the average of any dynamical variable B. Our
purpose now is to evaluate these changes. In the following, in order to
simplify the notation, we assume that the equilibrium average of B is zero
($\langle B \rangle_0 = 0$). To calculate $\langle B(t) \rangle$ up to the first order in powers of the applied
field, we will follow a procedure similar to that introduced by Kubo. The
time evolution of the *density matrix* is given by

$$\frac{\partial}{\partial t}\rho(t) = -i\mathcal{H}\rho(t) \tag{6.41a}$$

(see Chapter 2). As the Liouville operator $\hat{\mathcal{L}}$ depends linearly on the Hamiltonian, it is possible to separate it into an unperturbed part ($\hat{\mathcal{L}}_0$) and

a perturbation $(\mathcal{L}')$. We are interested in evaluating Eq. (6.41a) with $\rho(-\infty) = \rho_0 \propto \exp(-\beta\mathcal{H}_0)$. We look for solutions of the form

$$\rho(t) = \rho_0 + \Delta\rho(t) \tag{6.41b}$$

Introducing this into Eq. (6.41a), and retaining only first order contributions, gives

$$\frac{\partial}{\partial t}\Delta\rho(t) = -i\mathcal{L}_0\Delta\rho(t) - i\mathcal{L}'\rho_0(t)$$

$$= -\frac{i}{\hbar}[\mathcal{H}_0, \Delta\rho(t)] + \frac{i}{\hbar}[A, \rho_0]\mathcal{F}(t) \tag{6.42}$$

that can be formally integrated as

$$\Delta\rho(t) = \int_{-\infty}^{t} dt' e^{-i(t-t')\mathcal{L}_0} i\mathcal{H}'(t')\rho_0 \tag{6.43}$$

This result allows us to calculate $\langle B(t)\rangle$ up to the first order in the perturbation

$$\langle B(t)\rangle = \text{Tr}\{\Delta\rho(t)B\}$$

$$= \frac{i}{\hbar}\int_{-\infty}^{t} dt' \mathcal{F}(t')\text{Tr}\{e^{-i(t-t')\mathcal{H}_0}[A, \rho_0]B\} \tag{6.44}$$

Here we can use the cyclic properties of the trace, together with the definition of the Liouville operator, to obtain for arbitrary operators C and D:

$$\text{Tr}\{\mathcal{H}_0 CD\} = -\text{Tr}\{C\mathcal{H}_0 D\} \tag{6.45}$$

Using this in Eq. (6.44), with $C = [A, \rho_0]$ and $D = B$, and recalling also Eq. (6.3a), we obtain

$$\langle B(t)\rangle = \frac{i}{\hbar}\int_{-\infty}^{t} dt' \mathcal{F}(t')\text{Tr}\{[A, \rho_0]B(t - t')\}$$

$$= \frac{i}{\hbar}\int_{-\infty}^{t} dt' \mathcal{F}(t')\text{Tr}\{[\rho_0, B(t - t')]A\}$$

$$= \frac{i}{\hbar}\int_{-\infty}^{t} dt' \mathcal{F}(t')\text{Tr}\{\rho_0[B(t - t'), A]\} \tag{6.46}$$

Defining the so called *causal function* as

$$\theta_{BA}(t) = \frac{i}{\hbar}\text{Tr}\{\rho_0[B(t - t'), A]\} = \frac{i}{\hbar}\langle[B(t - t'), A]\rangle_0 \tag{6.47a}$$

the average value of B can be written as

$$\langle B(t) \rangle = \int_{-\infty}^{t} dt' \mathcal{F}(t') \theta_{BA}(t - t') \tag{6.47b}$$

From the previous expressions it is obvious the relation that exists among response and correlation functions. In the literature it is usual to take $\theta_{BA}(t) = 0$ for $t < 0$ (implying causality, and giving a meaning to the name of *causal* for the function defined above), which allows us to extend the integration to $t \to +\infty$ (including a step Heaviside time function in time). Then, $\langle B(t) \rangle$ is a superposition of the retarded effects of the external field acting at previous times.

Replacing now Eq. (6.40c) into Eq. (6.47b), and since $\theta_{BA}(t)$ is real (because it equals $i/\hbar$ times the thermal average of the commutator of two Hermitian operators), we get

$$\langle B(t) \rangle = \Re \left\{ \mathcal{F}_0 e^{-i(\omega + i\eta)t} \int_{-\infty}^{t} dt' e^{-i(\omega + i\eta)(t' - t)} \theta_{BA}(t - t') \right\}$$

$$= \Re \left\{ \mathcal{F}_0 e^{-i(\omega + i\eta)t} \int_{0}^{\infty} ds\, e^{-i(\omega + i\eta)s} \theta_{BA}(s) \right\}$$

$$= \Re \left\{ \mathcal{F}_0 e^{-i(\omega + i\eta)t} \theta_{BA}(\omega + i\eta) \right\} \tag{6.48a}$$

We can now define the *dynamical response function* (also called *dynamical susceptibility*. Recall that the *static susceptibility* describes the magnetic response -magnetization- of a substance to a static magnetic field, or the electric response -polarization- to a static electric field. In both cases a linear proportionality between the quantities is assumed. The present case is a natural generalization for time dependent situations).

$$\chi_{BA}(\omega) = \chi'_{BA}(\omega) + i\chi''_{BA}(\omega) = \lim_{\eta \to 0^+} \theta_{BA}(\omega + i\eta) \tag{6.48b}$$

which allows us to write Eq. (6.48a) as

$$\langle B(t) \rangle = \Re \left\{ \chi_{BA}(\omega) \mathcal{F}_0 e^{-i\omega t} \right\} \tag{6.48c}$$

In the case of space dependent external fields, the generalization of the above results is immediate. If the system is invariant under space translations in equilibrium, Eq. (6.47b) becomes

$$\langle B(\mathbf{r}, t) \rangle = \int_{-\infty}^{t} dt' \int d\mathbf{r}'\, \theta_{BA}(\mathbf{r} - \mathbf{r}', t - t') \mathcal{F}(\mathbf{r}', t') \tag{6.49a}$$

and the response function, or *generalized susceptibility*, is connected with the *non-local causal function* $\theta_{BA}(\mathbf{r}, t)$ through

$$\chi_{BA}(\mathbf{k}, \omega) = \lim_{\eta \to 0^+} \int_0^\infty dt \int d\mathbf{r} \theta_{BA}(\mathbf{r}, t) e^{i(\omega + i\eta)t - i\mathbf{k} \cdot \mathbf{r}} \qquad (6.49b)$$

This results in

$$\langle B(\mathbf{k}, t) \rangle = \Re \left\{ \chi_{BA}(\mathbf{k}, \omega) \mathcal{F}(\mathbf{k}) e^{-i\omega t} \right\} \qquad (6.49c)$$

In the last section of this chapter we shall give examples of application of these results.

6.4   Fluctuation–dissipation theorem, and properties of response functions

From the results indicated in Eqs. (6.47), it is clear that the linear response of a system to a (weak) external perturbation is completely described in terms of *equilibrium averages*. We shall now make more explicit the connection between $\theta_{BA}(\mathbf{r}, t)$ and the time correlation functions. From the Neumann equation

$$\dot{A} = -\frac{i}{\hbar}[\mathcal{H}_0, A] \qquad (6.50)$$

and recalling that $\rho_0 \propto e^{-\beta \mathcal{H}_0}$, it is easy to derive the *Kubo identity*

$$-\frac{i}{\hbar}[\rho_0, A] = \int_0^\beta d\lambda \rho_0 e^{\lambda \mathcal{H}_0} \dot{A} e^{-\lambda \mathcal{H}_0} = \int_0^\beta d\lambda \rho_0 \dot{A}(-i\hbar\lambda) \qquad (6.51)$$

Replacing this in Eqs. (6.46) and (6.47), we obtain

$$\theta_{BA}(t) = -\int_0^\beta d\lambda \langle \dot{B}(-i\hbar\lambda) A \rangle_0 = -\beta C_{BA}^{(c)}(t) = -\beta \frac{d}{dt} C_{BA}^{(c)}(t) \qquad (6.52a)$$

where $C_{BA}^{(c)}(t)$ is the canonical time correlation function of B and A, as defined in Eq. (6.13). In the classical limit ($\hbar \to 0$) we have

$$\theta_{BA}(t) = -\beta \langle \dot{B}(t) A \rangle_0 = \beta \langle B(t) \dot{A} \rangle_0 \qquad (6.52b)$$

We have thus found that the system response to a weak external perturbation is entirely determined by the equilibrium correlations of the fluctuating dynamical variables A and B. The expression in Eq. (6.52b), is the most general form of the *fluctuation-dissipation theorem*. We are now going to express it in a more conventional form, making its meaning clear.

According to the previous result and Eq. (6.49b), the response function is given by

$$\chi_{BA}(\omega) = -\beta \int_0^\infty dt C_{\dot{B}A}^{(c)}(t) e^{i\omega t}$$

$$= -\beta C_{\dot{B}A}^{(c)}(z = \omega) = \beta \{ C_{BA}^{(c)}(0) + i\omega C_{BA}^{(c)}(z = \omega) \} \qquad (6.53)$$

A very important situation corresponds to the case when $B = A$ (or $A^\dagger$), where we immediately conclude, from Eq. (6.21a), that the imaginary part of the susceptibility is

$$\chi_{AA}''(\omega) = \pi \beta \omega C_{AA}^{(c)}(\omega) \qquad (6.54)$$

This is, for this special case, a more compact form of the *fluctuation-dissipation theorem*. Moreover, we shall show that $\chi_{AA}''(\omega)$ is associated with the energy dissipated by the perturbed system, which originates the name of the theorem. First we discuss another simple relation among the Fourier-Laplace transform of the canonical time-dependent correlation functions, and the one-sided correlation functions,

$$C_{BA}^{(c)}(\omega) = \frac{1}{\beta \hbar \omega} \left[1 - e^{-\beta \hbar \omega} \right] C_{BA}(\omega) \qquad (6.55a)$$

and, for $B = A$, from Eq. (6.54),

$$\chi_{AA}''(\omega) = \frac{\pi}{\hbar} \left[1 - e^{-\beta \hbar \omega} \right] C_{AA}(\omega) \qquad (6.55b)$$

From Eq. (6.55b) and the positivity of the power spectrum for the self-correlation function, it follows that

$$\chi_{AA}''(\omega) \geq 0 \qquad (6.56a)$$

Let us see now a physical proof of this relation, showing that $\chi_{AA}''(\omega)$ is connected with energy dissipation. The rate of energy variation under the influence of the perturbation $\mathcal{H}'$ is

$$\frac{dE}{dt} = \frac{d}{dt} \text{Tr}\{\rho(t)\mathcal{H}'\} = -\frac{d}{dt} \mathcal{F}(t) \langle A(t) \rangle \qquad (6.56b)$$

The total energy variation over a time interval of $2T$, under the influence of an external monochromatic field, is

$$\Delta E = \int_{-T}^T \frac{dE}{dt} dt = T \mathcal{F}_0^2 \omega \chi_{AA}''(\omega) \geq 0 \qquad (6.56c)$$

where we have only used Eq. (6.48c), and then obtained that $\chi_{AA}''(\omega)$ is directly related to the energy dissipation.

For completeness, and similarly to what we have done for the correlation functions, we now look at some properties of the response functions:

(a) Properties of the *causal response function* arise from its definition, Eq. (6.47a) or from Eq. (6.53), and the properties of the time correlation functions previously discussed. In particular, $\theta_{BA}(t)$ is a real stationary function of t and, if ϵ_A and $\epsilon_B = -\epsilon_{\tilde{B}}$ are the signatures of the A and B operators under temporal inversion, from Eq. (6.11) we have

$$\theta_{BA}(-t) = -\epsilon_A \epsilon_B \theta_{BA}(t) \tag{6.57}$$

(b) From the definition Eq. (6.48b) for the response function, and due to $\theta_{BA}(t)$ being real, we find that over the real axis

$$\chi_{BA}(-\omega) = \chi^*_{BA}(\omega) = \chi'_{BA}(\omega) - i\chi''_{BA}(\omega) \tag{6.58}$$

Then, χ' is an *even* function of ω, meanwhile χ'' is an *odd* function of ω.

(c) We now introduce the Fourier transform of the causal function $\theta_{AB}(\omega)$ and the auxiliary function $\epsilon_{AB}(\omega)$, as

$$\epsilon_{AB}(\omega) = -i\pi\theta_{AB}(\omega) \tag{6.59a}$$

It turns out that $\chi_{BA}(z)$ is the Hilbert transform of $\epsilon_{AB}(\omega)$

$$\chi_{BA}(z) = \int_{-\infty}^{\infty} \frac{d\omega}{\pi} \epsilon_{A}\omega) \frac{1}{\omega - z} \tag{6.59b}$$

$$\chi_{BA}(\omega) = \lim_{\eta \to 0} \chi_{BA}(\omega + i\eta)$$

$$= \mathcal{P} \int_{-\infty}^{\infty} \frac{d\omega'}{\pi} \epsilon_{AB}(\omega') \frac{1}{\omega' - \omega} + i\epsilon_{AB}(\omega) \tag{6.59c}$$

$\mathcal{P}$ indicates the *principal part* of the integral.

(d) Now, we restrict ourselves to the case $B = A$. It includes the important case of the *density-density* correlation function ($B = A = \rho$). From $\theta_{AA}(-t) = -\theta_{AA}(t)$, it is easy to prove that

$$\epsilon_{AA}(\omega) = \chi''_{AA}(\omega) \tag{6.60a}$$

and, from Eq. (6.59c),

$$\chi'_{AA}(\omega) = \mathcal{P} \int_{-\infty}^{\infty} \frac{d\omega'}{\pi} \chi''_{AA}(\omega') \frac{1}{\omega' - \omega} \tag{6.60b}$$

The last equation, relating the real and the imaginary part of the response function, corresponds to one of the *Kramers–Kronig relations*, the other one being

$$\chi''_{AA}(\omega) = -\mathcal{P} \int_{-\infty}^{\infty} \frac{d\omega'}{\pi} \chi'_{AA}(\omega') \frac{1}{\omega' - \omega} \tag{6.60c}$$

In analogy with the dielectric constant, we see that the *dissipative* and *dispersive* parts of the response function are not independent. According to Eq. (6.55b), it is possible to measure the dissipative part directly, while the dispersive part is calculated from Eq. (6.59b).

(e) It is possible to do a high frequency expansion of the response function, corresponding to a short time expansion of the time correlation function. If we consider the general case including $\mathbf{k}$ dependence, from Eqs. (6.59a) and (6.60b) we obtain

$$\chi_{AA}(z) = -\sum_{n=1}^{\infty} \frac{a_{2n}}{z^{2n}} \tag{6.61a}$$

where

$$a_{2n} = \int_{-\infty}^{\infty} \frac{d\omega}{\pi} \chi_{AA}(\mathbf{k}, \omega)\omega^{2n-1} \tag{6.61b}$$

Only even powers appear, because χ''_{AA} is odd in ω. The expansion is *asymptotic*, in the sense that it is valid only if $|z|$ is large compared with all the characteristic frequencies of the system. In the classical limit ($\hbar \to 0$), from Eq. (6.55b) we get

$$\chi''_{AA}(\omega) = \pi\beta\omega C_{AA}(\omega) \tag{6.62a}$$

Then, using Eq. (6.26),

$$a_{2n} = \beta\langle\omega^{2n}\rangle_{AA} \tag{6.62b}$$

which clearly shows the connection with the short time expansion.

(f) From Eq. (6.59b) we deduce that the zero frequency limit of $\chi(\mathbf{k}, \omega)$, called the *static susceptibility* $\chi_{AA}(\mathbf{k})$, is

$$\chi_{AA}(\mathbf{k}) = \chi(\mathbf{k}, z = 0) = \int_{-\infty}^{\infty} \frac{d\omega}{\pi} \chi''_{AA}(\mathbf{k}, \omega)\omega^{-1}$$

$$= \beta \int_{-\infty}^{\infty} C_{AA}(\mathbf{k}, \omega)d\omega = \beta C_{AA}(\mathbf{k}, 0) \tag{6.63}$$

It is instructive to analyze the very important case of the density-density response function. Using the notation of the previous paragraph, from Eqs. (6.63) and (6.37b) we have

$$\chi_{\rho\rho}(\mathbf{k}) = \beta N S(\mathbf{k}) \tag{6.64a}$$

where $S(\mathbf{k})$ is the *structure factor* (familiar from solids, X–ray scattering, etc), discussed in relation with Eq. (6.39). It is a well known fact that

$$\lim_{k\to 0} S(\mathbf{k}) = \frac{X_T}{X_T^0} \tag{6.64b}$$

X_T being the *isothermal compressibility* of the system of interacting particles, and $X_T^0 = \rho k_B T$ the one corresponding to an ideal gas system with the same density and temperature. X_T can be considered as the macroscopic susceptibility of a fluid under external pressure changes. Due to the relation

$$\lim_{k \to 0} \chi(\mathbf{k}, z = 0) = N \rho X_T \qquad (6.65)$$

the dynamical susceptibility arises as a natural extension of the macroscopic susceptibility at finite frequency and wave length, and is usually called *generalized susceptibility*. Such a point of view was systematically extended when it was recognized that for a system perturbed by an external field, the linear response theory description in terms of temporal correlation functions must coincide with the one arising from the linearized equations of hydrodynamics, in the limit where all the relevant physical quantities vary slowly in space and time. Typical examples of this approach are found in [Forster (1975)] and [Kadanoff and Martin (1963)].

6.5   Some examples

Here we want to discuss how the formalism presented in the previous paragraphs works. As a few examples, we are going to show how to calculate the dielectric susceptibility of a liquid as well as some transport coefficients.

6.5.1   *Dielectric susceptibility*

Let us consider the dielectric response of a liquid to an external periodic electric field. We assume the liquid composed by N polar molecules, each one with a permanent dipole $\bar{\mu}_i$. The total dipolar moment is

$$\mathbf{M} = \sum_i \bar{\mu}_i \qquad (6.66a)$$

and the coupling with the external electric field $\mathbf{E}(t)$ is

$$\mathcal{H}'(t) = -\mathbf{M} \cdot \mathbf{E}(t) \qquad (6.66b)$$

It is clear that in equilibrium and without an external field, the mean value of the total dipolar moment will be zero. The relations (6.49a) and (6.49c) lead us to

$$\langle \mathbf{M} \rangle = \Re\{\chi_e(\omega)\mathbf{E}(t)\} \qquad (6.67a)$$

where, using Eqs. (6.48c) and (6.52b), the (classical) susceptibility is

$$\chi_e(\omega) = \frac{\beta}{3} \int_0^\infty dt \langle \dot{\mathbf{M}}(t)\mathbf{M}(0)\rangle e^{i\omega t} \tag{6.67b}$$

The special case of dilute solutions of non-polarizable molecules, with dipolar moments $\bar{\mu}$ gives, after neglecting correlations among the orientations of different dipoles,

$$\chi_e(\omega) = \beta \frac{\mu^2 N}{3} \int_0^\infty dt \langle \dot{\mathbf{u}}(t)\mathbf{u}(0)\rangle e^{i\omega t} \tag{6.67c}$$

where $\mathbf{u}$ is the unitary vector pointing in the direction of the dipole. Equation (67b) constitutes the starting point for the theory of dielectric relaxation. If the external electric field $\mathbf{E}(t)$ is independent of time, Eq. (6.67c) coincides with the known, static, result

$$\chi_e(\omega) = \beta \frac{\mu^2 N}{3} \langle \bar{\mu}^2 \rangle \tag{6.67d}$$

indicating that the average polarizability per molecule is inversely proportional to the temperature. This is a result that could be expected from an effect where the applied field must overcome the thermal fluctuation.

6.5.2 Transport coefficients

6.5.2.1 Electric conductivity

We consider a system of charged particles in an electric field $\mathbf{E}(t)$. The coupling of the system with this field (or perturbation) is

$$\mathcal{H}'(t) = -\sum_i e_i \mathbf{r}_i \cdot \mathbf{E}(t) \tag{6.68a}$$

$\mathbf{r}_i$ being the coordinate of the particle with charge e_i. The external electric field has the form

$$\mathbf{E}(t) = \mathbf{E}_0 e^{i(\omega_0 + i\nu)t} \tag{6.68b}$$

with ν a small (positive) parameter such that $\mathbf{E}(t) \to 0$ when $t \to -\infty$. The conductivity σ is determined from

$$\frac{\langle \bar{\mathbf{J}} \rangle}{V} = \bar{\sigma} \cdot \mathbf{E} \tag{6.69a}$$

where the current $\mathbf{J}$ is given by

$$\mathbf{J} = \sum_i e_i \dot{\mathbf{r}}_i = \sum_i \left(\frac{e_i}{m}\right) \mathbf{p}_i \tag{6.69b}$$

Using Eqs. (6.52a), (6.53), (6.44) and (6.49b), together with $A = \sum_i e_i \mathbf{r}_i$ and $B = \sum_i e_i \dot{\mathbf{r}}_i$, we find *Nakano's formula*

$$\bar{\sigma} = \mathrm{Tr} \int_{-\infty}^{t} V^{-1} e^{i\omega(t'-t)} dt' \int_0^{\beta} d\lambda \rho_0 \mathbf{J}(-i\hbar\lambda) \mathbf{J}(t-t') \qquad (6.70a)$$

Finally, the frequency dependent conductivity is obtained as

$$\bar{\sigma}(\omega) = V^{-1} \mathrm{Tr} \int_0^{\infty} e^{i(\omega+i\nu)\tau} d\tau \int_0^{\beta} d\lambda \rho_0 \mathbf{J}(-i\hbar\lambda) \mathbf{J}(\tau) \qquad (6.70b)$$

where ν is taken to be zero after the integration over τ. If we consider a time independent electric field, we shall recover the known result obtained in introductory courses on statistical mechanics within the framework of the relaxation time approximation for the Boltzmann equation.

6.5.2.2 *Diffusion constant*

We assume a gravitational or centrifugal field in the direction z, acting uniformly on all the particles of the system. In this case the perturbation is given by

$$\mathcal{H}'(t) = -\sum_i z_i \mathcal{F} \qquad (6.71a)$$

The force $\mathcal{F}$ is related with the average velocity v by

$$\mathcal{F} = \gamma v \qquad (6.71b)$$

where γ is the friction constant. It is related to the diffusion constant through the *Einstein relation*

$$\mathcal{D} = kT\gamma^{-1} \qquad (6.71c)$$

Then, we can obtain the diffusion constant evaluating the velocity average, per unit force, multiplied by kT

$$\mathcal{D} = kT\mathcal{F}^{-1} \langle m^{-1} \sum_i p_{z,i} \rangle = kT\mathcal{F}^{-1} \langle v_z \rangle \qquad (6.72a)$$

where $p_{z,i}$ is the z–component of the i-th particle momentum. Using the known expressions for $A = \sum_i z_i$ and $B = v_z$, we find

$$\mathcal{D} = kT \, \mathrm{Tr} \int_0^{\infty} d\tau \int_0^{\beta} d\lambda v_z(-i\hbar\lambda) v_z(\tau) \qquad (6.72b)$$

This last result is more general than the one found in the previous chapter in Eq. (5.19b).

6.5.2.3 *Viscosity coefficient*

We consider an incompressible, homogeneous Newtonian fluid, whose shear stress tensor is proportional to the tensor $\mathbb{Q}$, the velocity gradient, with the shear viscosity η as the proportionality constant:

$$s = 2\eta \mathbb{Q} \tag{6.73}$$

For a shear flow in the $x - y$ plane, the symmetrized tensor $\mathbb{Q}$ will be given by

$$\mathbb{Q}_{xy} = \mathbb{Q}_{yx} = \frac{1}{2}\left(\frac{\partial u_y}{\partial x} + \frac{\partial u_x}{\partial y}\right) \tag{6.74}$$

whereas all its other components vanish. In Eq. (6.74) u_x and u_y are the components of the average velocity of the system. The stress tensor is obtained considering the momentum transfer and a drift flow due to external potentials. We can write

$$s = s_K + s_\phi = V^{-1}\{\langle \mathbb{J}_K \rangle + \langle \mathbb{J}_\phi \rangle\} \tag{6.75a}$$

where

$$\mathbb{J}_K = -\sum_i (\mathbf{p}_i - m\mathbf{u})(\mathbf{p}_i - m\mathbf{u}) \tag{6.75b}$$

$$\mathbb{J}_\phi = \frac{1}{2}\sum_{i \neq j}\left[\frac{(\mathbf{q}_j - \mathbf{q}_i)(\mathbf{q}_j - \mathbf{q}_i)}{r_{ij}}\frac{\partial \phi_{ji}}{\partial r_{ij}}\right] \tag{6.75c}$$

Let us assume that $\mathbb{Q}$ has a sinusoidal oscillation in time. Then $\mathbb{Q}$ is given as

$$\mathbb{Q} = \dot{\mathbb{S}} \quad ; \quad \mathbb{S} = \mathbb{S}_0 e^{(i\omega+\nu)t} \tag{6.76a}$$

with $\nu > 0$. Note that

$$\int_{-\infty}^{t} dt = \int_{-\infty}^{t} 2\eta \dot{\mathbb{S}} dt = 2\eta \mathbb{S} = V^{-1}\langle \int_{-\infty}^{t} \mathbb{J} dt \rangle \tag{6.76b}$$

The last equality suggests a method for calculating the viscosity coefficient η, by evaluating the average of the physical quantity $B = \int_{-\infty}^{t} \mathbb{J} dt$. The perturbation term in the Hamiltonian could be taken as

$$\mathcal{H}'(t) = -(-\mathbb{J}_{xy})(2\mathbb{S}_{xy}) = -A\mathcal{F} \tag{6.77}$$

with $A = -\mathbb{J}_{xy}$ and $\mathcal{F} = 2\mathbb{S}_{xy}$. According to Eqs. (6.47b), (6.49b), (6.49c) and (6.52a), the response to this perturbation will be given by

$$\theta_{BA}(t) = \int_{0}^{\beta} d\lambda \text{Tr}\{\rho_0 \mathbb{J}(-i\hbar\lambda)\mathbb{J}(t)\} \tag{6.78a}$$

Then, we have

$$\eta \mathbb{S}_{xy} = V^{-1} \int_{-\infty}^{t} \theta_{BA}(t-t') \mathbb{S}_{xy}(t') dt' = V^{-1} \int_{0}^{\infty} \theta_{BA}(\tau) \mathbb{S}_{xy}(t-\tau) d\tau$$

$$= V^{-1} \mathbb{S}_{0xy} \int_{0}^{\infty} \theta_{BA}(\tau) e^{-(i\omega+\nu)\tau} d\tau \qquad (6.78b)$$

and we arrive to

$$\eta(\omega) = V^{-1} \lim_{\nu \to 0} \int_{0}^{\infty} d\tau e^{-(i\omega+\nu)\tau} \mathrm{Tr} \int_{0}^{\beta} d\lambda \{\rho_0 \mathbb{J}(-i\hbar\lambda) \mathbb{J}(-\tau)\} \qquad (6.79a)$$

The static viscosity becomes then

$$\eta = V^{-1} \int_{0}^{\infty} d\tau \mathrm{Tr} \int_{0}^{\beta} d\lambda \{\rho_0 \mathbb{J}(-i\hbar\lambda) \mathbb{J}(-\tau)\} \qquad (6.79b)$$

At this point we stop the discussion of this subject. It is worth remarking that the linear response theory so far discussed is applicable to systems out of equilibrium, but not too far from it. In the next two chapters we are going to discuss some aspects of the techniques employed to treat systems far from equilibrium.

PART 3

The mesoscopic approach far from equilibrium: Non-extended systems

... most natural phenomena are complex systems,
and regarding them as special cases is like referring to
the class of animals that are not elephants as non-elephants.
Stanislaw Ulam

If you are a young physics student, your textbook is full of linear
problems, and you become adept at solving them. When you're
confronted with a nonlinear problem, you're taught immediately to
linearize it; you make an approximation, use a special case.
But when you venture into the real world,
you realize that many problems are non-linear in an essential way
and cannot be linearized meaningfully. You would just lose the physics.
J. Doyne Farmer

The mesoscopic approach ... from equilibrium: Non-extended systems

Chapter 7

Stability of dissipative dynamical systems

I left for the various futures (but not all of them)
my garden of the bifurcating paths ...
Jorge Luis Borges

7.1 Introduction

In the forties, equilibrium thermodynamics was the only "serious" tool available for analyzing the behavior of a macroscopic system. Though there existed already a theory for systems *slightly* out of equilibrium (Onsager's theory, which we met in Chapter 5), there was no prescription to proceed in far-from-equilibrium situations. Furthermore, the discovery that far from equilibrium, matter acquires new properties—seen afterwards to be typical of non-equilibrium situations—came as a surprise to most people. Those cases often correspond to systems submitted to *strong* external constraints (e.g. strong fluxes of energy or chemical reactives) rather than being isolated or in contact with a *single* heat source at a time (or more, but diferring infinitesimally in the case of slight disequilibrium). Clearly, in order to understand the world around us, all those completely new properties must be studied. Examples of these phenomena in fields traditionally regarded as belonging to physics are the laser, the Bénard instability in convective flows and the Gunn effect in semiconductors. However, the occurrence of nonequilibrium phenomena largely transcends this realm, having widespread interest and implications for the understanding of cooperative phenomena in chemistry (e.g. the Belousov–Zhabotinskii reaction), biology and fields even more remote like sociology and economics.

The term *dissipative structures* was coined to name states exhibiting those new properties, as e.g. long range coherent organization. Another surprise was the possibility of *multiple* stationary states (for slight disequilibrium only one stationary state exists, that tends to equilibrium as temperature gradients tend to zero). This fact entails consequent transitions and *hysteresis* phenomena, all of them arising from the *nonlinear* character of the macroscopic laws governing matter under strong nonequilibrium regimes.

In this chapter, we shall introduce some basic tools needed for an adequate analysis of a system under those extreme conditions. As a starting point, we discuss the kind of behavior one can expect at the macroscopic level when an external *control parameter* (related to the above indicated external fluxes) is varied. We will show that the steady state solutions can be strongly affected, to the point that the macroscopic behavior can change drastically. Through some examples we will also introduce such notions as *attractors*, *limit cycles*, *bifurcations* and *symmetry breaking*. In the next chapters, we will continue analyzing the effect of *external fluctuations* on the macroscopic behavior. Contrarily to the common belief that fast fluctuations average out, we will find that their effect has deep consequences when they occur near one of the indicated instability points, giving rise to completely new behavior.

7.2   Phase plane analysis and linear stability theory

We want to discuss here how to analyze the stability (or the loss thereof) of a system subject to a variation of some external parameter(s). We recall what we have done in Chapter 3 in relation with van Kampen's expansion. To the highest order in $\Omega^{-1/2}$, this procedure yields the equation governing the macroscopic evolution of the system, see Eqs. (3.16) and (3.24). Up to the next order in $\Omega^{-1/2}$, it gives the Fokker–Planck equation governing the (Gaussian) fluctuations around the macroscopic trajectory, see Eqs. (3.18) and (3.25). In Sec. 3.3, we briefly discussed the limitations of this program and suggested the framework used to analyze the indicated stability. In order to introduce the necessary elements for such an analysis, we will study the stability of stationary solutions of nonlinear differential equations (NLDE). It is clear that this study is relevant insofar as the macroscopic equations (for instance those obtained through an Ω expansion), are, in general, NLDE. Here, and in order to introduce some basic notions of *linear stability theory*, we will only consider the space independent case. The space-dependent case will be treated in chapter 9 within a reaction–diffusion framework.

Consider a set of NLDE of arbitrary order. This system can be reduced to another set with a larger number of first order NLDE, that is easier to analyze. As an example of this reduction process, consider the equation:

$$\ddot{x} + \alpha \dot{x}^{\nu} + \beta x^{\mu} + \gamma = 0 \qquad (7.1a)$$

for the general case with ν and $\mu \neq 1$. This equation, might describe the dynamics of a unit mass particle subject to a coordinate and velocity dependent field of force. Introducing a new variable y, defined by $y = \dot{x}$, we obtain the equivalent system

$$\dot{x} = y$$

$$\dot{y} = -\alpha y^{\nu} - \beta x^{\mu} - \gamma \qquad (7.1b)$$

Hence, we can argue in general that it is enough to study the stability properties of solutions of sets of first order NLDE of the general form

$$\dot{x}_j = \mathcal{F}_j(x_1, \ldots, x_N), \qquad j = 1, \ldots, N \qquad (7.2)$$

where $\mathcal{F}_j(x_1, \ldots, x_N)$ are in general nonlinear functions. For simplicity, we will restrict ourselves to sets of two first order NLDE, corresponding to a general second order autonomous system, i.e.:

$$\frac{dx_1}{dt} \equiv \dot{x}_1 = f(x_1, x_2)$$

$$\frac{dx_2}{dt} \equiv \dot{x}_2 = g(x_1, x_2) \qquad (7.3)$$

Through the study of this system we will learn the necessary basic elements of stability theory for our present needs.

If, for certain values of the coordinates, say (x_1^0, x_2^0), the functions $f(x_1, x_2)$ and $g(x_1, x_2)$ satisfy the very general Lipschitz conditions, Eqs. (7.3) have a unique solution in the neighborhood of the point (x_1^0, x_2^0). In what follows, we shall assume that these conditions are satisfied.

Time dependent solutions of Eqs. (7.3) describe trajectories in the phase plane [that is the (x_1, x_2) plane] called *phase curves* or *phase trajectories*, that are solutions of the equation

$$\frac{dx_1}{dx_2} = \frac{f(x_1, x_2)}{g(x_1, x_2)} \qquad (7.4)$$

Through any point (x_1^0, x_2^0) there is a unique phase curve, with the exception of the *singular* or *fixed points* (x_1^s, x_2^s), where $\dot{x}_1 = \dot{x}_2 = 0$ or

$$f(x_1^s, x_2^s) = 0 \quad ; \quad g(x_1^s, x_2^s) = 0. \qquad (7.5)$$

A fixed point, corresponding to a steady state solution of Eqs. (7.3), can always be moved to the origin by the change of variables $x_1 \to x_1 - x_1^s$ and $x_2 \to x_2 - x_2^s$. Therefore, and without loss of generality, we shall assume that the singular point is located at the origin. We then consider a system described by Eqs. (7.3) which is in a steady state at $(x_1^s, x_2^s) = (0, 0)$.

If the system is at the steady state, it is of the maximum importance to know how the system will behave under the influence of a small perturbation. Here we face several possibilities. The system can leave this steady state and move to another one; it can remain in the neighborhood of the original steady state; or it can decay back to the original state.

In order to analyze the different possibilities we use a *linear stability analysis*. By this procedure we can say something regarding the stability of the system in the neighborhood of the steady state, but nothing about the global stability of the system. To discuss stability in the neighborhood of the steady state we write the solution in terms of the departure from the steady state, i.e.:

$$x_1 = x_1^s + \delta x_1 \quad ; \quad x_2 = x_2^s + \delta x_2 \tag{7.6}$$

inserting this into Eqs. (7.3), using that $x_1^s = x_2^s = 0$, and expanding up to first order in the departures $(\delta x_1, \delta x_2)$:

$$\dot{x}_1 = \delta \dot{x}_1 = \tag{7.7a}$$

$$= f(0,0) + \left(\frac{\partial f}{\partial x_1}\right)_{0,0} \delta x_1 + \left(\frac{\partial f}{\partial x_2}\right)_{0,0} \delta x_2 + \mathcal{O}(\delta x_1^2, \delta x_2^2)$$

$$\dot{x}_2 = \delta \dot{x}_2 = \tag{7.7b}$$

$$= g(0,0) + \left(\frac{\partial g}{\partial x_1}\right)_{0,0} \delta x_1 + \left(\frac{\partial g}{\partial x_2}\right)_{0,0} \delta x_2 + \mathcal{O}(\delta x_1^2, \delta x_2^2)$$

Keeping in mind Eq. (7.5), calling

$$\left(\frac{\partial f}{\partial x_1}\right)_{0,0} = a \quad ; \quad \left(\frac{\partial f}{\partial x_1}\right)_{0,0} = b;$$

$$\left(\frac{\partial f}{\partial x_1}\right)_{0,0} = c \quad ; \quad \left(\frac{\partial f}{\partial x_1}\right)_{0,0} = d$$

and considering very small values of the δx_j, so that we can neglect higher order terms, we reduce the problem to the analysis of the following linear system

$$\begin{pmatrix} \delta \dot{x}_1 \\ \delta \dot{x}_2 \end{pmatrix} = \begin{pmatrix} a & b \\ c & d \end{pmatrix} \begin{pmatrix} \delta x_1 \\ \delta x_2 \end{pmatrix} = \bar{\mathbb{A}} \begin{pmatrix} \delta x_1 \\ \delta x_2 \end{pmatrix} \tag{7.8}$$

The solutions of Eq. (7.8) give the parametric forms of the phase curves in the neighborhood of the steady state (at the origin), with time as the parameter.

The general form of the solution of Eq. (7.8) (except if $\lambda_1 = \lambda_2$) is

$$\begin{pmatrix} \delta x_1(t) \\ \delta x_2(t) \end{pmatrix} = \alpha \hat{C}_1 e^{-\lambda_1 t} + \beta \hat{C}_2 e^{-\lambda_2 t} \tag{7.9}$$

where α and β are arbitrary constants, $\hat{C}_1$ and $\hat{C}_2$ are the eigenvectors (the *normal modes*) of the matrix $\bar{\mathbb{A}}$, associated to the eigenvalues λ_1 and λ_2. These eigenvalues are determined from the relation

$$\det\left(\bar{\mathbb{A}} - \lambda \bar{\mathbb{I}}\right) = \det \begin{pmatrix} a - \lambda & b \\ c & d - \lambda \end{pmatrix} = 0 \tag{7.10a}$$

yielding

$$\lambda_{1,2} = \frac{1}{2}\left\{ (a + d) \pm \left[(a + d)^2 - 4(ad - bc) \right]^{1/2} \right\} \tag{7.10b}$$

It is then clear that the temporal behavior of the system, originally in the steady state $(x_1^s, x_2^s) = (0, 0)$, will depend, after applying a small perturbation, on the characteristics of the eigenvalues λ_j. We have the following possibilities:

(i) Both eigenvalues, λ_1 and λ_2, are real and negative ($\lambda_1 < \lambda_2 < 0$);
(ii) both eigenvalues are real and positive ($0 < \lambda_1 < \lambda_2$);
(iii) both eigenvalues are real, but $\lambda_1 < 0 < \lambda_2$;
(iv) both eigenvalues are pure imaginary;
(v) both eigenvalues are complex conjugates with $\Re(\lambda_1) = \Re(\lambda_2) < 0$;
(vi) both eigenvalues are complex conjugates with $\Re(\lambda_1) = \Re(\lambda_2) > 0$.

The different situations that could arise, according to the kind of eigenvalues we find, correspond to the phase trajectories depicted in Fig. 7.1. Case (i) corresponds to a solution that decays for increasing time, and is called *stable node*. Case (ii) is the opposite situation, and corresponds to an *unstable node*. Case (iii) is intermediate between both previous situations: in one direction the system is stable, and is unstable in the other, this corresponds to a *saddle point*. Case (iv) indicates a periodic behavior with a constant amplitude called *center*. Case (v) corresponds to periodic behavior, but with a decaying amplitude, this is a *stable focus*. Case (vi) is the opposite to the previous one and corresponds to an *unstable focus*.

Within this general scheme, even when we extend the results to a larger number of variables, it is possible to identify three basic situations:

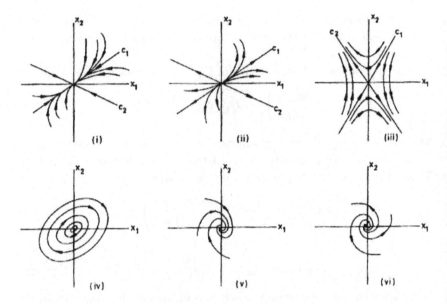

Fig. 7.1   The phase portrait for the different types of fixed points as indicated in the text.

(a) $\Re(\lambda_j) < 0$ for j;
(b) at least one $\Re(\lambda_j) > 0$;
(c) at least one $\Re(\lambda_j) = 0$.

The above analysis yields the following results (remember that this corresponds to *small perturbations!*):

(a) All $\Re(\lambda_j) < 0$: the steady state is called *asymptotically stable*. Whatever the form of the nonlinear terms in Eq. (7.3), after a small perturbation the normal modes decay. These types of solution are also called *attractors*, and the region of phase space including all the points such that any initial state finally tends to the attractor form its *basin of attraction*. In this context, an *equilibrium state* will correspond to a *universal attractor* (all initial states decay to it).
(b) At least one $\Re(\lambda_j) > 0$: the steady state is *unstable*, that is, the normal mode associated with this eigenvalue will grow with time.
   In either case (a) or (b), the behavior of the individual modes will be oscillatory if $\Im(\lambda_j) \neq 0$, and monotonic otherwise.

(c) At least one $\Re(\lambda_j) = 0$, all other $\Re(\lambda_k) < 0$: the steady state is *marginally stable* with respect to the mode having $\Re(\lambda_j) = 0$. As a solution of the linearized equation this mode will neither grow nor decay, but could oscillate if in addition it has $\Im(\lambda_j) \neq 0$. Here, the explicit form of the nonlinear terms will determine whether this marginally stable steady state is stable or unstable. If stable, it could correspond to a more complicated topological object than the zero-dimensional *attractors* we have discussed before, that we will not study here.

For the case when several fixed points coexist (that is there are several steady state solutions), the basin of attraction of each attractor is separated from the others by curves of *neutral* points, known as *separatrices*, as indicated in Fig. 7.2. However, this aspect is beyond the linear analysis we are considering.

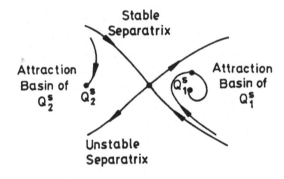

Fig. 7.2

7.3 Limit cycles, bifurcations, symmetry breaking

Besides the cases we have just analyzed, for nonconservative nonlinear equations it is also possible to find a new, very important, kind of steady solution, called *limit cycle*, corresponding to stable (and also unstable) *periodic* solutions. If such a periodic solution is stable, all the solutions in its neighborhood will decay to it for long times.

To exemplify this behavior consider a system described by the following set of equations

$$\dot{x}_1 = x_2 + x_1 R^{-1/2}(1 - R) \tag{7.11a}$$

$$\dot{x}_2 = -x_1 + x_2 R^{-1/2}(1 - R) \tag{7.11b}$$

with $R = x_1^2 + x_2^2$. Changing to polar coordinates $x_1 = \rho \cos\varphi$ and $x_2 = \rho \sin\varphi$, the set of equations transforms into

$$\dot{\rho} = 1 - \rho^2 \tag{7.12a}$$

$$\dot{\varphi} = -1 \tag{7.12b}$$

Equation (7.12b) fixes $\dot{\varphi}$, and it is then possible to solve Eq. (7.12a) and obtain

$$\int_{\rho_0}^{\rho(t} d\rho \left(1 - \rho^2\right)^{-1} = \int_0^t dt = t = \frac{1}{2}\ln\left|\frac{1+\rho}{1-\rho}\right|_{\rho_0}^{\rho(t)} \tag{7.13}$$

We can rewrite this as

$$\rho(t) = \frac{Ae^{2t} - 1}{Ae^{2t} + 1} \tag{7.14}$$

with $A = (1 + \rho_0)/(1 - \rho_0)$. The last equation shows that, regardless of the value of ρ_0, we have $r(t) \to 1$ for $t \to \infty$. A particular case will be the value $\rho_0 = 1$, in which case $r(t) = 1$ for all t. Hence, we have shown that a system described by Eqs. (7.11), (7.12) has the kind of behavior corresponding to a stable limit cycle. The phase portrait of this attractor, for the more general non-circular case, is shown in Fig. 7.3.

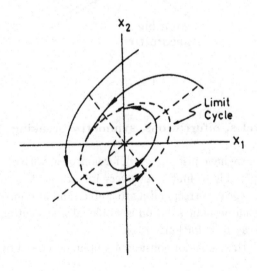

Fig. 7.3

We now want to analyze another useful example of limit cycle behavior, the *Van der Pol oscillator*. This general type of equation arises in several different systems: onset of coherent radiation in lasers, self-excitation in electric circuits, nonlinear mechanics, self-organization in chemical reactions, indicating again the relevance of limit cycle behavior. The approach we will use is typical of a kind of special techniques called *singular perturbation techniques*. These are often necessary in cases where the small quantity in which we expand, appears in the equations in such a way that, in order to obtain the zero order contribution, the neglect of the contribution of the term where it arises changes the order of the differential equation. A very well known example is the $\hbar$ expansion through the WKB method in quantum mechanics. We will discuss one of these techniques working it out in the context of the indicated nonlinear oscillator.

The NLDE that describes the Van der Pol oscillator has the form

$$\ddot{x} + \omega_0^2 x - \gamma(1 - x^2)\dot{x} = 0 \qquad (7.15)$$

with $\gamma > 0$. For $\gamma = 0$ this equation reduces to the simple harmonic oscillator, whose solution is

$$x(t) = \alpha \sin(\omega_0 t + \varphi) \qquad (7.16a)$$

$$\dot{x}(t) = \alpha\omega_0 \cos(\omega_0 t + \varphi) \qquad (7.16b)$$

For the case when γ is very small, we are interested in solutions that retain the form of Eqs. (16), but where α (the amplitude) and φ (the phase) become functions of time to be determined. In order that such an Ansatz be consistent, we look for a condition by differentiating the expression of $x(t)$ with respect to time

$$\dot{x}(t) = \dot{\alpha}(t) \sin[\omega_0 t + \varphi(t)] + \alpha(t)\omega_0 \cos[\omega_0 t + \varphi(t)]$$

$$+ \alpha(t)\dot{\varphi}(t) \cos[\omega_0 t + \varphi(t)] \qquad (7.17)$$

Using the second of Eqs. (16), we obtain

$$\dot{\alpha}(t) \sin[\omega_0 t + \varphi(t)] + \alpha(t)\dot{\varphi}(t) \cos[\omega_0 t + \varphi(t)] = 0 \qquad (7.18)$$

If we now differentiate Eq. (16b), and replace the expressions for x, $\dot{x}$ and $\ddot{x}$, we find the pair of differential equations

$$\dot{\alpha}(t) = -\gamma F(t) \cos[\omega_0 t + \varphi(t)] \qquad (7.19a)$$

$$\dot{\varphi}(t) = \frac{\gamma}{\alpha(t)} F(t) \cos[\omega_0 t + \varphi(t)] \qquad (7.19b)$$

where

$$F(t) = \left\{ \alpha(t)^3 \sin^2[\omega_0 t + \varphi(t)] \cos[\omega_0 t + \varphi(t)] - \alpha(t) \cos[\omega_0 t + \varphi(t)] \right\}$$

If—as we have assumed before—γ is a very small parameter, we could also suppose that both the amplitude $\alpha(t)$ and the phase $\varphi(t)$ are so slowly varying functions of t that during a period $\tau = 2\pi/\omega_0$, $\alpha(t)$ and $\dot{\varphi}(t)$ remain constant. We determine their (constant) values by *averaging* over one period (in what is called *first Krilov–Bogoliubov approximation*), obtaining

$$\alpha = \frac{\gamma \alpha}{2} \left(1 - \frac{\alpha^2}{4} \right) \tag{7.20a}$$

$$\dot{\varphi} = 0 \tag{7.20b}$$

Eq. (20a) can be rewritten as

$$\frac{d}{dt}\alpha^2 = \gamma \alpha^2 \left(1 - \frac{\alpha^2}{4} \right) \tag{7.21}$$

This equation can be integrated yielding

$$\alpha^2(t) = \alpha_0^2 e^{\gamma t} \left[1 + \frac{\alpha_0^2}{4} \left(e^{\gamma t} - 1 \right) \right]^{-1} \tag{7.22}$$

When we replace this into the expression for $x(t)$, we get

$$x(t) = \alpha_0 e^{\gamma t/2} \left[1 + \frac{\alpha_0^2}{4} \left(e^{\gamma t} - 1 \right) \right]^{-1/2} \sin[\omega_0 t + \varphi_0] \tag{7.23}$$

This result indicates that, for any arbitrary initial state $(x_0, \dot{x}_0)$, defining an amplitude

$$\alpha_0 = \left[x_0^2 + \left(\frac{\dot{x}}{\omega_0} \right)^2 \right]^{1/2} > 0 \tag{7.24}$$

the (approximate) solution of Eq. (15) given by the trajectory in Eq. (23), spirals toward a circle of radius 2, corresponding to a limit cycle. For large values of γ, this first order approximation fails, and one can improve it including higher-order harmonics. However, such approximation converge very slowly, and it is better to resort to an *adiabatic elimination procedure* for the velocity variable (i.e. assuming $\ddot{x} \approx 0$).

According to the above classification one might be tempted to conclude that a given system can only be described by a particular kind of fixed point or attractor. But this is not the case. The most interesting aspects of nonequilibrium phenomena arise from the fact that the same system can show a variety of behaviors, each one corresponding to a different attractor. The change from a given state to another is produced by the variation of some of the external constraints (or control parameters) acting on the system, so that the original (or *reference*) state becomes unstable, and subsequently a *bifurcation* to new branches of states occurs.

We will analyze two kinds of instabilities which may lead to a stable limit cycle from a fixed point. For our discussion we will refer to Fig. 7.4. In part (i) of the figure we depict the variation of the eigenvalue λ associated with the unstable original mode. This is usually called *the thermodynamic branch* as it is the direct extrapolation of the equilibrium states, sharing

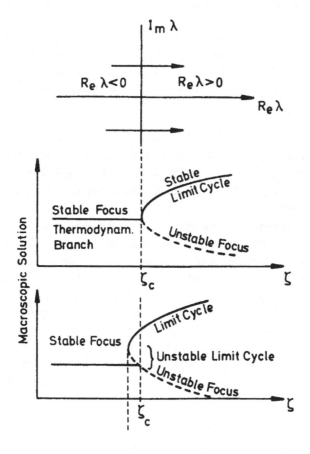

Fig. 7.4

with them the important property of asymptotic stability, since in this range the system is able to damp internal fluctuations or external disturbances, and we can still describe the behavior of the system, essentially, within a thermodynamic approach. The horizontal axis indicates the real part of the eigenvalue and the vertical axis the imaginary part. The real part of λ crosses the imaginary axis, from the negative to the positive values (left to right), as the control parameter ζ takes a critical value $\zeta = \zeta_c$. In parts (ii) and (iii) of the figure, the horizontal axis represents the variation of the control parameter ζ, and the vertical axis schematically indicates a steady state solution of the NLDE describing the system and may represent several different physical or chemical properties (i.e.a concentration of some reactive for a chemical system, an amplitude of oscillation for a mode in a fluid, etc).

As the parameter ζ is varied from left to right (in parts (ii) and (iii) of the figure), a pair of complex eigenvalues λ cross the imaginary axis (part (i)). Consider the case where, before crossing, the steady state solution (x_1^0, x_2^0) is a *stable focus*. As soon as $\Re(\lambda)$ goes through zero and becomes positive for $\zeta = \zeta_c$, the solution may:

(a) bifurcate into an unstable *focus* and a stable *limit cycle*. Beyond the bifurcation point ζ_c, the limit cycle is the only stable solution. This kind of transition, where the limit cycle arises continuously for $\zeta > \zeta_c$, is called a *soft self-excitation*. A bifurcation to the right is called a *supercritical* one.

(b) The bifurcation to a limit cycle may also be *subcritical*, that is, it may occur at the left of $\zeta = \zeta_c$ as indicated in part (iii). The limit cycle towards which the system bifurcates at ζ_c is unstable, and a stable limit cycle may be reached for $\zeta_a < \zeta < \zeta_c$ but only in response to a finite perturbation that exceeds a certain threshold. For a smaller perturbation, the system will return to the stable steady state. But if the perturbation exceeds the threshold (as indicated in the figure) then it will continue to grow until the system reaches a stable limit cycle. Due to the existence of a threshold this is called a *hard self-excitation*. For small perturbations the system will remain in the stable steady state until $\zeta > \zeta_c$, where the steady state becomes unstable and the system jumps *abruptly* to the limit cycle, in contrast to the *continuous* transition of the previous case. Mathematically both types of instabilities are classified as *Hopf's bifurcations*.

We can understand why this phenomenon of bifurcation should be associated with a kind of catastrophical change. Indeed, the instant of the transition (neighborhood of $\zeta = \zeta_c$) is crucial as the system has to make a critical choice there. For instance, in the Bénard convection it is associated with the appearance of right or left-handed cells in a given space region (i.e., branches 1 or 2 in the figure). There is nothing in the set up of the system that allows us to predict which state will arise. It is only *chance* that could decide through the effect of *fluctuations*. The fluctuations will allow "exploring" the "landscape" of the system, make some initial unsuccessful attempts and finally a particular fluctuation will take over. It is within this framework that the interplay between chance and constraint, or fluctuations and irreversibility, that underlies all instability phenomena, is clearly seen.

When discussing the kind of transitions associated with these nonequilibrium instabilities, it is usual to adopt the language of equilibrium thermodynamic phase transitions and critical phenomena. For instance, the supercritical bifurcation is analogous to a second-order phase transition, while the subcritical resembles a first-order one.

Now, and in order to introduce some notions related with the concept of *symmetry breaking* as well as with *global stability*, we will work out a useful mechanical analogy. Let us analyze the example of an *overdamped anharmonic oscillator*. The classical equation of motion of such a system is

$$m\frac{dv}{dt} = -\gamma v + F(x) \tag{7.25a}$$

where x is the position and is the velocity of a particle of mass m, and γ is the friction coefficient. Considering that $v = dx/dt$, Eq. (25a) can be rewritten as

$$m\ddot{x} + \gamma\dot{x} = F(x) \tag{7.25b}$$

We will concentrate on the particular case in which the particle is light, that is its mass m is very small, while the friction coefficient γ is very large. This corresponds to *overdamped motion*, in which the first term on the left hand side, when compared with the second one, can be neglected (that is: we assume $\ddot{x} \approx 0$). This approximation is the prototype of the procedure called *adiabatic elimination*. Now we can make a change of time scale according to $t \to \gamma t$, and in this way eliminate the constant γ from the equation, which finally reads

$$\dot{x} = F(x) \tag{7.26}$$

It has the same form as the equation we have analyzed before (i.e. Eq. (2)). Now, for a one dimensional problem, we always have that the force $F(x)$ can be derived from a potential $V(x)$, according to

$$F(x) = -\frac{\partial}{\partial x}V(x) \tag{7.27}$$

For the harmonic case $V(x) = 1\,2k_0x^2$. However, we are interested in the nonharmonic case. We assume a force that has, besides a harmonic linear dependence on the coordinate, a cubic term:

$$F(x) = -k_0x - k_1x^3 \tag{7.28a}$$

from a quartic potential

$$V(x) = \frac{1}{2}k_0x^2 + \frac{1}{4}k_1x^4 \tag{7.28b}$$

The form of the potential is depicted in Fig. 7.5. In part (i) we show the case $k_0 > 0$, while the case $k_0 < 0$ is shown in part (ii). The equilibrium points will be determined from $F(x) = 0$. From the figure it is clear that in each of these two cases we have a completely different situation.

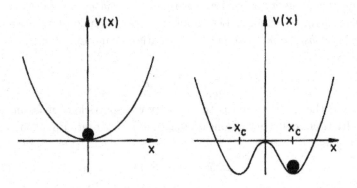

Fig. 7.5

In the first case, for $k_0 > 0$ and $k_1 > 0$, the unique solution is $x = 0$, and is stable; whereas in the second, for $k_0 < 0$ and $k_1 > 0$, we have three solutions, namely, $x = 0$ which is unstable, and two stable symmetric solutions $x = \pm x_c$ (where $x_c = [|k_0|/k_1]^{1/2}$). Here we meet again the *bifurcation* phenomenon discussed above. The bifurcation diagram will have the form indicated in Fig. 7.6.

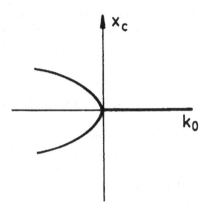

Fig. 7.6

It is easy to prove, within linear stability analysis, that both solutions $x = \pm x_c$, are stable. Also, it is simply proven that Eq. (26) with $F(x)$ given by Eq. (28a) is invariant under the transformation $x \to -x$, that is, Eq. (26) is symmetric with respect to this transformation. Also the potential in Eq. (28b) remains invariant under such a transformation. Although the problem, as described by Eqs. (26), (28a) is completely symmetric under inversion, the symmetry is now broken as the system will adopt one of the two possible solutions. We then have that, when we change k_0 slowly from positive to negative values, we reach $k_0 = 0$ where the stable equilibrium solution $x = 0$ becomes unstable. This phenomenon is usually described as a *symmetry-breaking instability*.

Consider now the case with $k_0 < 0$ and $k_1 < 0$. The only steady state solution of $F(x) = 0$ is again $x = 0$, that now results to be unstable.

We can extend these results to higher dimensional situations to conclude that, when there is a potential function $V(x_1, \ldots, x_n)$ from which we can derive the forces $F_j(x_1, \ldots, x_n) = -\partial V / \partial x_j$, we can discuss the stability of the steady state solutions just by looking at the form of the potential. In other words, we have a global stability criterion. However, there are too many systems which do not have a potential. It is in these cases where a theorem due to *Lyapunov* comes to help us. This theorem states that, if certain conditions are fulfilled, there exists a function which has the desirable properties of the potential and makes it possible to discuss global stability, but is not based on the requirement that the forces be derived from a potential. Even though this is a beautiful theory, its applicability is rather limited because there are very few examples of systems where such a *Lyapunov function* can be determined. A typical example of a Lyapunov function is Boltzmann's $\mathcal{H}$ function discussed in Sec. 4.8.

7.4   Notions about the "nonequilibrium potential"

From the equilibrium phenomena point of view, both classical mechanics and reversible equilibrium thermodynamics (the two most prominent branches of macroscopic physics) are characterized by **extremum principles**. In classical mechanics it is the principle of stationary action that determines the classical trajectory, while equilibrium thermodynamics is characterized by the maximum entropy principle in closed systems. Both extremum principles originate as the macroscopic limit of more fundamental theories (quantum mechanics and statistical mechanics respectively) where those principles are violated through the occurrence of fluctuations: quantum fluctuations on one hand (that assign finite probability amplitudes to nonclassical trajectories) and thermal fluctuations (assigning nonzero probabilities to states with less than the maximum entropy) on the other. It is this deep connection between fluctuation phenomena and extremum principles that allows the system to explore a neighborhood of the extremum state and thereby identifying it.

The evolution equations governing the nonequilibrium phenomena belong neither to the realm of equilibrium thermodynamics (therefore thermodynamic extremum principles are not applicable) nor to the realm of classical mechanics (implying that the principle of least action is also not applicable). Hence, an extremum principle allowing to characterize time-dependent or time-independent solutions of such evolution equations is not readily available. However, classical fluctuations are also present in nonequilibrium systems, just from thermal origin or due to some stochastic perturbations of general nature. Just as in equilibrium thermodynamics, these fluctuations allow to explore not only the deterministic nonequilibrium trajectory, but also its neighborhood. Hence, it is expected that some extremum principle must also hold when one is able to identify the minimized potential function, enabling us to characterize stable steady states (or attractors) of the macroscopic evolution equations (as well as the unstable ones or separatrices) by some extremum condition. This is what one seeks in order to understand pattern selection in self-organizing systems and other related phenomena.

The main goal of this section is to offer an introductory view to the application of the nonequilibrium potential picture into reaction-diffusion systems. We can summarize the main idea behind the nonequilibrium potential approach by saying that the selection of a stationary state or pattern is strongly influenced by fluctuations. It is only *chance* which could decide, through the effect of *fluctuations*. The fluctuations will enable us to explore the *landscape* of the system, and after making some initial unsuccessful attempts finally a particular fluctuation will take over. It is within this framework that the interplay between chance and constraint, or fluctuations and irreversibility, underlying all instability phenomena, is clearly seen [Nicolis and Prigogine (1977); Haken (1978); Prigogine (1980); Malchow and Schimansky-Geier (1985); Nicolis (1989); Mikhailov (1990); Wio (1994); Nicolis (1995)].

7.4.1 *Lyapunov function and global stability*

We can extend the previous results to higher dimensional situations to conclude that, when there is a potential function $V(x_1, \ldots, x_n)$ from which we can derive the forces

$$F_j(x_1, \ldots, x_n) = -\frac{\partial}{\partial x_j} V(x_1, \ldots, x_n), \qquad (7.29)$$

we can discuss the stability of the steady state solutions just by looking at the form of the potential. In other words, we have a *global stability criterion*. However, there are a large majority of systems which do not have such a potential. There is a theorem due to *Lyapunov* stating that, if certain conditions are fulfilled, there exists a function which has the desirable properties making it possible to discuss global stability. Such a function is not based on the requirement that the forces be derived from a potential, i.e. the system could be *non-variational*, which means that Eq. (7.29) is not fulfilled.

To fix ideas, in the two variable case indicated in Eq. (7.3), Lyapunov's theorem states that: if there exists a function $V(x_1, x_2)$ such that it has a minimum at the fixed point $(\bar{x}_1, \bar{x}_2)$, in the neighborhood of this fixed point and both $V(x_1, x_2) > 0$ and $dV(x_1, x_2)/dt \leq 0$, then such a fixed point will be asymptotically stable. [Reichl (1980); Kreuzer (1984); Nicolis (1995)]

At this point, and with the idea of the Lyapunov function in mind, it is worth to make a brief classification of the different possibilities for the flow in the phase space. [Hohenberg and Halperin (1977); Graham (1978, 1987); Cross and Hohenberg (1993); San Miguel and Toral (2000)]

(i) *Relaxational Gradient Flow*: If there is a potential function $V(x_1, \ldots, x_n)$ such that $F_j(x_1, \ldots, x_n)$ fulfills Eq. (7.29) (i.e., it is a variational system), implying

$$\dot{x}_j = -\frac{\partial}{\partial x_j} V(x_1, \ldots, x_n), \tag{7.30}$$

where the fixed points will correspond to the extrema of $V(x_1, \ldots, x_n)$, the phase space flow will correspond to what is called a *relaxational gradient flow*, and the system will evolve towards the minimum of $V(x_1, \ldots, x_n)$ following trajectories that correspond to the line of *steepest descent*. Clearly $V(x_1, \ldots, x_n)$ is a Lyapunov functional as it also fulfills

$$\frac{dV}{dt} = \sum \frac{\partial V}{\partial x_j} \frac{dx_j}{dt} = -\sum \left(\frac{\partial V}{\partial x_j} \right)^2 \leq 0. \tag{7.31}$$

This behaviour is depicted in part (a) of Fig. 7.7.

(ii) *Relaxational non-Gradient Flow*: Consider a system governed by the equation

$$\dot{x}_j = -\sum (\mathbb{T})_{jl} \frac{\partial V}{\partial x_l}, \tag{7.32}$$

where $\mathbb{T}$ is a real, symmetric, positive definite matrix. The fixed points of the system will still correspond to the extrema of V. However, the trajectories in phase space that will evolve towards the minima of V, will not follow the steepest descent lines. This means that the transient dynamics will not be governed just by V.

A by now classical example of this situation is the Cahn–Hilliard equation for spinodal decomposition, [Langer (1987); Hohenberg and Halperin (1977); Gunton and Droz (1983); San Miguel (1985); Kirkaldy (1992); Cross and Hohenberg (1993)] where $(\mathbb{T})_{jl} = -\nabla^2$. It is clear that V is still a Lyapunov functional as

$$\frac{dV}{dt} = \sum (\mathbb{T})_{jl} \frac{\partial V}{\partial x_j} \frac{\partial V}{\partial x_l} \leq 0. \tag{7.33}$$

This behavior is depicted in part (b) of Fig. 7.7.

(iii) *Non-Relaxational Potential Flow*: Here we can consider two situations:

(a) In the first case we assume

$$\dot{x}_j = -\sum (\mathbb{K})_{jl} \frac{\partial V}{\partial x_l}, \tag{7.34}$$

Where $\mathbb{K}$ is an arbitrary, positive definite matrix. We can separate it into a symmetric ($\mathbb{S}$) and an antisymmetric ($\mathbb{F}$) part

$$\mathbb{K} = \mathbb{S} + \mathbb{F}$$

$$\mathbb{S} = \frac{1}{2}(\mathbb{K} + \mathbb{K}^T) \quad \mathbb{S} = \mathbb{S}^T$$

$$\mathbb{F} = \frac{1}{2}(\mathbb{K} - \mathbb{K}^T) \quad \mathbb{F} = -\mathbb{F}^T. \tag{7.35}$$

The fixed points are again given by the extrema of V. On the other hand we have that V also fulfills

$$\frac{dV}{dt} = \sum (\mathbb{S})_{jl} \frac{\partial V}{\partial x_j} \frac{\partial V}{\partial x_l} - \sum (\mathbb{F})_{jl} \frac{\partial V}{\partial x_j} \frac{\partial V}{\partial x_l} \leq 0, \tag{7.36}$$

as, clearly, the first term on the rhs is ≤ 0, while the second one is $= 0$. Hence V is again a Lyapunov function. The later result implies that the antisymmetric part of $\mathbb{K}$ induces a flow in the system that keeps the Lyapunov functional constant (that is *without cost*). A typical situation is depicted in part (c) of Fig. 7.7.

(b) In the second case we consider

$$\dot{x}_j = f_j = -\sum (\mathbb{T})_{jl} \frac{\partial V}{\partial x_l} + w_j, \tag{7.37}$$

with $(\mathbb{T})_{jl}$ as in (ii) and w_j an arbitrary function. In the present case, $V(x_1, x_2, ...)$ will be a Lyapunov functional if the second term on the rhs of

$$\frac{dV}{dt} = -\sum (\mathbb{T})_{jl} \frac{\partial V}{\partial x_j} \frac{\partial V}{\partial x_l} + \sum w_j \frac{\partial V}{\partial x_j} \tag{7.38}$$

is zero. For this to be true, the following orthogonality condition must be fulfilled

$$\sum_l \left(f_l + \sum_j (\mathbb{T})_{jl} \frac{\partial V}{\partial x_j} \right) \frac{\partial V}{\partial x_l} = 0, \tag{7.39}$$

or

$$\left(\vec{F} + \mathbb{T} \nabla V \right) \cdot \nabla V = 0, \tag{7.40}$$

which is analogous to a Hamilton–Jacobi equation. In such a case we have $\frac{dV}{dt} \leq 0$, and V is a Lyapunov functional.

A more general discussion on such a classification of dynamical flows for complex fields can be found in [San Miguel and Toral (2000)].

Fig. 7.7   (a) Case (i); (b) case (ii) and (c) case (iii).

7.4.2   *Nonequilibrium potential*

At this point it is worth asking about the effect of noise or fluctuations on the dynamical equation—(7.30), (7.32), (7.32) or (7.37), according to the case—as well as on the stability of the fixed points. We start writing

$$\frac{dx_j}{dt} = F_j(\vec{x}) + \sum_l \vec{g}_{jl}\xi_l(t), \qquad (7.41)$$

where $\xi_l(t)$ are white noise sources of zero mean and correlations $\langle \xi_j(t)\xi_l(t')\rangle = 2\gamma\delta_{jl}\delta(t - t')$. The associated FPE for $P((\vec{x}),t) = P(x_1,\ldots,x_n,t)$, will be a generalization of the FPE discussed in Chapter 2, given by

$$\frac{\partial}{\partial t}P(\vec{x},t) = \sum_j \frac{\partial}{\partial x_j}\left(-[F_j(\vec{x})P(\vec{x},t)] + \gamma\sum_l \frac{\partial}{\partial x_l}[\vec{G}_{jl}P(\vec{x},t)]\right) \quad (7.42)$$

with $\vec{G} = \vec{g}\vec{g}^T$. If we can write that $F_j(\vec{x}) = \sum_l(\mathbb{S})_{jl}\frac{\partial}{\partial x_l}V(\vec{x})$, we find again the situation studied before in Sec. 7.4.1 where, from a deterministic point of view, we have a *relaxational flow*. Hence, if $\vec{G} = \mathbb{S}$, it is possible to derive the expression

$$P_{st}(\vec{x}) \simeq e^{-V(\vec{x})/\gamma}. \qquad (7.43)$$

Clearly, if $g_{jl} = \delta_{jl}$, the result is trivial.

Let us now consider the case where $F_j(\vec{x}) = -\sum_l(\mathbb{S})_{jl}\frac{\partial V}{\partial x_l} + w_j$, that in analogy with Sec. 7.4.1 corresponds to a *nonrelaxational flow*. We assume that $\mathbb{S} = \vec{g}.\vec{g}^T$. Hence we will have that the Hamilton–Jacobi-like equation (7.39) that now reads

$$\sum_l \left(F_l + \sum_j(\mathbb{S})_{jl}\frac{\partial V}{\partial x_j}\right)\frac{\partial V}{\partial x_l} = 0, \qquad (7.44)$$

yields, in the deterministic case, a function $V(\vec{x})$ that is a Lyapunov functional of the problem. However, it has been proved by Graham and collaborators [Graham (1978, 1987)] that in such a case and in the weak noise limit ($\gamma \to 0$), the stationary solution of the multivariate FPE (7.42) associated to the set of SDE (7.41) is given by

$$P_{\mathrm{st}}(\vec{x}) \sim \exp\left[-\frac{V(\vec{x})}{\gamma} + \mathcal{O}(\gamma)\right]. \qquad (7.45)$$

with $V(\vec{x})$ the solution of (7.44). This functional corresponds to the *nonequilibrium* or *Graham's potential* [Graham (1978, 1987)].

As discussed in the introduction, we can interpret the effect of noise saying that it induces fluctuations in the system around one of the minima of $V(\vec{x})$, fluctuations that allow the system to explore the neighborhood of such a point and in this way identify it as a minimum.

The knowledge of the potential $V(\vec{x})$ offers us, at least in principle, the possibility of getting information about

(i) fixed points (extreme of $V(\vec{x})$),
(ii) local stability of the fixed points,
(iii) global stability of the minima (attractors),
(iv) barrier heights between different minima,
(v) decay times of metastable states.

It is the last point, decay of metastable states, one of the aspects that will be discussed in the following chapters.

Chapter 8

Noise-induced phenomena in non-extended dynamical systems

From that day on, I never moved the pieces on the chessboard.
Jorge Luis Borges

8.1 Introduction

Fluctuations or noise have had a changing role in the history of science. Historically, we can identify three views of noise. In the first, until the end of 19th century, noise was considered a nuisance to be avoided or eliminated. A second stage started at the beginning of the 20th century, where through the work of Einstein on Brownian motion, and the study of fluctuations via Onsager relations and fluctuation–dissipation relations it became possible to actually obtain information about a physical system from its fluctuations. The third stage started about four decades ago, and is marked by the recognition that noise can actually play a driving role that induces new phenomena. Examples where noise in fact induces order include noise induced phase transitions, noise induced transport, and stochastic resonance. In this chapter we will discuss some of such noise induced phenomena. While many of these fluctuation-induced phenomena involve temporal fluctuations, spatial fluctuations (disorder) can also play a similar role. It is natural to speculate on the relation of these phenomena to some physical, biological, and chemical examples.

We will start discussing the possibility of transitions among different attractors of a dynamical system driven by fluctuations, that is discussing some elementary aspects of passage time dynamics.

8.2   Bistability, escape times, critical phenomena

In this section we will return to van Kampen's expansion procedure to discuss the very general situation of *bistability*, that is when there are simultaneously two stable steady states.

With the background of Chapters 3 and 7, we now reformulate the stability criterion. We have assumed that the system's macroscopic variable $\mathcal{Z}$ can be written as $\mathcal{Z} = \Omega\phi(t)+\Omega^{1/2}\xi$. We call $\alpha_0(\phi)$ the nonlinear function in the macroscopic equation, i.e. Eqs. (3.16) and (3.24) or (3.12). If there is a positive constant ϵ such that for all ϕ in the neighborhood of ϕ_0 (this being the steady state solution of the macroscopic equation)

$$\frac{\partial}{\partial\phi}\alpha_0(\phi) \leq -\epsilon < 0 \tag{8.1}$$

holds, there is only one stationary macrostate ϕ_0. Associated with it we also have the *mesostate* (that is a stationary stable solution of the Master Equation) described by the stationary probability distribution $P^s(X)$, meaning that this distribution has a sharp peak at $\mathcal{Z} = \Omega\phi_0$, of width $\Omega^{1/2}$. In the limit $\Omega \to \infty$, this peak becomes a delta function.

What happens when the mesostate does not have only one peak, and hence it is not possible to relate it with a unique macrostate? Consider the case in which we can approximately write $P(\mathcal{Z})$ as

$$P(\mathcal{Z}) \cong p_1\delta(\mathcal{Z} - \mathcal{Z}_1) + p_2\delta(\mathcal{Z} - \mathcal{Z}_2) \tag{8.2}$$

with $\mathcal{Z}_1 \neq \mathcal{Z}_2$, p_j is the weight of the macrostate $\mathcal{Z}_j$, and $p_1 + p_2 = 1$. We then have two mesoscopic states, each one related with one of the macroscopic states, i.e. $\phi_1 = \mathcal{Z}_1/\Omega$ and $\phi_2 = \mathcal{Z}_2/\Omega$.

We will assume that the nonlinear function $\alpha_0(\phi)$ has the *bistable* shape indicated in Fig. 8.1. According to the notation in the figure, the steady state solution ϕ_0 is *unstable* since $\alpha_0'(\phi_0) > 0$. If the system is prepared in this state, any small perturbation will move the system to one of the states ϕ_1 or ϕ_2. This *bistable* behavior is typical of several physical and chemical systems (lasers, nucleation phenomena, etc). As an example consider *Schlögl's model* describing the following system of catalytic reactions

$$A + 2X \underset{\leftarrow}{\overset{\rightarrow}{\rightleftarrows}} 3X \tag{8.3a}$$

$$X \underset{\leftarrow}{\overset{\rightarrow}{\rightleftarrows}} B \tag{8.3b}$$

whose macroscopic equation is

$$\dot{\phi} = \kappa_1\alpha\phi^2 - \kappa_1'\phi^3 - \kappa_2\phi + \kappa_2'\beta \tag{8.4}$$

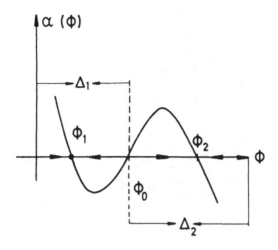

Fig. 8.1

where ϕ is the density of the reactive X, while α and β are the (fixed densities) of the reactives A and B, and κ_j indicates the different reaction rates. In this case, α and β, could be used as control parameters whose variations produce nonequilibrium phase transitions.

Coming back to Fig. 8.1, and considering the situation where the steady state solutions are well separated, if we prepare the system in region Δ_1, that corresponds to the basin of attraction of ϕ_1, due to the effect of some *giant fluctuations* (that could have a low, but still finite, probability), the system can jump to the region Δ_2, that corresponds to the basin of attraction of ϕ_2. Clearly, the inverse process, the transition from Δ_2 to Δ_1, shall also happen. Then ϕ_1 and ϕ_2 turn out to be *metastable states*. If such giant fluctuations through ϕ_0 are rare, we can distinguish two well separated *time scales*:

(a) A short time scale, during which the **small** fluctuations around ϕ_1 (ϕ_2) lead to equilibrium inside Δ_1 (Δ_2). This is indicated in Fig. 8.2.
(b) A long time scale, large fluctuations take over, producing the transition.

This last timescale can be estimated to be of the order of the height of the stationary density at ϕ_0: $\tau_{gf} \propto P^s(\phi_0)$. In chemistry, this estimate is written as

$$\tau_{gf}^{-1} \propto \exp\left[-V(\phi_0)/kT\right],$$

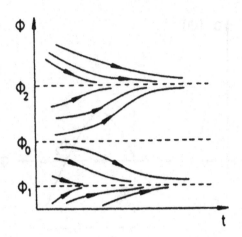

Fig. 8.2

and is known as *Arrhenius' law* (as before, we indicate by $V(\phi) = -\partial\alpha_0(\phi)/\partial\phi$ the "potential" associated with the nonlinear "force").

Near the critical point of the transition, that is when the nonlinear function $\alpha_0(\phi)$ is too flat (see part (a) of Fig. 8.3) and ϕ_1, ϕ_0 and ϕ_2 coalesce, there is no difference between both time scales. The different possibilities according to the form of the potential are indicated in part (b) of Fig. 8.3.

Hence, near a macroscopic instability point the fluctuations generate macroscopic effects, making it impossible to distinguish between a macroscopic contribution and fluctuations. We can then conclude that there is no mesostate associated with ϕ_0.

If we consider distributions initially centered at ϕ_0, the system will evolve according to the following stages:

stage 1) the distribution widens very fast, fluctuations through ϕ_0 are negligible.

stage 2) two peaks, located at ϕ_1 and ϕ_2, start to develop, separated by a "valley" in ϕ_0. There is practically a zero probability flux through ϕ_0, and each peak has a probability (weight) given by

$$p_1 = \int_{-\infty}^{0} d\mathcal{Z} P(\mathcal{Z}, t) \quad ; \quad p_2 = \int_{0}^{\infty} d\mathcal{Z} P(\mathcal{Z}, t) \qquad (8.5)$$

Both are almost constant in t, and essentially determined by the initial distribution.

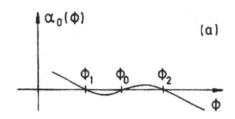

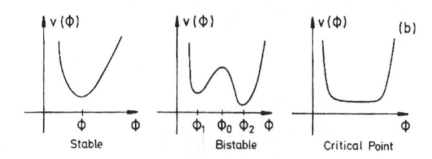

Fig. 8.3

stage 3) the peaks reach their local equilibrium around ϕ_1 and ϕ_2. The resulting mesostate does not correspond to a single macrostate, but to a pair of them. It is a *metastable mesostate* because in the longer time scale there is transfer of probability from ϕ_1 to ϕ_2 and viceversa.

In Fig. 8.4 we depicted the qualitative form of the "potential" and the stationary probability distribution that will result.

We are then able to estimate, recalling that $p_1 + p_2 = 1$, that the equation for the evolution of p_1 and p_2 has the form:

$$\dot{p}_1 = -\tau_{21}^{-1}p_1 + \tau_{12}^{-1}p_2 = -p_2 \qquad (8.6)$$

with τ_{21}^{-1} (τ_{12}^{-1}) the probability per unit time that the system in ϕ_2 (ϕ_1) fluctuates through ϕ_0 and makes a transition to ϕ_1 (ϕ_2). The stationary values fulfill

$$\tau_{21}^{-1}p_1^s = \tau_{12}^{-1}p_2^s \qquad (8.7)$$

and after a time longer than τ_{21} or τ_{12} has elapsed, p_1 and p_2 reach their stationary values

$$p_1^s = \tau_{12}[\tau_{21} + \tau_{12}]^{-1} \quad ; \quad p_2^s = \tau_{21}[\tau_{21} + \tau_{12}]^{-1} \qquad (8.8)$$

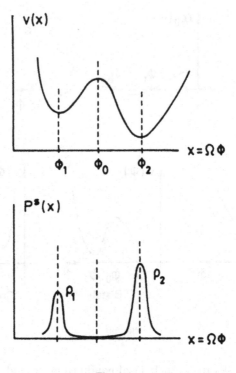

Fig. 8.4

Then, $P(\mathcal{Z}, t)$ has also reached its equilibrium value $P^s(\mathcal{Z})$, showing two peaks near ϕ_1 and ϕ_2. As the integrals of each are p_1 and p_2, respectively, we can make the Gaussian approximation

$$P^s(\mathcal{Z}) \approx p_1^s \left[V''(\phi_1)/2\pi\mathcal{D}\right]^{1/2} \exp\left[-V''(\phi_1)(\mathcal{Z} - \phi_1\Omega)^2/2\mathcal{D}\right]$$
$$+ p_1^s \left[V''(\phi_2)/2\pi\mathcal{D}\right]^{1/2} \exp\left[-V''(\phi_1)(\mathcal{Z} - \phi_2\Omega)^2/2\mathcal{D}\right] \quad (8.9)$$

where $\mathcal{D}$ is a measure of the noise intensity (in other words, the diffusion coefficient). The ratio of the time that the particle spends near ϕ_1 or ϕ_2, is given by the ratio p_1^s/p_2^s, which is also equal to τ_{12}/τ_{21}. The use of Arrhenius' expression provides a rough estimate

$$\frac{p_1^s}{p_2^s} = \frac{\tau_{12}}{\tau_{21}} \sim \exp\left[-\frac{V(\phi_1) - V(\phi_2)}{\mathcal{D}}\right] \quad (8.10)$$

On the other hand, as the ratio is given by $P^s(X)$, also the prefactor is known

$$\frac{p_1^s}{p_2^s} = \left(\frac{V''(\phi_2)}{V''(\phi_1)}\right)^{1/2} \exp\left[-\frac{V(\phi_1) - V(\phi_2)}{\mathcal{D}}\right] \tag{8.11}$$

However, this still does not give us the explicit value of τ_{21}. The above sketched procedure configures a *first passage time* evaluation.

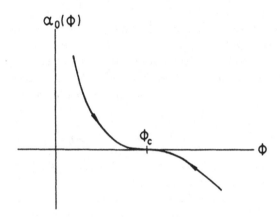

Fig. 8.5

To conclude this section we will briefly discuss the case of *critical fluctuations*. Let ϕ_c be a stationary solution of the macroscopic equation, i.e. $\alpha_0(\phi_c) = 0$. We have seen that if $\alpha_0(\phi_c)' < 0$ the steady state is stable, and is unstable if $\alpha_0(\phi_c)' > 0$. If it happens that $\alpha_0(\phi_c)' = 0$, the state is in general unstable, but it might be stable. For instance, if $\alpha_0(\phi_c) = \alpha_0(\phi_c)' = \alpha_0(\phi_c)^{(2)} = 0$, but $\alpha_0(\phi_c)^{(3)} < 0$, such a state has the typical characteristics of a *critical point*, as indicated in Fig. 8.5.

However, the stability of a critical point is much weaker than that of the points ϕ_1 and ϕ_2 in Fig. 8.1. For instance, in order to discuss its stability, we must include higher order contributions to the departure from the steady state $(\delta\phi = \phi - \phi_c)$ in the local method of analysis we have been using. We will find the following equation for $\delta\phi$:

$$\frac{d}{dt}\delta\phi = -\frac{1}{6}|\alpha_0^3(\phi_c)|\delta\phi^3 + \mathcal{O}(\delta\phi^4) \tag{8.12}$$

If the initial departure $\delta\phi(t = t_0) = \delta\phi_0$ is small enough, such that the last contribution is negligible, we have the solution

$$\delta\phi(t) = \delta\phi_0 \left[1 + \frac{1}{3} |\alpha_0^3(\phi_c)| (t - t_0) \right]^{-1/2} \tag{8.13}$$

indicating that $\delta\phi$ goes to zero, but only as $t^{-1/2}$ instead of exponentially as before. This fact corresponds to the *critical slowing-down* of the macroscopic approach to equilibrium near a critical point. A well known example of this phenomenon is *critical opalescence* in a fluid near a transition.

Now, before starting to study the effect of some external noise on the macroscopic description, we will discuss another interesting phenomenon that is also typical of far from equilibrium phase transitions; that of *hysteresis*. We again consider the *Schlögl model* of Eqs. (8.3)–(8.4). As indicated after Fig. 8.1, the macroscopic equation governing the time behavior ϕ, the density of the reactive X, is

$$\dot{\phi} = \omega\phi^2 - \kappa_1'\phi^3 - \kappa_2\phi + \eta \tag{8.14}$$

where $\omega = \kappa_1\alpha$, and $\eta = \kappa_2'\beta$; α and β are the (fixed) densities of the reactives A and B, and the κ_j indicates the different reaction rates. As before, ω and η (or α and β), could be used as control parameters whose variations produce nonequilibrium phase transitions, and again the steady state solutions of this equation are determined putting the r.h.s. of Eq. (8.14) equal to zero. When the coefficient $\omega < \omega_c = [3\kappa_1'\kappa_2]^{1/2}$, we have only one solution for each value of the parameter η. For $\omega > \omega_c$, there is a range of values of η such that, for $\eta \in [\eta_0, \eta_1]$ three solutions are possible, while for $\eta \notin [\eta_0, \eta_1]$, we find again only one solution. These behaviors are depicted in Fig. 8.6. In part (a), we show the situations when $\omega > \omega_c$ and $\omega = \omega_c$. In part (b), the case of $\omega > \omega_c$, for different values of η (that is $\eta \in [\eta_0, \eta_1]$ or $\eta \notin [\eta_0, \eta_1]$) are shown.

As before, we denote the three steady states respectively by ϕ_1, ϕ_0 and ϕ_2 ($\phi_1 < \phi_0 < \phi_2$). For $\omega > \omega_c$, the steady (homogeneous) states are usually classified as:

(a) a *low density* phase when $\eta < \eta_1, \phi < \phi_1$;
(b) a *high density* phase when $\eta > \eta_0, \phi < \phi_2$;
(c) an *intermediate* region for $\eta \in [\eta_0, \eta_1]$ and $\phi = \phi_1, \phi_0$ or ϕ_2.

In Fig. 8.7 we depicted the values of ϕ for which the r.h.s. of Eq. (8.14) is zero, as function of η and for three cases: $\omega_0 > \omega_c, \omega_1 > \omega_c$ and $\omega = \omega_c$. When $\omega = \omega_0 = \omega_c$, if one starts with $\eta < \eta_0$, the system has only one steady state corresponding to the low density phase. As η increases beyond

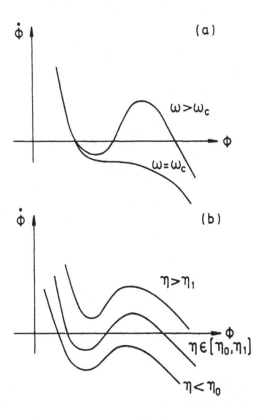

Fig. 8.6

η_0, three states become possible, and beyond η_1, again only one state is possible that corresponds to the high density phase. Hence, when η sweeps through the region $[\eta_0, \eta_1]$, the system undergoes a (discontinuous) phase transition from a low to a high density phase. However, if $\omega < \omega_c$, as η is varied, the system undergoes a continuous transition.

As is shown in Fig. 8.7, through a procedure similar to that used in equilibrium phase transitions (the so called *Maxwell's construction*), we find a *hysteretic* behavior. Following the terminology introduced before, we have a *supercritical bifurcation* occurring at η_0 and ϕ_2, and a *subcritical bifurcation* when η_1 and ϕ_1. All these characteristics are indicated in the figure.

We stop the discussion of these aspects here and turn now to study the effect of external noise on the macroscopic description.

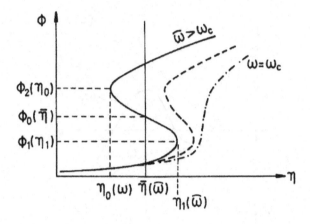

Fig. 8.7

8.3   Noise-induced transitions

In this section we will be concerned with a somewhat different problem. We will study here the influence of external noise on the behavior of the equations governing the macroscopic evolution of a system, particularly regarding the nonequilibrium instabilities indicated above. In most of these transitions the *bifurcation parameter* is some externally controlled parameter, that in general will be subject to fluctuations. Near the bifurcation point these fluctuations, even when characterized by a small variance (weak intensity), can deeply influence the macroscopic behavior of the system.

We will start from the macroscopic NLDE and associate to it a stochastic differential equation (SDE) by assuming that the external parameter is defined as a Gaussian white noise whose mean value is given by the corresponding deterministic value, and a certain variance around this mean. The mathematical tools we will use here are the same that we have considered in Chap. 1. We shall present this phenomenon by discussing a couple of examples. Another very interesting problem, that we will not consider here because it is beyond the scope of this book, is the interplay between the external noise and the intrinsic (internal) fluctuations of the system.

As the first example we will study the *Malthus–Verhulst model*, which was originally proposed within the field of population dynamics, in order to describe the evolution of a biological population. Let describe the number (or density) of individuals of a certain population. This number will change according to the growth rate g minus the death rate d

$$\dot{n} = g - d \tag{8.15a}$$

The birth and death rates depend on the number of individuals present, then as the simplest form it is assumed that

$$g = \gamma n \quad ; \quad d = \delta n \tag{8.15b}$$

Calling $\alpha = \gamma - \delta$, the evolution equation for has the form

$$\dot{n} = \alpha n \tag{8.15c}$$

and allows for either an exponentially growing or decaying population. Using the linear stability analysis we have been discussing, the particular case $\alpha = 0$, is to be unstable against small perturbations. Hence, it is necessary to consider that γ or δ (or both) must depend on , among other reasons due to a limited food supply. In order to correct for this fact another term, sometimes described as the *struggle for life*, is included, resulting in the equation

$$\dot{n} = \alpha n - \beta n^2 \tag{8.15d}$$

By an adequate scaling of the variable ($\rightarrow q = \beta$) we rewrite the last equation as

$$\dot{q} = \alpha q - q^2 \tag{8.16}$$

If we call q_0 the initial value of the (scaled) population, the solution of Eq. (8.16) is

$$q(t) = q_0 e^{\alpha t} \left[1 + \alpha^{-1} q_0 \left(e^{\alpha t} - 1\right)\right]^{-1} \tag{8.17}$$

Let us analyze this solution. According to the stability analysis we have discussed in previous section, for $\alpha < 0$, Eq. (8.16) has only one stationary state solution, $q = 0$, which is stable. At $\alpha = 0$ we have a bifurcation, this solution becomes unstable and a new stable steady state branch arises with $q = \alpha$. As this branch emerges in a continuous but non-differentiable way, we say that the system undergoes a second-order phase transition at $\alpha = 0$. This is depicted in Fig. 8.8.

As we indicated before, we assume that the parameter α is subject to fluctuations, indicating that one (or all) of the original parameters γ, δ or β, fluctuates. Here we take α to be a Gaussian white noise with mean α_0 and variance σ^2. Hence, we will consider the stochastic differential equation (SDE) associated with Eq. (8.16) which is

$$dq = \left(-q^2 + \alpha q\right) dt + \sigma q dW \tag{8.18a}$$

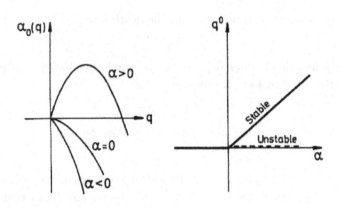

Fig. 8.8

where W is the Wiener process, discussed in Chapter 1 and in sections 2.2, 2.3. The last equation can be formally written as

$$dq = f(q)dt + g(q)dW \qquad (8.18\text{b})$$

As in the deterministic case, and in order to analyze the possible instabilities, we want to compute the stationary solutions of Eqs. (8.18). The relevant stationary solutions are the so called *stationary points* of the SDE, that correspond to those points q_0 for which

$$f(q) = g(q) = 0 \qquad (8.19)$$

One such point is $q_0 = 0$ (indicating that extinction is always possible for a system described by Eqs. (8.19)) and it is the only one. Now we will investigate the stability of such a point. The discussion of the stability for SDE is a somewhat more complicated task than for the case of NLDE; hence we will use some results without demonstration. Since, due to the physical meaning of the variable q (a population number or density), we always have $q \geq 0$, we are only interested in the stability of $q_0 = 0$ from the right. It is possible to show that this point is stable if

$$\mathbb{J} = \int_0^z du \exp\left[\int_u^z d\epsilon 2f\epsilon)g(\epsilon)^{-2}\right] \qquad (8.20\text{a})$$

is finite (with $z > 0$). In our case we have

$$\mathbb{J} = C \int_0^z du\, u^{-(2\alpha/\sigma^2)-1} e^{2u/\sigma^2} \qquad (8.20\text{b})$$

And, because it is possible to set the bounds

$$C \int_0^z du u^{-(2\alpha/\sigma^2)-1} \leq \mathbb{J} \leq C \int_0^z du u^{-(2\alpha/\sigma^2)-1} e^{2z/\sigma^2}$$

we find that $\mathbb{J}$ is finite if and only if $\alpha < 0$. That is, the stationary point $q_0 = 0$ is stable for $\alpha < 0$ and unstable otherwise, which is in agreement with the results of the deterministic case. The question that naturally arises is whether, in addition to the stationary point $q_0 = 0$, the SDE in Eqs. (8.18) has another stationary solution. The simplest form to answer this question is by analyzing the Fokker–Planck equation (FPE) associated with the SDE Eq. (8.18). As discussed in section 2.3, the corresponding FPE will have the form

$$\frac{\partial}{\partial t} P(q,t) = -\frac{\partial}{\partial q} \left[(\alpha q - q^2) P(q,t) \right] + \frac{\sigma^2}{2} \frac{\partial^2}{\partial q^2} \left[q^2 P(q,t) \right]. \tag{8.21}$$

We are particularly interested in the stationary solution of this equation, that is the probability distribution $P_{st}(q)$ such that $\partial P_{st}(q)/\partial t = 0$ and satisfying the equation

$$0 = -\frac{\partial}{\partial q} \left[(\alpha q - q^2) P_{st}(q) \right] + \frac{\sigma^2}{2} \frac{\partial^2}{\partial q^2} \left[q^2 P_{st}(q) \right] \tag{8.22}$$

The solution of this equation, assuming natural boundary conditions (due to the fact that the processes never reach the boundaries and then the probability flux at the boundaries is zero) is

$$P_{st}(q) = N g(q)^{-2} \exp \left[\int^q d\epsilon 2 f(\epsilon) g(\epsilon)^{-2} \right] \tag{8.23}$$

$P_{st}(q)$ will be considered a probability density if and only if it is normalizable, that is if its integral over the range $[0, \infty]$ is finite. According to this, we will say that the stationary solution of Eq. (8.21) exists if this condition is fulfilled. In the one dimensional case it can be shown that, if the stationary solution of the FPE exists, the functional defined as

$$\Psi[P(q,t)] = \int dz P(z,t) \ln \left[\frac{P(z,t)}{P_{st}(z)} \right] + \varphi \tag{8.24}$$

is a Lyapunov functional of the FPE, from which the stability of the $P_{st}(q)$ follows. In our case the form of the stationary solution is

$$P_{st}(q) = N q^{(2\alpha/\sigma^2)-1} e^{2q/\sigma^2} \tag{8.25}$$

This function is found to be integrable over $[0, \infty]$ only if $\alpha > 0$, yielding

$$N^{-1} \int_0^\infty dq P_{st}(q) = \int_0^\infty dq q^{(2\alpha/\sigma^2)-1} e^{2q/\sigma^2} = \frac{(2/\sigma^2)^{-(2\alpha/\sigma^2}}{\Gamma(2\alpha/\sigma^2)} \tag{8.26}$$

where, in order to normalize $P_{st}(q)$ we must choose

$$N = \frac{(2/\sigma^2)^{-(2\alpha/\sigma^2}}{\Gamma(2\alpha/\sigma^2)}.$$

The stationary solution we have obtained was derived under the assumption that we have *natural* boundaries (that is, the process never reaches the boundaries, or equivalently its density flux goes to zero at those boundaries). We will assume that this is true without proof. Hence, our stationary probability distribution is (for $\alpha > 0$)

$$P_{st}(q) = \frac{(2/\sigma^2)^{-(2\alpha/\sigma^2}}{\Gamma(2\alpha/\sigma^2)} q^{(2\alpha/\sigma^2)-1} e^{2q/\sigma^2} \qquad (8.27)$$

The important point here is to notice the drastic change in the character of this stationary distribution for $\alpha = \sigma^2/2$: if $0 < \alpha < \sigma^2/2$, $P_{st}(q)$ is divergent for $q = 0$; while for $\alpha > \sigma^2/2$, $P_{st}(q = 0) = 0$.

Summarizing, by taking into account the influence of an external noise on the control parameter, we can predict the following behavior of the system.

(a) for $\alpha < 0$ the stationary point $q = 0$ is stable, and we can visualize the stationary probability distribution as a Dirac-delta function concentrated at the origin;

(b) the value $\alpha = 0$ is a kind of *transition point*, since the point $q = 0$ is unstable for $\alpha > 0$, and a new $P_{st}(q)$ arises;

(c) for $0 < \alpha < \sigma^2/2$, $P_{st}(q)$ is divergent for $q = 0$, keeping part of the property of a delta function. Even though $q = 0$ is no longer stable, it remains the most probable value. We can interpret this saying that the delta-function *starts to leak* to the right as α crosses the point $\alpha = 0$;

(d) for $\alpha > \sigma^2/2$, there is again a change in the character of $P_{st}(q)$, the value $\alpha = \sigma^2/2$ becomes a second order transition point produced only by the external noise. According to the classification made in the previous section at $\alpha = 0$ we have a *soft transition*, while for $\alpha = \sigma^2/2$ there is a *hard transition*: the divergence of $P_{st}(q)$ for $q = 0$ not only disappears, but we also have $P_{st}(q = 0) = 0$.

All these results for the probability density $P_{st}(q)$ are shown in Fig. 8.9. There, curve (i) shows the delta-like behavior at the origin for $\alpha = 0$; curve (ii) indicates the behavior for $0 < \alpha < \sigma^2/2$, with the divergence for

$q = 0$; finally, curve (iii) indicates the new qualitative behavior of $P_{st}(q)$ for $\alpha > \sigma^2/2$, with zero value at the origin.

Another traditional way to look at the behavior of the probability distribution (as an indicator for a transition in the steady state behavior) is to study the extrema of $P_{st}(q)$, as well as its mean value and variance. From Eq. (8.22) it follows that

$$0 = -\left(\alpha q_m - q_m^2\right) P_{st}(q_m) + \frac{1}{2}\sigma^2 \partial/\partial q \left(q_m^2 P_{st}(q_m)\right) \qquad (8.28a)$$

where q_m indicates the coordinate values for the extrema of the probability density. Since $\partial P_{st}(q_m)/\partial q = 0$, this reduces to

$$0 = -\left(\alpha q_m - q_m^2\right) + \sigma^2 q_m, \qquad (8.28b)$$

if $P_{st}(q_m) \neq 0$. This yields $q_{m1} = 0$ and $q_{m2} = \alpha - \sigma^2/2$ (if $\alpha > \sigma^2/2$). Further analysis of these values indicates that q_{m2} is always a maximum, while q_{m1} is a maximum only for $0 < \alpha < \sigma^2/2$. Hence, we see that the Verhulst model in a fluctuating environment is influenced in such a way that, instead of having only one transition point as in the deterministic case, it has two transition points: one at $\alpha = 0$ that corresponds to the transition in the nature of the boundary at the origin, and the other at $\alpha = \sigma^2/2$. The latter corresponds to an abrupt change in the shape of the probability distribution, whose maximum occurs now at a finite value of q.

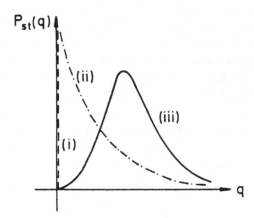

Fig. 8.9

Let us now look at the mean value and the variance of the stationary density. They are given by

$$\langle q \rangle = \int_0^\infty dq\, q P_{st}(q) = \alpha \tag{8.29a}$$

$$\langle q^2 \rangle = \int_0^\infty dq\, q^2 P_{st}(q) = \alpha^2 + \alpha\frac{\sigma^2}{2} \tag{8.29b}$$

$$\langle \Delta q^2 \rangle = \langle q^2 \rangle - \langle q \rangle^2 = \alpha\frac{\sigma^2}{2} \tag{8.29c}$$

Thus, although we have seen that the character of the stationary probability distribution changes at $\alpha = \sigma^2/2$, this fact is not reflected at all in the mean value or variance. This indicates that the study of those moments is not enough for a characterization of these noise induced phenomena. The curve $\langle q \rangle$ vs α is the same as in the deterministic case. It is possible to understand the appearance of the hard transition point from the relation between $\langle \Delta q^2 \rangle$ and $\langle q \rangle$. We have that

$$(\langle \Delta q^2 \rangle)^{-1/2} = \left(\frac{\sigma^2}{2\alpha}\right)^{-1/2} \langle q \rangle \tag{8.30}$$

We can then interpret that for $0 < \alpha < \sigma^2/2$ the fluctuations dominate over the deterministic autocatalytic growth of the population and extinction, even though is not certain, is the most probable outcome. The transition point $\alpha = \sigma^2/2$ is characterized by $\langle \Delta q^2 \rangle = \langle q \rangle^2$ indicating that fluctuations have the same importance as the cooperative effects, while for $\alpha = \sigma^2/2$ autocatalytic growth prevails over fluctuations.

This kind of noise-induced transition [Horsthemke and Lefever (1984)], which essentially corresponds to a shift of the transition phenomena already present in the deterministic bifurcation diagram, can occur for arbitrarily small values of the fluctuations intensity, if the system is close enough to the deterministic instability point. This kind of phenomenon is expected to occur typically in the neighborhood of instability points of systems subject to a multiplicative noise. We will analyze now another example showing still deeper modifications of the macroscopic behavior of nonlinear systems.

The model we will study now is described by the following macroscopic equation

$$\dot{q} = \alpha - q + \beta q(1 - q) \tag{8.31}$$

where $q \in [0, 1]$, and β is a parameter that couples to the environment. This model was initially introduced on a purely theoretical basis. However it is not artificial and has a realistic interpretation in the field of population genetics. Although we will not interpret or justify it in such a context, we will refer to it for short as the *genetic model*, and consider only a possible chemical realization of it, according to the following reaction scheme

$$Q \overset{\overset{k_1}{\rightarrow}}{\underset{k_2}{\leftarrow}} Y$$

$$A + Q + Y \overset{k_3}{\rightarrow} 2Y + A^*$$

$$B + Q + Y \overset{k_4}{\rightarrow} 2Q + B^* \qquad (8.32)$$

From these chemical equations it is clear that the reactions conserve the total number of Q and Y particles (whose densities we indicate with the same letters)

$$Q + Y = N = \text{const.} \qquad (8.33)$$

Hence, defining the fraction $q = Q/N$, and considering the following definition of parameters

$$\alpha = k_2(k_1 + k_2)^{-1}$$

and

$$\beta = \frac{(k_3 B + k_4 B^* - k_1 A - k_2 A^*)}{(k_2 A^* + k_4 B^*)}$$

We find that Eq. (8.31) governs the evolution of q. In the above indicated reaction scheme, the reactives A, B, A^* and B^* play the role of catalyzers. In order to simplify the expressions, and with a minimal loss of generality in our analysis we will adopt $\alpha = 1/2$, yielding a symmetric behavior

$$\dot{q} = \frac{1}{2} - q + \beta q(1 - q) \qquad (8.34)$$

The physically meaningful steady state value is then

$$q_s = \frac{\left[\beta - 1 + (\beta^2 + 1)^{1/2}\right]}{(2\beta)} \qquad (8.35)$$

corresponding to a one to one mapping between the intervals $(-\infty, +\infty)$ for β and $[0, 1]$ for q, as indicated by the curve labelled "0" in Fig. 8.10. It is possible to show that the *eigenvalue* $\tau_r = -(1 + \beta^2)^{1/2}$, arising within the linear stability analysis, is always negative regardless of the sign of β. Furthermore, as $\dot{q} < 0$ for $q > q_s$ and $\dot{q} > 0$ for $q < q_s$, these stationary states are asymptotically (globally) stable. From a thermodynamical

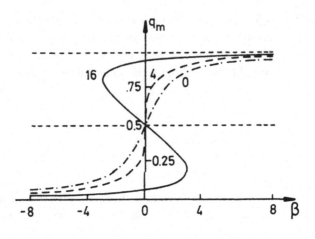

Fig. 8.10

point of view, this is the unique possible stable (*thermodynamic*) branch of stationary states under deterministic environmental conditions. Even in the case when the actual value of the ratio (A^*B^*/AB) differs considerably from the well known chemical equilibrium mass action value

$$\left(\frac{A^*B^*}{AB} \right)_{eq} = \frac{k_1 k_2}{k_3 k_4}$$

the system always evolves in time towards states belonging to the thermodynamic branch. Hence, under deterministic environmental conditions no instabilities can occur, and any possible transition phenomena that could happen in a fluctuating environment will then be purely a noise effect.

Let us assume now that the system is coupled to a noise environment which is reflected in the fact that the parameter β fluctuates. For instance, in a chemical context, we can assume that the densities of the reactives A and B are fluctuating quantities (and A^* and B^* are large enough to neglect their fluctuations). Then we write $\beta = \beta_0 + \sigma\xi$, with ξ a white noise. Keeping $\alpha = 1/2$, the SDE associated with Eq. (8.34) is

$$dq = \left[\frac{1}{2} - q + \beta_0 q(1-q) \right] dt + \sigma q(1-q)dW. \qquad (8.36)$$

As in the previous example, the boundaries 0 and 1 of the state space are intrinsic boundaries of the diffusion process q induced by the Wiener process W. Also in this case both boundaries are natural for the whole range of values of β and σ.

The FPE that corresponds to the SDE Eq. (8.36) is

$$\frac{\partial}{\partial t} P(q,t) = -\frac{\partial}{\partial q} \left\{ \left[\frac{1}{2} - q + \beta_0 q(1-q) \right] P(q,t) \right\}$$

$$+ \frac{\sigma^2}{2} \frac{\partial^2}{\partial q^2} \left\{ [q(1-q)]^2 P(q,t) \right\}. \qquad (8.37)$$

As we have discussed in the previous model, identifying which are the functions (q) and (q), the expression of the stationary probability distribution will be given by Eq. (49), yielding in this case

$$P_{st}(q) = N[q(1-q)]^2 \exp \left\{ -\frac{2}{\sigma^2} \left[[2q(1-q)]^{-1} + \beta_0 \ln \left(\frac{1-q}{q} \right) \right] \right\} \quad (8.38)$$

For simplicity we will discuss the case $\beta_0 = 0$, where the stationary solution in Eq. (8.35) reduces to $q_s = 1/2$. As before we can find the extrema of the distribution in Eq. (8.38) obtaining the equation

$$\frac{1}{2} - q_m + \beta_0 q_m (1 - q_m) - \sigma^2 q_m (1 - q_m)(1 - 2q_m) = 0 \qquad (8.39)$$

whose roots are

$$q_{m1} = 1/2 \text{ and } q_{m\pm} = \frac{1}{2} \left[1 \pm \left(\frac{2}{\sigma^2} \right)^{1/2} \right] \qquad (8.40)$$

If $\sigma^2 < 2$, there is only one real root $q_{m1} = 1/2$. But when $\sigma^2 > 2$, the stationary probability distribution has three extrema. Since $P_{st}(q)$ tends to zero for $q \to 0$ or $q \to 1$, as indicated in Fig. 8.10, we have the following situation:

(a) for $\sigma^2 < 2$, $q_{m1} = 1/2$ is a maximum;
(b) for $\sigma^2 > 2$, $q_{m1} = 1/2$ becomes a minimum and two maxima arise at $q_{m\pm}$, that tend to 0 and to 1 as σ^2 tends to infinity.

For the asymmetric case ($\beta_0 \neq 0$), the situation is qualitatively the same, with only a shift along the β-axis.

The behavior of the stationary probability distribution is depicted in Fig. 8.11, and corresponds to the following. Calling $\sigma_c^2 = 2$ the critical value of the noise variance, we have that for $\sigma^2 < \sigma_c^2$, $P_{st}(q)$ has a unimodal form; for $\sigma^2 = \sigma_c^2$ the maximum of the probability distribution becomes flat; and for $\sigma^2 > \sigma_c^2$, a transition to a bimodal behavior occurs.

This unimodal–bimodal transition is triggered solely by external noise. For the asymmetric case, even if the deterministic stationary solution is near to either of the boundaries (0 or 1), when the noise intensity becomes

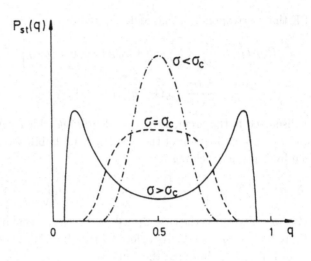

Fig. 8.11

larger than some critical value, the stationary probability distribution will always become bimodal. If we keep σ^2 fixed but larger than σ_c^2, and move along the β-axis, due to the "s" shape of the curves (see Fig. 8.9), the behavior will resemble that occurring in first order phase transitions.

The qualitative change in the steady state behavior of the system can be traced back to the fact that the degree of the polynomial in Eq. (8.39), giving the extrema of the stationary probability density, is one order higher than the one in Eq. (8.34) that (for = 0) gives the deterministic steady states. An interesting way to visualize this is through the *nonequilibrium potential*, the kind of potential we have discussed in Sec. 7.3 [see Eq. (3.27)]. In Fig. 8.12 we depicted the form of such a potential as the parameter σ^2 is varied from $\sigma^2 < \sigma_c^2$ to $\sigma^2 > \sigma_c^2$. In the first case the potential has only one minimum, in correspondence with the existence of only one stable steady state. For $\sigma^2 = \sigma_c^2$ the bottom of the minimum becomes flat, a fact reflected in the probability distribution that also has a flat maximum. Finally, for $\sigma^2 > \sigma_c^2$, the potential develops two minima, and correspondingly the probability distribution acquires the bimodal behavior.

The examples we have just studied show that it is necessary to revise some traditional notions when dealing with systems interacting with fluctuating environments. In particular we must abandon the common belief that due to the rapidity of those fluctuations we can always average them out. If this were true, the system would always remain in its *equilibrium*

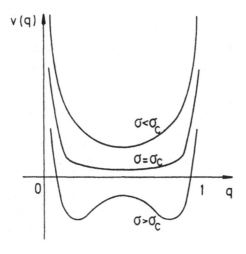

Fig. 8.12

state and no transition would occur. However, the previous examples, as well as several experimental situations, clearly indicate that this is not what happens in reality.

8.4 Notions of stochastic resonance

One of the most fascinating cooperative effects arising out of the interplay between deterministic and random dynamics in a nonlinear system is the phenomenon of *stochastic resonance* (SR). This phenomenon is characterized by the enhancement of the signal-to-noise ratio (SNR) caused by injection of noise into a periodically modulated nonlinear system. The increase in the noise intensity from small initial values induces an increase in the SNR ratio until it reaches a maximum, beyond which there is a decay of SNR for large noise values. Some recent reviews and conference proceedings clearly show the wide interest of this phenomenon and the state of the art [Moss (1992); Gingl *et al.* (1993, 1995); Bulsara *et al.* (1995); Lindner *et al.* (1995, 1996)].

The basic picture of SR has been illustrated by means of a mechanical analogy. Consider a particle moving in a double well potential like the one in Fig. 8.13 and subject to friction. Consider a weak signal that periodically modulates the potential alternatively raising and lowering the wells relative to the barrier. Here weak implies that the modulation is too small to deterministically excite the particle over the barrier. Besides modulation, we also consider the effect of noise, that alone is enough to induce irregular switchings between the wells. In the high friction limit the dynamics can be modeled by

$$\dot{x}(t) = -\frac{dU_0(x)}{dx} + \xi(t) + A\cos\Omega_0 t, \qquad (8.41)$$

where $U_0(x)$ is the bare potential, $A\cos\Omega_0 t$ is the *signal* or modulation and $\xi(t)$ is the random contribution. The phenomenon of SR is the nonlinear cooperative effect whereby the small signal entrains the noise inducing hopping in such a way that the transitions becomes surprisingly regular. Even more, the regularity can improve with the addition of more noise, at least up to a point: it is optimally sensitive at some non-zero level of input noise.

The two essential features of SR in the bistable potential are: that it is a threshold phenomenon, and that its statistical properties are nonstationary. Consider the quartic potential

$$U(x) = U_0(x) + cx = -\frac{a}{2}x^2 + \frac{b}{4}x^4 + cx, \qquad (8.42)$$

with $a, b > 0$ and $c = A\cos\Omega_0 t$. Regarding the threshold feature, the threshold c_{th} is the value of c for which the deterministic switching becomes possible, i.e. the value of c at which the bistability is destroyed ($c_{th} = \pm[4a^3/27b]^{1/2}$). Hence, weak modulation requires $A < c_{th}$ implying that no deterministic switching can occur with the signal alone. The nonstationarity becomes evident when noise is added, and the potential becomes

$$U(x) = -\frac{a}{2}x^2 + \frac{b}{4}x^4 + x[A\cos\Omega_0 t + \xi(t)].$$

The LE that drives the motion of the particle is

$$\dot{x}(t) = ax - bx^3 + [A\cos\Omega_0 t + \xi(t)], \qquad (8.43)$$

and non-stationarity means that the probability density is a (periodic) function of time [Jung (1993)].

Within the indicated picture, the only important dynamical events are the well-to-well switching transitions that can occur whenever

$$|A\cos\Omega_0 t + \xi(t)| \geq c_{th}, \qquad (8.44)$$

indicating that SR is, fundamentally, a threshold phenomenon.

To further clarify the mechanism, let us simplify the picture even more and assume that the modulation is such that during the first half period the left well is kept fixed below the right one, while the situation is reversed during the second half period. Hence, considering the Kramers formula (8.11), it is clear that, during the first half period, the average decay time for jumping from the right well to the left one will be shorter than the reverse transition. The situation is reversed during the second half period. If the noise intensity is such that this decay time is of the order of half the period (while for the reverse less probable transition it is larger), we will meet a *tuning* condition between the random jumps and the modulation that corresponds to the SR phenomenon.

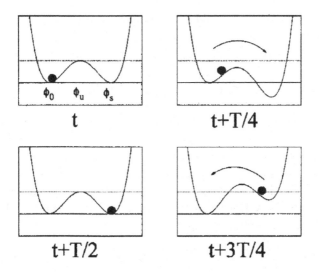

Fig. 8.13   Schematic picture of the transition cycle

In order to make a more quantitative description of the phenomenon it is necessary to evaluate the power spectrum of the particle motion in the indicated generic bistable potential (8.42). To do that, we will follow here McNamara and Wiesenfeld's (MNW) calculation [McNamara and Wiesenfeld (1989)]. We start defining two discrete variables $x_\pm$ describing the position of the particle in either the right $(+)$ or left $(-)$ potential well (for instance, in the indicated bistable potential $x_\pm = \pm[a/b]^{1/2}$), and the corresponding probabilities $\Xi_\pm(t)(\Xi_+(t) = 1 - \Xi_-(t))$. We can then write a rate (or master) equation in terms of $W_\pm$, the transition rates out of the $\pm$ states

$$\frac{d}{dt}\Xi_+(t) = -\frac{d}{dt}\Xi_-(t) = W_-\Xi_-(t) - W_+\Xi_+(t). \tag{8.45}$$

Clearly, the only dynamical variables are the particle populations (or probabilities) within the wells, as corresponds to a this approximate two-state dynamics, analogous to the discussion in the previous subsection.

In order to solve (8.45) at least some approximate form for the $W_\pm$ is required. Following MNW we use an expansion in terms of a small parameter $\eta_0 \cos \Omega_0 t$, where $\eta_0 = A/\gamma$ (γ being the noise intensity)

$$W_\pm = \frac{1}{2}\left(\alpha_0 \pm \alpha_1 \eta_0 \cos \Omega_0 t + \ldots\right), \tag{8.46}$$

where α_0 and $\alpha_1 \eta_0$ are treated as parameters of the system. According to what was discussed previously, α_0 and α_1 can be related to the unperturbed Kramers rate (with $r_K \simeq \tau_K^{-1}$) in an adiabatic-like approximation

$$r_K \simeq r_{K,0}\left(1 + \frac{A\,|\,x_\pm\,|}{\gamma}\cos \Omega_0 t\right). \tag{8.47}$$

This allows us to express $W_\pm$ (within a phase factor) in (8.45), and to solve it finding a solution $\Xi_+(t)$. From this solution we can construct the (conditional) autocorrelation function $\langle x(t)x(t+\tau) \mid x_0, t_0\rangle$, that in the asymptotic limit ($t_0 \to -\infty$) yields the desired correlation function $\langle x(t)x(t+\tau)\rangle$. From this last quantity we can obtain the power spectrum through the Wiener–Kintchine theorem [van Kampen (1982); Gardiner (1985); Wio (1994)] yielding

$$S(\omega) = \left(1 - \frac{(\alpha_1 \eta_0)^2}{2[\alpha_0^2 + \Omega_0^2]}\right)\left(\frac{4\alpha_0\langle x^2\rangle}{[\alpha_0^2 + \omega^2]}\right) + \pi\frac{\langle x^2\rangle(\alpha_1\eta_0)^2}{2[\alpha_0^2 + \Omega_0^2]}\delta(\omega - \Omega_0). \tag{8.48}$$

This result makes two notable predictions, both borne out by experiment:

(i) the shape of the power spectrum is a delta contribution arising from the modulation, riding on a Lorentzian noise background;
(ii) the total power (signal plus noise) is a constant.

The latter property (that is strictly true only for the bistable model) means that the power in the signal part of the response grows at the expense of the noise power. This result demonstrates that, in such a bistable system, the proper application of noise at the input, can result in more order at the output. This could not be possible with a linear system. Moreover, the nonlinear system must be out of equilibrium.

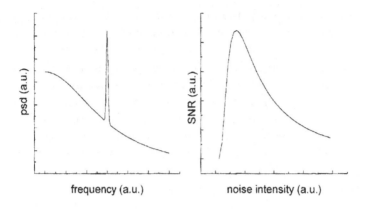

Fig. 8.14   (i) psd *vs.* ω and (ii) SNR *vs.* γ

A quantity that typically has been used to quantify this phenomenon is the *signal-to-noise ratio* (SNR) defined as the ratio between the power from the signal (second term on the r.h.s. of (8.48)) to the noise power (first term on the r.h.s. of (8.48), evaluated at $\omega = \Omega_0$). Using the form of the Kramers rate $r_{K,0} = \tau_{K,0}^{-1}$, it is possible to obtain (for the bistable potential) the approximate result

$$\mathcal{SNR} \simeq \left(\frac{A\Delta U_0}{\gamma} \right)^2 e^{-\Delta U_0/\gamma}, \tag{8.49}$$

where ΔU_0 is the barrier high: $\Delta U_0 = a^2/4b$. The qualitative form of the power spectral density (psd) and the SNR as function of the noise intensity γ is depicted in Fig. 8.14.

The maximum of the curve of SNR results for a value of γ that makes the Kramers time (τ_K) roughly equal to half the period of the modulation. For more details on the SR phenomenon we refer to [Moss (1992); Gingl *et al.* (1993, 1995); Bulsara *et al.* (1995); Lindner *et al.* (1995, 1996)].

8.5   Brownian motors

Einstein relation is a first example of the intimate relation between thermal noise and dissipation that characterizes thermal equilibrium, the fluctuation–dissipation theorem, put on firm ground only much later. It is just this overdamped Brownian noise which we attempt to harvest with the concept of a Brownian motor [Maruyama *et al.* (2002)]. The question is: can one extract energy from Brownian (quantum or classical) particles in asymmetric set-ups in order to perform useful work against an external load? If true, then it would be possible to rectify thermal Brownian motion so as to separate, shuttle or pump particles on a micro- or even nanoscale. In view of the laws of thermodynamics, in particular the second law, the answer is obviously a strong no. If we could indeed succeed, then such a device would constitute a Maxwell demon perpetuum mobile of the second kind [Maruyama *et al.* (2009)]. The possibility is thus to depart from thermal equilibrium, so that the constraints of thermodynamic laws no longer apply, leading us to study non-equilibrium statistical mechanics in asymmetric systems. There, the symmetry is broken either

(i) by the system characteristics, such as an asymmetric periodic potential (or substrate) which lacks reflection symmetry, called ratchet-like potentials, or

(ii) the dynamics itself that may break the symmetry in the time domain.

Clearly, noise-induced, directed transport in the presence of a static bias is trivial. It is also an everyday experience that macroscopic, unbiased disturbances can cause directed motion. The example of a self winding wrist watch, or even windmills are well known proofs. The challenge becomes rather intricate when we consider motion on the micro-scale. There, the subtle interplay of thermal noise, nonlinearity, asymmetry, and unbiased driving of either stochastic, or chaotic, or deterministic origin can indeed induce a rectification of the noise, resulting in directed motion of Brownian particles [Maruyama *et al.* (2002)]. As a consequence, new possibilities open up to optimize and control transport on the micro- and/or nano-scale, including novel applications in physics, nanochemistry, materials science, nanoelectronics and in biological systems such as in molecular motors [Maruyama *et al.* (2002)].

In order to quantify and develop further these considerations, we focus on a Brownian particle in one dimension with coordinate $x(t)$ and mass m, which is governed by a Newtonian equation of motion of the form

$$m\ddot{x}(t) + V'(x(t)) = -\eta\dot{x}(t) + \xi(t), \qquad (8.50)$$

where $V(x)$ is a periodic potential with period L, $V(x + L) = V(x)$, which exhibits broken spatial symmetry (a ratchet potential). A typical example is

$$V(x) = V_0[\sin(2\pi x/L) + 0.25\sin(4\pi x/L)], \qquad (8.51)$$

which is depicted in Fig. 8.15.

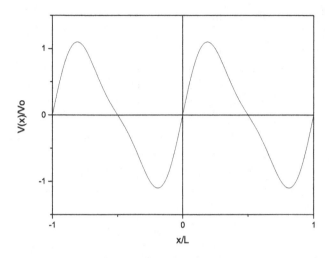

Fig. 8.15

The left-hand side in (8.50) represents the deterministic, conservative part of the particle dynamics, while the right-hand side accounts for the effects of the thermal environment. These are energy dissipation, modeled as viscous friction with a friction coefficient η, and randomly fluctuating forces in the form of the thermal noise $\xi(t)$. These two terms are not independent of each other, because they both have the same origin: the interaction of the particle $x(t)$ with the huge number of microscopic degrees of freedom of the environment. In fact, our assumption that the environment is an equilibrium heat bath at temperature T, whose effect on the system can be modeled in the form of the phenomenological Ansatz appearing on the right-hand side of (51), completely fixes all statistical properties of the fluctuations $\xi(t)$ without the need to refer to any microscopic details of the environment [Maruyama *et al.* (2002)]. That is, in order not to allow for a perpetuum mobile of the second kind, the fluctuations $\xi(t)$ are bound to be a Gaussian white noise of zero mean,

$$\langle \xi(t) \rangle = 0, \tag{8.52a}$$

satisfying the fluctuation–dissipation relation (see Chapters 2 and 6)

$$\langle \xi(t)\xi(t') \rangle = 2\eta k_B T \delta(t - t'), \tag{8.52b}$$

where k_B is Boltzmann's constant, i.e. the noise $\xi(t)$ is uncorrelated in time.

The state variable $x(t)$ in (8.50) will usually be referred to as a *Brownian particle* and may represent some type of collective degree of freedom or other relevant (slow) state variable of the system under investigation. Examples include a chemical reaction coordinate, geometric configuration coordinates, or some internal degrees of freedom of cellular transport enzymes as they occur in the modeling of molecular motors and pumps, the Josephson phase in a superconducting quantum interference device, etc. In these typically very small systems the fluctuation dynamics is often, to a good approximation, governed by an overdamped Langevin dynamics, that is, the inertia term $m\ddot{x}(t)$ then becomes negligible (see [Maruyama *et al.* (2002)] and Chapter 2). We thus arrive at the following minimal equilibrium ratchet model

$$\eta\dot{x}(t) = -V'(x(t)) + \xi(t). \tag{8.53}$$

The quantity of foremost interest in the context of transport in periodic systems is the average particle current in the long-time limit (after initial transients have died out), i.e.

$$\langle\dot{x}(t)\rangle = \langle\lim_{t\to\infty}\frac{x(t) - x(0)}{t}\rangle. \tag{8.54}$$

It is intuitively plausible, and it can also be readily confirmed by a more rigorous formal calculation, that—as far as the velocity $\dot{x}(t)$ is concerned—the infinitely extended state space in (8.53) can be substituted by a circle, i.e. we can identify $x+L$ with x. Accordingly, the probability density $P(x,t)$ associated with a statistical ensemble of independently sampled random processes (8.53) also has the spatial periodicity L and is normalized in $[0, L]$. Moreover, one can infer that the average particle current (8.54) can be rewritten in the form

$$\langle\dot{x}(t)\rangle = -\lim_{t\to\infty}\frac{1}{t}\int_0^t dt'\frac{V'(x(t))}{\eta}P(x,t'). \tag{8.55}$$

The time evolution of $P(x,t)$ is quantitatively described by the associated Fokker–Planck equation. For the present dynamics in (8.53), this equation yields in the long-time limit—as one would have expected—a Boltzmann-type, steady-state solution $P_{st}(x)$ of the form

$$P_{st}(x) = \frac{\exp\left[-V(x)/k_BT\right]}{\int_0^L dy\exp\left[-V(y)/k_BT\right]}. \tag{8.56}$$

With (8.55) this implies for the particle current the result

$$\langle\dot{x}(t)\rangle = 0. \tag{8.57}$$

In other words, we find once again that at thermal equilibrium, in a spatially periodic potential there arises—in spite of the system's intrinsic asymmetry—no preferential direction of the random Brownian motion.

Finally, we complement our minimal ratchet model (8.53) by an additional homogeneous, static force F:

$$\eta \dot{x}(t) = -V'(x(t)) + F + \xi(t). \tag{8.58}$$

It is instructive to incorporate the ratchet potential $V(x)$ and the force F into a single effective potential $V_{\text{eff}}(x) = V(x) - xF$. For example, for a negative force, $F < 0$, pulling the particles to the left, the effective potential will be like in Fig. 8.15, but now tilted to the left. In view of the result (8.57) for $F = 0$, it is suggested that in such a tilted potential the particles will move on average "downhill", i.e. $\langle \dot{x} \rangle < 0$ for $F < 0$, and similarly $\langle \dot{x} \rangle > 0$ for $F > 0$. This conclusion can also be confirmed with detailed quantitative calculations along similar lines to the above discussed case for $F = 0$. Here, we content ourselves with the remark that detailed balance symmetry is broken when $F \neq 0$, suggesting according to Curie's principle the emergence of a non-zero current. Such a current can then only point in the same direction as F, in order not to yield a perpetuum mobile of the second kind.

The appearance of a non-vanishing current $\langle \dot{x} \rangle$ furthermore signals that the ratchet system is driven away from thermal equilibrium by the static force F in (8.58) with the corresponding possibility for a motor action.

We now turn to the central issue of this section, namely the phenomenon of noise-induced, directed transport in a spatially periodic, asymmetric system away from thermal equilibrium: the so-called "ratchet effect", often illustrated by the example of an on–off ratchet model (see Fig. 8.17).

Here, we will elucidate the prominent physics with a different example, namely the so-called *temperature ratchet model*. This Brownian motor model, however, actually turns out to be closely related to the on–off ratchet model. This example is particularly suitable for the purpose of illustrating—besides the ratchet effect per se—also several other basic physical concepts.

In order to illuminate the mechanism of a Brownian motor, we consider as an extension of the model in (8.58), the situation where the noise strength, as represented by the temperature T of the Gaussian white noise $\xi(t)$ in (5), is subjected to periodic, temporal modulations with period $\mathcal{T}$, i.e.

$$\langle \xi(t)\xi(t') \rangle = 2\eta k_B T(t)\delta(t - t') \qquad (8.59a)$$

$$T(t) = T(t + \mathcal{T}). \qquad (8.59b)$$

Two typical examples are

$$T(t) = \bar{T}\left(1 + A\, s[\sin(\frac{2\pi t}{\mathcal{T}})]\right) \qquad (8.60a)$$

$$T(t) = \bar{T}\left(1 + A\sin(\frac{2\pi t}{\mathcal{T}})\right)^2, \qquad (8.60b)$$

where $s[x]$ indicates the sign function, and $A < 1$. With the first case the temperature thus jumps between $T(t) = \bar{T}[1 + A]$ and $T(t) = \bar{T}[1 - A]$ at every half period, $\mathcal{T}/2$. Due to these permanent changes of the temperature, $T(t)$, the system approaches a periodic long-time asymptotics, which in general can only be handled numerically.

We next come to the pivotal feature of the temperature ratchet model (8.58)–(8.59): In the case of the statically tilted model with a time-independent temperature T we have seen above that for a given force, say $F < 0$, the particle will move *downhill* on average, i.e. $\langle \dot{x} \rangle < 0$. This fact holds true for any fixed (non-zero) value of T. Returning to the temperature ratchet with T being now subjected to periodic temporal variations, one therefore should expect that the particles still move *downhill* on average. The numerically calculated *load curve* depicted in Fig. 8.16 demonstrates [Maruyama *et al.* (2002)] that the opposite is true within an entire interval of negative F values. Surprisingly indeed, the particles are climbing *uphill* on average, thereby performing work against the load force, F.

This upward directed motion is apparently triggered by no other source than the white thermal noise $\xi(t)$. A conversion (or rectification) of random fluctuations into useful work as exemplified above is termed the *ratchet effect*. For a setup of this type, the names thermal ratchet, Brownian motor, Brownian rectifier, stochastic ratchet, or simply ratchet are in use [Maruyama *et al.* (2002)]. Because the average particle current $\langle \dot{x} \rangle$ usually depends continuously on the load force F it is for a qualitative analysis sufficient to consider the case $F = 0$: the occurrence of the ratchet effect is then equivalent to a finite current, $\langle \dot{x} \rangle \neq 0$ for $F = 0$, i.e. the unbiased Brownian motor implements a "particle pump".

In order to understand the basic physical mechanism behind the ratchet effect at $F = 0$, we focus on strong, i.e. $|A| \leq 1$, dichotomous periodic temperature modulations from (8.60a). During an initial time interval, say

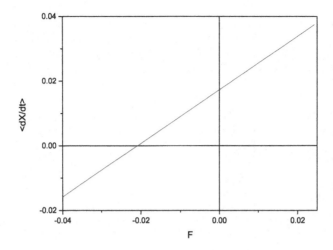

Fig. 8.16

$t \in [\frac{T}{2}, T]$, the thermal energy $k_B T(t)$ is kept at a constant value, $k_B \bar{T}[1 - A]$, much smaller than the potential barrier $\triangle V$ between two neighboring local minima of $V(x)$. Thus, all of the particles will have accumulated in the close vicinity of the potential minima at the end of this time interval, as sketched in the lower panel of Fig. 8.17. Then, the thermal energy jumps to a value $k_B \bar{T}[1 + A]$ much larger than $\triangle V$ and remains there during another half period, say $t \in [T, \frac{3T}{2}]$. Because the particles then hardly feel the potential any more in comparison to the violent thermal noise, they spread out in a manner which is typical for the case of free thermal diffusion (upper panel in Fig. 8.17). Finally, $T(t)$ jumps back to its original low value of $\bar{T}[1 - A]$, and the particles slide downhill towards the respective closest local minima of $V(x)$. Due to the asymmetry of $V(x)$, the original population of one given minimum is redistributed asymmetrically, yielding a net average displacement after one temporal period T.

In the case where $V(x)$ has exactly one minimum and maximum per period, L (as in Fig. 8.17), it is quite obvious that if the local minimum is closer to its adjacent maximum to the right (Fig. 8.17) a positive particle current, $\langle \dot{x} \rangle > 0$, will arise. Put differently, upon inspection of the lower part of Fig. 8.17, it is intuitively clear that during the cool-down cycle the particles must diffuse a long distance to the left and only a short distance to the right, yielding a net transport against the steep hill towards the right.

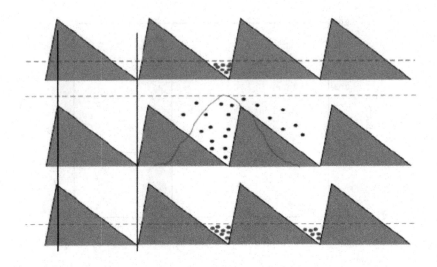

Fig. 8.17

All these predictions rely on our assumptions that $\bar{T}[1-A]$ and $\bar{T}[1+A]$ are much smaller and larger than $\triangle V$, respectively, and that the time period $\mathcal{T}$ is large enough. For more general temperature modulations, the direction of the current becomes much less obvious to predict.

We stop the discussion on noise induced phenomena in 0-dimension here, and will consider, in a next chapter, the role played by the instabilities we have been studying, when we include space variables, to build up nonequilibrium spatial structures or *patterns* out of homogeneous states.

PART 4

The mesoscopic approach far from equilibrium: Extended systems

We must confess in all humility that,
while number is a product of our mind alone,
space has a reality beyond the mind
whose rules we cannot completely prescribe.
Carl F. Gauss

. . . I often say that when you can measure what you are speaking about,
and express it in numbers, you know something about it;
but when you cannot measure it, when you cannot express it in
numbers, your knowledge is of a meager and unsatisfactory kind:
it may be the beginning of knowledge, but you have scarcely in your
thoughts advanced to the stage of Science, whatever the matter may be.
Lord Kelvin

Chapter 9

Formation and propagation of patterns in far-from-equilibrium systems

....
The detail of the pattern is movement,
....
The knowledge imposes a pattern, and falsifies,
for the pattern is new in every moment
....

T. S. Eliot

9.1 Introduction

In the previous chapter we have began discussing how to treat far from equilibrium situations, involving systems that, far from being isolated, are submitted to strong external constraints such as energy or chemical reactive fluxes. There, we only considered the case of spatially homogeneous systems, where the time behavior is the relevant one. However, all around us we find examples where the break of spatial homogeneity leads to the formation or propagation of *spatial structures* or *patterns*. This happens at all length scales, from the very small (cellular structures in biology, propagation of nerve signals along the neuron axon), to intermediate ranges (spiral propagation of electric signals in cardiac tissue, pacemakers as well as spirals in the Belousov–Zhabotinskii reaction, Bénard convection in fluids, vortex structure in superconductors, sea waves), to even larger scales (convective motion in the ocean, cloud patterns in the Earth as well as other planetary atmospheres), to the very large space scales (star clouds, nebulae, galaxy clusters).

As was discussed before, the study of these nonequilibrium phenomena leading to space-time or *dissipative structures*, has a widespread interest due to their implications for the understanding of cooperative phenomena in physics, chemistry and biology. However, the study of all these different cases, seems a formidable task. Notwithstanding, there are some general underlying principles that may be learnt through the study of simple model examples.

Among the latter, one picture that has become very useful in the description of pattern formation and propagation is that called *active* (or *excitable*) *media*. A distributed active medium can be viewed as a set of active elements (that is, each element being a system with two or more possible steady states) representing small parts of a (continuous) system, interacting among each other, for instance, through some transport mechanism (typically a diffusion process). The interplay of the internal nonlinearities of each element with the coupling among them, together with the effect of external control parameters, are at the origin of the space structures.

In this chapter we present some underlying principles, by focussing our discussion on the *reaction–diffusion model*. We will study one- and two-component cases, where this can be done, to a higher degree, in an analytical way. We will not only discuss the formation of static patterns, but also a few principles governing their propagation.

9.2 Reaction–diffusion descriptions and pattern formation

Let us consider a distributed active medium, viewed as a set of active elements interacting with each other. We will assume that the interactions between the different elements that compose the *active medium* are local in time and also that the variation in space is not too sharp. This implies that we can neglect memory effects, as well as space derivatives of order higher than two. Within the formalism discussed in the last chapter, the general form of the macroscopic equation, for the case of only one relevant macroscopic variable ϕ, will then be

$$\dot{\phi} = \mathcal{F}(\phi; \frac{\partial \phi}{\partial \mathbf{r}}; \frac{\partial^2 \phi}{\partial \mathbf{r}^2}; \ldots)\} \tag{9.1}$$

Expanding this in terms of gradients, etc, we obtain

$$\dot{\phi} = \mathcal{F}(\phi) + \bar{A}\frac{\partial \phi}{\partial \mathbf{r}} + B(\frac{\partial \phi}{\partial \mathbf{r}})^2 + D\frac{\partial^2 \phi}{\partial \mathbf{r}^2} + \ldots \tag{9.2}$$

When the medium is isotropic, the second term on the r.h.s. disappears. Also, typically, the term involving gradients is absent, and neglecting higher order derivatives, the equation reduces to

$$\frac{\partial \phi}{\partial t} = \mathcal{F}(\phi) + D\frac{\partial^2 \phi}{\partial \mathbf{r}^2} \tag{9.3a}$$

This equation conforms the so called *reaction–diffusion model*. We will not consider a more rigorous derivation of this equation, and adopt it as a kind of phenomenological approach. However, it is possible to guess how to get it more rigorously. Consider the system to be divided in cells, and described through a Master Equation composed of two kinds of contribution. The first describes the behavior within each cell, corresponding to one of the active elements (usually called the *reactive part*). The second corresponds to the interaction of the different cells with each other. By an adequate limiting procedure, it is possible to achieve a continuous space description that, in some approximation, has the form of a reaction-diffusion scheme. For details of this approach we refer the reader to van Kampen's book.

Clearly, the reaction-diffusion equation for one macroscopic variable shown in Eq. (9.3a), can be easily extended to several macroscopic variables $\{\phi_1, \phi_2, \ldots, \phi_n\}$. In such a case we have a system of coupled reaction–diffusion equations that, when the matrix of diffusion coefficients is diagonal, reads

$$\frac{\partial}{\partial t}\phi_1 = D_1\frac{\partial^2}{\partial \mathbf{r}^2}\phi_1 + \mathcal{F}_1(\phi_1, \phi_2, \ldots)$$

$$\ldots\ldots\ldots\ldots\ldots\ldots\ldots\ldots\ldots\ldots\ldots$$

$$\frac{\partial}{\partial t}\phi_n = D_n\frac{\partial^2}{\partial \mathbf{r}^2}\phi_n + \mathcal{F}_n(\phi_1, \phi_2, \ldots) \tag{9.3b}$$

However, and in order to proceed with the analysis of the reaction–diffusion model we will start focussing on the one variable case, as indicated in Eq. (9.3a). We also consider initially a one dimensional system, i.e. $\partial^2/\partial \mathbf{r}^2 \to \partial^2/\partial x^2$.

The first step is to look for stationary solutions, that is to consider $\partial \phi/\partial t = 0$. Our equation reduces to

$$0 = D\frac{d^2}{dx^2}\phi + \mathcal{F}(\phi) \tag{9.4a}$$

Now, for a one dimensional problem, we have seen earlier that the *reactive term* (or force) $\mathcal{F}(\phi)$ can always be derived from a *potential* $V(\phi)$, according to

$$\mathcal{F}(\phi) = \frac{\partial}{\partial \phi}V(\phi) \tag{9.4b}$$

In order to fix ideas we resort again to an example considered in the previous chapter, the *Schlögl model*. It describes the following reaction system

$$A + 2X \rightleftarrows 3X;$$

$$X \rightleftarrows B$$

and has the associated macroscopic reactive term

$$\mathcal{F}(\phi) = \omega\phi^2 - \kappa\phi^3 - \kappa_2\phi + \eta \tag{9.5}$$

where, as seen in the last chapter, ϕ is the density of the reactant X, $\omega = \kappa_1\alpha$ and $\eta = \kappa\beta$ (α and β are the fixed densities of the reactants A and B), while κ_j indicates the different reaction rates. The densities α and β (ω or η), are used as control parameters. Let us consider the bounded domain case: $x \in [-L, L], 2L$ being the system length. In principle, we can consider two different *boundary conditions*

(i) Dirichlet boundary conditions: $\phi(-L) = 0$, and $\phi(L) = 0$,
(ii) Neumann boundary conditions: $d\phi(x = -L)/dx = d\phi(x = L)/dx = 0$,
   with the physical meaning of zero flux at the boundary.

Another, more general, boundary condition that includes both previous cases as limiting ones, is the *albedo* boundary condition. It describes the situation with partial reflectivity at the boundary, which we will discuss latter.

It is clear that the search for homogeneous (space independent) solutions will give the same result as in the previous chapter. Hence, we will focus on the more interesting case of inhomogeneous solutions. Eq. (9.4b) can written as

$$V(\phi) = \int_0^\phi \mathcal{F}(\phi')d\phi' \quad ; \quad V(0) = 0 \tag{9.6a}$$

yielding for the potential

$$V(\phi) = \frac{1}{3}\omega\phi^3 - \frac{1}{4}\kappa_1'\phi^4 - \frac{1}{2}\kappa_2\phi^2 + \eta\phi \tag{9.6b}$$

The form of Eq. (9.4a), together with the mechanical analogy introduced in the last chapter—see Eqs. (7.26)–(7.26)—suggest its interpretation as describing a particle of mass D, moving under the influence of the potential $V(\phi)$. In order to do this we need to assimilate the coordinate x to a time variable (varying from $-L$ to L), and ϕ to a *spatial* coordinate. The first integral of motion is then

$$\frac{D}{2}\left(\frac{d\phi}{dx}\right)^2 + V(\phi) = E, \tag{9.7}$$

with E a constant. The last equation indicates the conservation of E, the analogous of the *total mechanical energy*. Exploiting this mechanical analogy, the following features of the solutions of Eq. (9.4a) (for the potential given in Eq. (9.6)) can be easily seen:

(i) The stationary homogeneous solutions ϕ_1, ϕ_0, ϕ_2, found at the end of Sec. 8.2 (when discussing the hysteresis effect in the Schlögl model), correspond to the extrema of the potential $V(\phi)$. The linear stability analysis indicated that ϕ_1 and ϕ_2, associated with maxima of $V(\phi)$, are stable solutions, while ϕ_0, that corresponds to a minimum of $V(\phi)$, is unstable.

(ii) If we do not impose the Neumann boundary conditions indicated above, then every value of E corresponds to a solution of Eq. (9.4a) in the range of ϕ where $E > V(\phi)$ (for given values of $\phi(-L)$ and $\phi(L)$).

(iii) When we impose Neumann boundary conditions, we require that $\phi(-L)$ and $\phi(L)$ became turning points of the *trajectory*, that is $E = V[\phi(-L)] = V[\phi(L)]$. This imposes a constraint on the acceptable solutions, restricting them to those confined to the *valley* between ϕ_1 and ϕ_2, as indicated in Fig. 9.1 by the values ϕ_m and ϕ_M. According to the discussion at the end of Sec. 8.2, such valleys exist only if $\omega > \omega_c$ and $\eta \in [\eta_0, \eta_1]$. The other case is indicated with the trajectory that starts at $\phi_m = 0$, bounces at $\phi_M = \phi_M^*$ and returns to the origin.

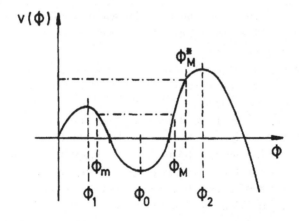

Fig. 9.1

Furthermore, in order that $\phi(x)$ be an acceptable solution it must satisfy $\phi(x) < \phi_2$. Also, for an infinite system (that is considering the limit $L \to \infty$), every value of E lower than the smaller of the two maxima $[V(\phi_1)$ and $V(\phi_2)]$ corresponds to an acceptable solution. For a finite system, we also have periodic solutions of period $2L/n$, with $n = 1, 2, \ldots$, each value of n corresponding to a trajectory bouncing n-times between the turning points ϕ_m and ϕ_M, in the "time" interval $x = -L$ to $x = L$. Clearly, for any given $V(\phi)$ of the form indicated above, not all values of E will yield solutions satisfying the boundary conditions. Resorting to known results from classical mechanics, the possible trajectories in phase space, parametrized with E, can be obtained from

$$\frac{d\phi}{dx} = \pm\sqrt{\frac{2}{D}}[E - V(\phi)]^{1/2} \tag{9.8}$$

Integrating the last equation, the solution is obtained as an inverse function according to

$$x(\phi) = \pm\sqrt{\frac{2}{D}} \int_0^\phi \frac{d\phi'}{[E - V(\phi')]^{-1/2}} \tag{9.9a}$$

If we now look for a solution with period $2L/n$, then, in the interval $x = -L/n$ to $x = L/n$, ϕ varies from $\phi_m = \phi(-L/n)$ to $\phi_M = \phi_n = \phi(L/n)$, with $V(\phi_m) = V(\phi_n)$. In this periodic case, the last equation reads

$$\frac{L}{n} = \pm\sqrt{\frac{2}{D}} \int_{\phi_m}^{\phi_n} \frac{d\phi'}{[E - V(\phi')]^{-1/2}} \tag{9.9b}$$

The above procedure is familiar from classical mechanics, and corresponds to the method of quadratures used to integrate the equation of motion for conservative systems.

In general, it is a difficult (if not impossible) task to find explicit solutions either for the stationary problem indicated in Eq. (9.4a), or for the (complete) time dependent one in Eq. (9.3a). However, there are situations where one is satisfied just with a qualitative analysis of the behavior of such solutions. Clearly then, the study of the stability becomes of primary importance. In this context, the methods developed in the previous chapter are of relevance. In the present case, we linearize the problem about the inhomogeneous stationary solution (say $\phi_s(x)$), considering a small time dependent perturbation, and obtain in this way linear differential equations that contain the needed information. Hence, we consider perturbed solutions of the form

$$\phi(x, t) = \phi_s(x) + \varphi(x)e^{\lambda t} \tag{9.10}$$

Replacing this into Eq. (9.3a), and linearizing in $\varphi(x)$, leads to the following eigenvalue equation

$$D\frac{d^2}{dx^2}\varphi(x) + \left[\frac{\partial}{\partial\phi}\mathcal{F}(\phi)\right]_{\phi=\phi_s}\varphi(x) = -\lambda\varphi(x) \qquad (9.11)$$

whose form, and the Neumann boundary conditions, suggest solutions of the type

$$\varphi_n(x) \approx \cos\left(\frac{n\pi x}{2L}\right) \qquad (9.12a)$$

provided that

$$\left[\frac{\partial}{\partial\phi}\mathcal{F}(\phi)\right]_{\phi=\phi_s} - \lambda = D\left(\frac{n\pi}{2L}\right)^2 \qquad (9.12b)$$

The last equation shows that there is a tight connection between the eigenvalue λ and the wave vector $k = n\pi/2L$ associated to the perturbation. Hence, it is possible to have cases such that, for a certain range of values of the wave length of the perturbation the system is stable, while for other ranges it becomes unstable.

In order to discuss the emergence of an instability, we will consider the scheme from a more general viewpoint, valid for a wider class of systems than those described by the Schlögl model. Let us start from Eq. (9.3a) for a general (infinite) problem, with a stationary homogeneous solution ϕ_s that is stable. The stability of this solution means that our earlier linear stability analysis will give (for a multi-component system) a set of eigenvalues λ, all having a negative real part [i.e. $\Re(\lambda) < 0$]. We focus on the one with the largest real part, denoted by $\lambda(k)$ to make explicit its dependence on the perturbation wave vector. Now suppose that there is a control parameter ϵ, whose variation could change the stability of the solution. That is, for $\epsilon < \epsilon_c$ we have $\Re[\lambda(k)] < 0$ (for all k); while for $\epsilon = \epsilon_c$, $\Re[\lambda(k_0)] > 0$ for some $k = k_0$. Here ϵ_c is the *critical value* of the parameter ϵ.

We introduce now, for $\epsilon_c \neq 0$, the *reduced* control parameter

$$\eta = \frac{\epsilon - \epsilon_c}{\epsilon_c} \qquad (9.13)$$

and show, in Fig. 9.2, the dependence of $\Re[\lambda(k)]$ with η. In part (a), for $\eta < 0$, the reference ϕ_s state is stable and $\Re(\lambda) < 0$, but it becomes unstable for $\eta \geq 0$. For $\eta = 0$, the instability sets in at the wave vector $k = k_0$: $\Re[\lambda(k_0)] = 0$. For $\eta > 0$, there is a band of wave vectors $(k_1 < k < k_2)$ for which the uniform state is unstable. For this situation, when $\eta = 0$, we can

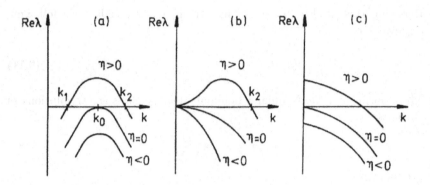

Fig. 9.2

have two kinds of instabilities: stationary if $\Im(\lambda) = 0$, or oscillatory when $\Im(\lambda) \neq 0$.

If for some reason (usually a conservation law) it happens that $\Re[\lambda(k = 0)] = 0$ for all values of η, another form of instability occurs. It is depicted in part (b) of Fig. 9.2. Here, $k_0 = 0$ is the critical wave vector, and for $\eta > 0$, the unstable band is $0 = k_1 < k < k_2$. It is possible to show that in general $k_2 \approx \eta^{1/2}$, and this indicates that the arising pattern occurs on a long length scale near the threshold $\eta = 0$. Once again we can find steady or oscillatory cases associated with $\Im(\lambda) = 0$ or $\neq 0$.

Finally, in part (c) of the figure, we depict a case where both the instability and the maximum growth rate, occur at $k_0 = 0$. This indicates that there is no intrinsic length scale. For this reason the pattern will presumably occur on a scale determined by the system size or by the dynamics. Once again we find steady or oscillatory cases associated with $\Im(\lambda) = 0$ or $\Im(\lambda) \neq 0$.

Another very interesting situation occurs, if we have a system of at least a two components, when there are two real roots and one of them becomes positive at some critical value of the control parameter. This is a situation leading to a spatially nonuniform steady state that is called a *Turing instability/bifurcation* (as opposed to the *Hopf instability/bifurcation* discussed in the previous chapter). This name is due to A. Turing, who was the first to note, in a now classic paper on morphogenesis, in the early fifties, the possibility of such a bifurcation in chemical kinetics.

One of the most commonly used models exemplifying all the characteristics we have so far discussed is the *Brusselator*. This model, introduced by Prigogine and Lefever, is a simplified version of a more elaborate model (the *Oregonator*) showing, qualitatively, a behavior similar to those observed in experiments related with the *Belousov–Zhabotinskii reaction*, in particular the existence of a transition to a limit cycle. Among the books quoted in the bibliography, those of Nicolis and Prigogine, Haken, Reichl, van Kampen, include nice discussions about the *Brusselator*. Here, we will choose a different (one component) model to exemplify the formation of spatial patterns, associated with an electrothermal instability.

9.3 Example: An electrothermal instability

In order to fix some of the ideas discussed in the previous section, we will analyze a solvable model of a physical system called the *ballast resistor*. This system consists essentially of a thin straight metal wire immersed in a gas, with both the temperature of the gas and the current that flux along the wire externally controlled. Depending on the values of these parameters, the temperature profile on the wire will be either homogeneous, or inhomogeneous (regions with different temperatures coexist along the wire). This device has been known and used as a current stabilizer for a very long time.

Here we will adopt a form of the model that has been used in some experiments on superconducting microbridges, called the *hot-spot* model. We consider a thin wire of length L, along which an electric current I is flowing. The wire is immersed in a heat bath with constant temperature T_B. The law of conservation of internal energy per unit length of the wire $u(x,t)$ is given by

$$\frac{\partial}{\partial t}u(x,t) = -\frac{\partial}{\partial x}\left[J(x,t) + h(x,t)I(x,t)\right] - Q(x,t) + I(x,t)E(x,t) \quad (9.14)$$

Here x is the position along the wire ($-L \leq x \leq L$), J is the heat current, h is the enthalpy per unit unit of charge carrier and unit length, Q is the energy flow dissipated into the gas per unit length, E is the electric field along the wire and IE is the heat generated by the current per unit length. As the Coulomb forces between the charges are very strong, we assume electro-neutrality of the wire, yielding

$$\frac{\partial}{\partial x}I(x,t) = 0 \Leftarrow I(x,t) = I(t) \quad (9.15)$$

However, the assumption of electro-neutrality will only be valid if one considers a range of time variation that is short when compared with the inverse of a typical plasma frequency of the electrons.

The quantities J, E and Q obey the phenomenological linear laws

$$J(x,t) = -\lambda \frac{\partial}{\partial x} T(x,t) + \Pi I(t), \quad \lambda > 0 \tag{9.16a}$$

$$E(x,t) = -\eta \frac{\partial}{\partial x} T(x,t) + RI(t), \quad R > 0 \tag{9.16b}$$

$$Q(x,t) = q\left[T(x,t) - T_B\right], \quad q > 0 \tag{9.16c}$$

Here $T(x,t)$ is the local temperature field. The parameters introduced in these equations are: λ the heat conductivity of the wire, R the isothermal resistivity per unit length and η the differential thermoelectric power of the wire which is related to the Peltier coefficient Π through an Onsager relation

$$T\eta = -\Pi$$

The last coefficient, q, is related to the energy dissipated into the gas due to the difference in temperature between the wire and the gas. All these coefficients may, in principle, depend on the local temperature of the wire, while q may also depend on T_B. The internal energy of the wire u is a function of the temperature only, so that

$$du(x,t) = cdT(x,t) \tag{9.17}$$

with c the heat capacity per unit length. Substituting Eqs. (9.16) and (9.17) into Eq. (9.14), we obtain the equation for the temperature profile of the wire

$$c\frac{\partial}{\partial t}T(x,t) = \frac{\partial}{\partial x}\lambda\frac{\partial}{\partial x}T + \sigma_t I\frac{\partial}{\partial x}T - q(T - T_B) + RI^2 \tag{9.18}$$

where Eq. (9.15) was used and we introduced the Thompson coefficient σ_t that corresponds to the heat effect due to the simultaneous presence of an electric current and a temperature gradient, and is given by

$$\sigma_t = -\eta - \frac{\partial \Pi}{\partial T} - \frac{\partial h}{\partial T} = T\frac{\partial \eta}{\partial T} - \frac{\partial h}{\partial T} \tag{9.19}$$

using the Onsager relations.

In order to simplify the equation further for the temperature profile, we assume that σ_t as defined above is approximately zero. We also assume that the specific heat c, the heat conductivity λ and the heat transfer coefficient q are all constant along the wire.

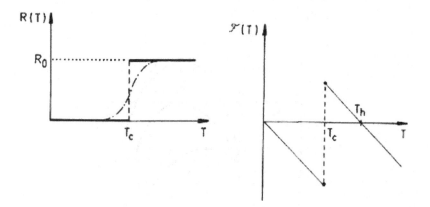

Fig. 9.3

As discussed earlier, we are interested in stationary solutions for the temperature field distribution. Hence, our equation has the form

$$\frac{\partial}{\partial x}\lambda\frac{\partial}{\partial x}T - q(T - T_B) + RI^2 = 0 \tag{9.20}$$

For the resistivity R we will adopt a piecewise-linear approximation of a realistic one, see the l.h.s. of Fig. 9.3, according to

$$R(T) = R_0\theta\left[T(x,t) - T_c\right] \tag{9.21}$$

with $\theta(z)$ the step Heaviside function $[\theta(z) = 1$ for $z > 0$, $\theta(z) = 0$ for $z < 0]$. The assumption of such a form for the resistance, means that for $T < T_c$ the wire is superconducting while it has a constant resistivity for $T > T_c$. Without loss of generality we may take the zero of the temperature scale at the heat bath temperature T_B (that is we take $T_B = 0$). We also introduce scaling to make the coordinate dimensionless as follows

$$y \equiv \left(\frac{q}{\lambda^2}\right)x \text{ with } y_L = \left(\frac{q}{\lambda^2}\right)L \tag{9.22a}$$

with $-y_L \leq y \leq y_L$. In our discussion we assume that the current I is fixed (the voltage difference depending on I), and we define the following effective temperature

$$T_h \equiv \frac{I^2 R_0}{q} \tag{9.22b}$$

With all these assumptions, Eq. (9.20) for T adopts the final form

$$\frac{\partial^2}{\partial y^2}T(y) - T + T_h\theta[T(y) - T_c] = \frac{d^2}{dy^2}T(y) + \frac{d}{dT}V(T) = 0 \tag{9.23}$$

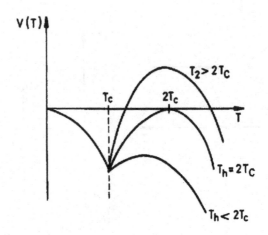

Fig. 9.4

where the potential $V(T)$ is defined, accordingly to Eq. (9.6a) as

$$V(T) = \int_0^T dT' \mathcal{F}(T') = \int_0^T dT' \{T_h\theta[T(y) - T_c] - T'\} \qquad (9.24)$$

So that

$$V(T) = T_h(T - T_c)\theta(T - T_c) - \frac{1}{2T^2} = \begin{cases} -\frac{1}{2T^2} & \text{for } T < T_c \\ V_h - \frac{1}{2T^2} & \text{for } T > T_c \end{cases} \qquad (9.25)$$

where $V_h \equiv T_h^2 Z/2$, and $Z = (1 - 2T_c/T_h)$. The shape of the potential for different values of the parameters is shown in Fig. 9.4. It has two parabolic branches, one for $T < T_c$ and the other for $T > T_c$. The point $T = T_c$ is common to both. If $T_c > T_h$, the potential has only one maximum, while for $T_c < T_h$ it has two maxima, one at the origin and the other at $T = T_h$, and one minimum at $T = T_c$. For $2T_c = T_h$, both maxima are coincident. The possibility of having two maxima for some values of the current I $(T_h = I^2 R_0/q)$ plays an essential role in the analysis of stationary structures. On the r.h.s. of Fig. 9.3 we show the form of the nonlinear function $\mathcal{F}(T)$. We can compare this with the corresponding term in the Schlögl model (Fig. 8.12), and see that the present form is a *mimic* of the other.

To complete the model, we need to specify the boundary conditions at both borders: $x = -L$ and $x = L$ (or $y = -y_l$ and $y = y_L$). As commented above, we will only consider Dirichlet or Neumann boundary conditions, given by

$$\frac{d}{dy}T\bigg|_{y=-y_L} = \frac{d}{dy}T\bigg|_{y=y_L} = 0 \qquad \text{Neumann b.c.} \qquad (9.26a)$$

$$T(-y_L) = T(y_L) = 0 \qquad \text{Dirichlet b.c.} \qquad (9.26b)$$

Clearly, for the discussion of the different structures that arise according to the boundary conditions imposed, we can exploit the mechanical analogy introduced in relation with Fig. 9.1.

Here as before—Eq. (9.7)—there is a constant of motion (associated with the conservation of the analogue of the total mechanical energy), the quantity

$$\frac{1}{2}\left(\frac{dT}{dx}\right)^2 + V(T) = E \qquad (9.27)$$

that, as in all one dimensional problems, has the characteristics of a Lyapunov functional.

To find the form of the stationary solutions one may distinguish two different regions

Cold regions, where $T(y) < T_c$, and Eq. (9.23) reduces to

$$\frac{\partial^2}{\partial y^2}T(y) - T = 0 \qquad (9.28a)$$

with solutions that have the general form

$$T(y) = \mathcal{A}_c e^y + \mathcal{B}_c e^{-y} \qquad (9.28b)$$

Hot regions, where $T(y) > T_c$, and Eq. (9.23) reduces to

$$\frac{\partial^2}{\partial y^2}T(y) - T + T_h = 0 \qquad (9.29a)$$

and with general solutions of the form

$$T(y) = \mathcal{A}_h e^y + \mathcal{B}_h e^{-y} + T_h \qquad (9.29b)$$

The parameters $\mathcal{A}_c$, $\mathcal{B}_c$, $\mathcal{A}_h$ and $\mathcal{B}_h$ are determined after imposing the boundary conditions. Furthermore, if we have a cold region on the left and a hot region on the right of a certain position coordinate y_c (or vice versa), both solutions should be joined together in such a way that Eq. (9.23) is satisfied at the transition point. This is the case if both T and dT/dy are continuous at $y = y_c$. Using these conditions, it is clear that

$$T(y_c) = T_c \qquad (9.30)$$

Note that, if the parameters in Eqs. (9.29) have been chosen in such a way that the corresponding temperature profiles give the same value of E, the continuity of T at $y = y_c$ automatically gives the continuity of dT/dy at this point.

Let us now analyze the stationary states. We first consider the homogeneous case. For Neumann b.c. the potential $V(T)$ must have a maximum. This leads to $T(y) = 0$ for all values of T_h (and therefore of the current I). However, if $T_h > T_c$, there is an additional homogeneous solution $T(y) = T_h$. Clearly both satisfy the Neumann boundary conditions. For Dirichlet b.c., there is only one possible homogeneous solution $T(y) = 0$.

We now turn to inhomogeneous stationary temperature profiles. Using our mechanical analogy, it is possible to find inhomogeneous solutions corresponding to several *bounces* between the *turning points* of the potential. Here, we will consider spatial temperature distributions having only one maximum, with two *cold* regions for $-y_L < y < -y_c$ and $y_c < y < y_L$, and one *hot* region for $-y_c < y < y_c$ (with two transition points at $y = \pm y_c$ due to symmetry). Through a linear stability analysis one can prove that solutions with several maxima (or several *bounces*) are always unstable.

Imposing the boundary conditions of Eqs. (9.26) on the solutions of the form indicated in Eqs. (9.29), the different constants are determined yielding the following. For Neumann b.c. we get

$$T^s(y) = T_h \begin{cases} -\dfrac{\sinh(y_c)\cosh(y_L+y)}{\sinh(y_L)} & -y_L < y < -y_c \\[2mm] 1 - \dfrac{\cosh(y)\sinh(y_L-y_c)}{\sinh(y_L)} & -y_c < y < y_c \\[2mm] -\dfrac{\sinh(y_c)\cosh(y_L-y)}{\sinh(y_L)} & y_c < y < y_L \end{cases} \qquad (9.31a)$$

while for Dirichlet b.c. one gets

$$T^s(y) = T_h \begin{cases} -\dfrac{\sinh(y_c)\sinh(y_L+y)}{\cosh(y_L)} & -y_L < y < -y_c \\[2mm] 1 - \dfrac{\cosh(y)\cosh(y_L-y_c)}{\cosh(y_L)} & -y_c < y < y_c \\[2mm] -\dfrac{\sinh(y_c)\cosh(y_L-y)}{\cosh(y_L)} & y_c < y < y_L \end{cases} \qquad (9.31b)$$

The points $\pm y_c$ are determined by the matching conditions at $y = \pm y_c$ (that is continuity of the function T and its derivative), resulting in the different equations as follows:

Neumann b.c.:

$$y_c = \frac{1}{2}\left\{ y_L - \ln\left[Z\cosh(y_L) - \sqrt{Z^2\cosh(y_L)^2 + 1} \right] \right\} \qquad (9.32a)$$

Dirichlet b.c.:

$$y_c^{(\pm)} = \frac{1}{2}\left\{y_L - \ln\left[Z\cosh(y_L) \pm \sqrt{Z^2\cosh(y_L)^2 + 1}\right]\right\} \qquad (9.32b)$$

Remember that here $Z = (1 - 2T_c/T_h)$.

The typical shapes for the patterns in the cases of Dirichlet and Neumann b.c. are shown in parts (a) and (b) of Fig. 9.5 respectively. The case of Dirichlet b.c. shows the possibility of existence of two solutions (that is two possible roots $y_c^{(\pm)}$, depending on the value of Z), while there is only one for Neumann b.c.

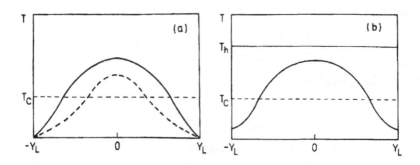

Fig. 9.5

We now analyze the stability of the structures that we have found. Following our previous discussion, we propose after Eq. (9.10), a perturbation of the following form

$$T(y,t) = T^s(y) + \varphi(y,t) \qquad (9.33)$$

Also, as discussed earlier, see Eq. (9.11), we substitute this into Eq. (9.23) and linearize in $\varphi(y,t)$, leading to an eigenvalue equation for α, which is

$$\frac{\partial}{\partial t}\varphi(y,t) = \left\{\frac{\partial^2}{\partial y^2}T^s(y) - T^s(y) + T_h\theta[T^s(y) - T_c]\right\}$$
$$+ \frac{\partial^2}{\partial y^2}2\varphi(y,t) - \varphi(y,t) - \kappa_0\sum_j \delta(y - y_j)\varphi(y,t) \qquad (9.34)$$

The term within the large parentheses is zero, as it must be from the stationary condition, reducing Eq. (9.34) to an equation for $\varphi(y)$. The parameter κ_0 arises from the discontinuities (due to the presence of the step $\theta(y)$ function) at $y_j = \pm y_c$, in the case of the structures given by Eqs. (9.31). This parameter is given by

$$\kappa_0 = T_h \left| \frac{dT^s(y)}{dy} \right|_{y=y_j}^{-1} \tag{9.35}$$

and with the specific b.c. its form is

Neumann:

$$\kappa_0 = T_h \left| \frac{dT^s(y)}{dy} \right|_{y=y_c}^{-1} = \frac{\sinh(y_L)}{\sinh(y_c)\sinh(y_L - y_c)} \tag{9.36a}$$

Dirichlet:

$$\kappa_0 = T_h \left| \frac{dT^s(y)}{dy} \right|_{y=y_c^{(\pm)}}^{-1} = \frac{\cosh(y_L)}{\sinh(y_c^{(\pm)})\cosh(y_L - y_c^{(\pm)})} \tag{9.36b}$$

We again propose $\varphi(y,t) = \varphi_0(y)e^{-\alpha t}$, leading in each region to solutions of the form

$$\varphi_0(y) = ae^{\lambda y} + be^{-\lambda y} \tag{9.37}$$

where the coefficients a and b depend on the region and are determined, as indicated before, by the boundary conditions. Writing the eigenvalue as $\alpha = 1 - \lambda^2$, replacing the form of Eq. (9.37) into the equation for $\varphi(y)$, finding the parameters and for each region, and after some algebra, we find the following equations for λ in the Neumann case

$$\lambda \left\{ \tanh(\lambda y_c) + \frac{\sinh[\lambda(y_L - y_c)]}{\cosh[\lambda(y_L - y_c)]} \right\} = \kappa_0 \tag{9.38a}$$

while for the Dirichlet case we get

$$\lambda \left\{ \tanh(\lambda y_c^{(\pm)}) + \frac{\cosh[\lambda(y_L - y_c^{(\pm)})]}{\sinh[\lambda(y_L - y_c^{(\pm)})]} \right\} = \kappa_0^{(\pm)} \tag{9.38b}$$

Let us consider the Dirichlet case, and call $f^{(\pm)}(\lambda)$ the l.h.s. of Eq. (9.38b). It is easy to prove that

$$\frac{\partial}{\partial \lambda} f^{(\pm)}(\lambda) > 0 \text{ for } \lambda > 0 \tag{9.39}$$

It follows that Eq. (9.38b) has at most one solution. Furthermore, one has that, for

$$\frac{2}{y_L} \ln \left[Z \sinh(y_L) \pm \sqrt{Z^2 \sinh(y_L)^2 - 1} \right] \neq 0 \tag{9.40}$$

it is

$$f^{(+)}(1) < \lambda \ln \left\{ \tanh(\lambda y_c^{(+)}) + \frac{\cosh[\lambda(y_L - y_c^{(+)})]}{\sinh[\lambda(y_L - y_c^{(+)})]} \right\} = f^{(+)}(\lambda^+) \quad (9.41a)$$

Since f is a monotonically increasing function of λ, it follows that a solution λ^+ exists and that, when Eq. (9.40) holds,

$$\lambda^+ > 0 \Rightarrow \alpha_{min}^+ < 0 \quad (9.41b)$$

indicating that the corresponding stationary states are unstable. For the negative solution, again assuming Eq. (9.40), we get

$$f^{(-)}(1) < \lambda \ln \left\{ \tanh(\lambda y_c^{(-)}) + \frac{\cosh[\lambda(y_L - y_c^{(-)})]}{\sinh[\lambda(y_L - y_c^{(-)})]} \right\} = f^{(-)}(\lambda^-) \quad (9.42a)$$

It follows that, if a solution exists

$$\lambda^- < 0 \Rightarrow \alpha_{min}^- > 0 \quad (9.42b)$$

and consequently the corresponding stationary states are stable.

We conclude that, for Dirichlet b.c., from the pair of simultaneous solutions, the one with the larger dissipation (i.e. the larger hot region) is stable while the other is unstable.

A similar analysis for the case of Neumann b.c., indicates that the homogeneous stationary solutions are stable, while inhomogeneous structures are always unstable.

Two final remarks to close this section. One concerns the boundary conditions, and the other the applicability of the piecewise linear approximation of the nonlinear "reactive" term.

The inhomogeneous solutions corresponding to the Neumann or Dirichlet b.c. can both be obtained (simultaneously) by considering a more general form of boundary condition called *albedo* b.c. (or *radiation conditions*), from which the two previous b.c. can be obtained as limiting cases, by changing the value of an appropriate parameter (on which the condition depends). The physical meaning of these different b.c. is the following:

Dirichlet b.c. imply that the density at the boundary is zero, indicating the presence of a perfect absorber or zero reflectivity,

Neumann b.c. correspond to a zero current at the boundary, indicating a perfect reflectiveness,

Albedo b.c. correspond to an intermediate situation, indicating a partially reflecting boundary.

The mathematical form of the albedo b.c. is

$$\frac{d}{dy}T\bigg|_{y=-y_L} = \kappa T(-y_L) \quad ; \quad \frac{d}{dy}T\bigg|_{y=y_L} = -\kappa T(y_L)$$

Here κ is the parameter related to the reflectivity of the boundary: if $\kappa \to 0$ this gives the Neumann b.c. (total reflectivity), while $\kappa \to \infty$ corresponds to Dirichlet b.c. (zero reflectivity). For a complete discussion of this boundary condition, we refer the reader to [Duderstadt and Martin (1979)].

The possibility of applying a piecewise linear approximation to the non-linear "reactive" term is not restricted to the one component case. Such an approximation has also been used for two component models of the *activator–inhibitor* type, by making a mimic of the nullclines by means of piecewise linear approximations. We will return to this point in Sec. 9.5.

9.4   Pattern propagation: (a) one component systems

In this section we discuss how to describe the propagation of structures in one component systems. For this reason we consider Eq. (9.3a) once more in its complete form, that is

$$\frac{\partial \phi}{\partial t} = D\frac{\partial^2 \phi}{\partial x^2} + \mathcal{F}(\phi) \tag{9.43}$$

We assume a one dimensional, infinite, system. To complete the description, we need to include some boundary conditions at infinity. Clearly, for a very general form of $\mathcal{F}(\phi)$ it is not easy to find an arbitrary solution of Eq. (9.43) fulfilling the choosen b.c.

However, there is a particular kind of solutions of great interest called *solitary waves* on which we will focus our attention. These waves are functions of the spatial (x) and temporal (t) coordinates, not independently, but through the following combination

$$\xi = x - ct \tag{9.44}$$

In terms of the new variable ξ, Eq. (9.43) adopts the form

$$D\frac{\partial^2 \phi}{\partial \xi^2} + c\frac{\partial \phi}{\partial \xi} + \mathcal{F}(\phi) = 0 \tag{9.45}$$

where $\partial/\partial t = -c\partial/\partial \xi$ and $\partial^2/\partial x^2 = \partial^2/\partial \xi^2$.

Here we can resort once more to the mechanical analogy we used earlier. We again interpret ϕ as the spatial coordinate of a particle of mass D moving in the force field $\mathcal{F}(\phi)$ [derived from the potential $V(\phi)$, see Eq.

(9.6a), i.e. $\mathcal{F}(\phi) = \partial V/\partial\phi]$, but now in the presence of a friction force proportional to the velocity of the particle, i.e. $\partial\phi/\partial\xi$. In such an analogy, c plays the role of the friction coefficient.

Let us concentrate on the situation where the potential $V(\phi)$ has a bistable form as indicated in Fig. 9.1, and ask for solutions of Eq. (9.45) with the boundary conditions

$$\phi \to \phi_2 \text{ for } \xi \to -\infty$$

$$\phi \to \phi_1 \text{ for } \xi \to \infty \tag{9.46}$$

The resulting wave, or moving pattern, is called a *trigger wave* (or *front*), because its propagation triggers the transition from one stationary state of the system (say ϕ_2) to the other (ϕ_1). This kind of waves has been observed, for instance, in chemically reacting media or as electrical activity that propagates without attenuation along the axonal membrane.

In order to analyze qualitatively the behavior of the system, we assume that the potential $V(\phi)$ has the form indicated in Fig. 9.1, that is $V(\phi_2) > V(\phi_1)$. This implies that the quantity ϑ, defined as

$$\vartheta = \int_{\phi_1}^{\phi_2} d\phi \mathcal{F}(\phi) \tag{9.47}$$

is positive.

When there is no friction present (i.e. $c = 0$), we have seen—as indicated in Eq. (9.7)—that the *total mechanical energy* is conserved. Hence, if we release a particle from the maximum ϕ_2, with a vanishingly small velocity, it reaches the lower maximum at ϕ_1 after a finite "time" ξ ($= x$, as we have $c = 0$). Due to conservation of energy, the particle moves further in the direction of negative values of ϕ. Clearly, in this case, we cannot fulfill the boundary conditions indicated in Eq. (9.46). In the opposite situation, that is if the friction is too large, the particle motion results overdamped and cannot reach the "point" ϕ_1. Instead, for $\xi \to \infty$ it stays at ϕ_0, where the potential has its minimum.

We can conclude that for this potential, there is only one value of the friction coefficient c for which the dissipation of mechanical energy is exactly that corresponding to the difference $\Delta V = \vartheta = V(\phi_2) - V(\phi_1)$. Hence, after starting at ϕ_2, the particle crosses over the minimum at ϕ_0, and after an infinite interval of time (that is $\xi \to \infty$) finally arrives at ϕ_1, with zero velocity. Such a particular value of c, say c_0, corresponds to the propagation speed of the trigger wave (or front). This indicates that both the propagation speed of the front as well as its profile, are determined univocally by the properties of the medium, and are the same for all trigger waves in this medium independently of the conditions that originated them.

Within the same picture we can see that when ΔV is reduced, c_0 is also reduced, and finally for $\Delta V = 0$ one has $c_0 = 0$. Hence, in the case when both maxima have the same height we have no propagation (only a stationary pattern is possible). However, for $\Delta V < 0$ we will find that the previous situation is reversed, and the propagation of the trigger wave (motion of the front connecting the states ϕ_1 and ϕ_2) is also reversed. That means the velocity c will have the opposite sign.

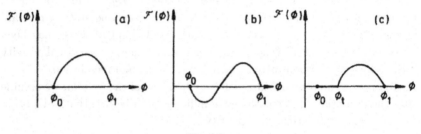

Fig. 9.6

Before analyzing these aspects in detail in a concrete example, we want to categorize a few typical situations of wave fronts according to the properties of the reactive term $\mathcal{F}(\phi)$. The different forms are indicated in Fig. 9.6.

Fisher model: $\mathcal{F}(\phi) \sim (\phi - \phi_0)(\phi_1 - \phi)$. This case was studied by Fisher in connection with a problem of population genetics. There is a $c^* > 0$, such that for every velocity $c > c^*$ a front exists. These fronts turn out to be less stable than in the next bistable case.

Bistable case: This is essentially the case we have discussed so far. As we have seen, there is a unique wave front connecting both stationary stable solutions ϕ_0 and ϕ_1. The stability analysis indicates that these are very stable structures.

Ignition case: In this case, fronts connecting ϕ_0 and ϕ_1 are also unique, with a unique velocity, and have a limited degree of stability. The front starts to propagate after a certain threshold value ϕ_t has been reached.

Let us now discuss the ballast resistor model introduced in Sec. 9.3. As we concentrate on the case of *solitary waves* (for the infinite system), through the change of variables indicated in Eq. (9.44), Eq. (9.23) adopts the form

$$\frac{\partial^2}{\partial \xi^2} T(\xi) + c \frac{\partial}{\partial \xi} T(\xi) - T(\xi) + T_h \theta[T(\xi) - T_c] = 0 \qquad (9.48)$$

As indicated in Fig. 9.7, we consider solutions with b.c.

$$T \to 0 \ \text{ for } \ \xi \to -\infty$$

$$T \to T_h \ \text{ for } \ \xi \to \infty \qquad (9.49)$$

In the notation used in the general discussion, we have $\phi_1 = T(-\infty) = 0$, $\phi_0 = T_c$ and $\phi_2 = T(\infty) = T_h$. There is a point $\xi = \xi_c$ at which we match the temperature profile as well as its derivative, and where $T(\xi_c) = T_c$. Due to translational symmetry we can choose $\xi_c = 0$, without loss of generality.

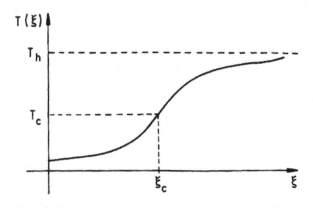

Fig. 9.7

The form of the solution is

$$T(\xi) = \begin{cases} T_c e^{\alpha_1 \xi} & \xi < 0 \\ (T_c - T_h) e^{\alpha_2 \xi} + T_h & \xi > 0 \end{cases} \qquad (9.50)$$

where

$$\alpha_{1,2} = -\frac{c}{2} \pm \sqrt{1 + \frac{c^2}{4}} \qquad (9.51)$$

From the discussion of the ballast model in Sec. 9.3 we know that, if $T_h > 2T_c$, $\phi_2 = T_h$ is the *dominant state* (where the potential V has its highest maximum), while for $T_h < 2T_c$ the situation is reversed. The quantity ϑ, given by Eq. (9.47), gives for the propagation velocity

$$c_0 = (2T_c - T_h)\left[T_c(T_h - T_c)\right]^{-1/2} \tag{9.52}$$

This result is in clear agreement with the previous discussion, that is for $T_h > 2T_c$ the propagation velocity is negative, and the front moves to the left, while for $T_h < 2T_c$ the propagation velocity is positive and the front moves to the right. Finally for $T_h = 2T_c$, the velocity is zero, and there is no propagation.

In Fig. 9.8 we present some numerical results for the propagation of a pair of patterns in the ballast model. Case (a) corresponds to the propagation of a temperature profile that has an initial step form, and shows the evolution for a few consecutive times. It is clear that the form indicated by Eq. (9.50) is reached asymptotically.

To analyze the stability of moving patterns, we can apply the same earlier ideas. If $T^0(\xi)$ is the solution, we propose

$$T(\xi, t) = T^0(\xi) + \varphi(\xi, t) \tag{9.53}$$

Replacing this form into Eq. (9.48), and linearizing in $\varphi(\xi, t)$, the following equation results

$$\frac{\partial}{\partial t}\varphi(\xi, t) = \frac{\partial^2}{\partial \xi^2}2\varphi(\xi, t) - \varphi(\xi, t) - \kappa_0\delta(\xi - \xi_c)\varphi(\xi, t) \tag{9.54}$$

where

$$\kappa_0 = T_h \left|\frac{dT^0(\xi)}{d\xi}\right|^{-1}_{\xi = \xi_c}. \tag{9.55}$$

As usual we propose $\varphi(\xi, t) = \varphi_0(\xi)e^{-\alpha t}$, leading in each region to solutions of the form

$$\varphi_0(\xi) = ae^{\lambda\xi} + be^{-\lambda\xi} \tag{9.56}$$

and the analysis follows the same lines as before. However, one of the possible eigenvalues results to be zero, indicating the existence of translational symmetry.

Returning to Fig. 9.8, the case (b) corresponds to the propagation of a different pattern that has two fronts, corresponding to a *bubble* like structure. In this case, we can make an analogy with a nucleation process: if the size of the bubble is larger than the critical one and the structure grows, producing a phase transition to a homogeneous state with $T = T_h$. In case (c) we have again the case of a bubble, but now the initial size is smaller than the critical one and the structure collapses.

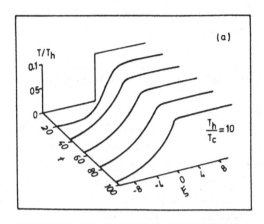

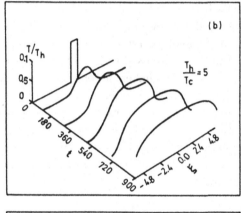

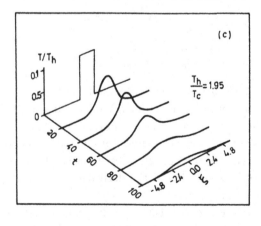

Fig. 9.8

We now turn to discuss the propagation phenomenon in systems with more than one component.

9.5   Pattern propagation: (b) two component systems

In order to make a realistic description, for the theoretical representation of travelling waves of chemical, physical or biological activity commonly observed in spatially distributed excitable media, we need to resort to models with more than one component. All excitable media share certain characteristic features. They have a stable rest state, and small perturbations are rapidly damped out. However, perturbations larger than a certain threshold trigger an abrupt and substantial response. After the indicated fast response, the media is typically refractory to further stimulation for some characteristic time until it recovers its full excitability. It is clear that such a sequence of events cannot be represented by a simple one component model of the kind we have discussed so far. On the other hand, the analysis of a model with a large number of components quickly becomes too cumbersome. Notwithstanding, experience has shown that it is enough to resort to two component models in order to be able to qualitatively (and sometimes quantitatively) reproduce several characteristics of real systems.

The set of equations corresponding to a model describing a typical two component system, with densities $u(x,t)$ and $v(x,t)$, according to Eq. (9.3b), has the general form

$$\frac{\partial u}{\partial t} = D_u \frac{\partial^2 u}{\partial x^2} + f(u,v) \tag{9.57a}$$

$$\frac{\partial v}{\partial t} = D_v \frac{\partial^2 v}{\partial x^2} + g(u,v) \tag{9.57b}$$

Depending on the actual form of the nonlinear terms $f(u,v)$ and $g(u,v)$, even such an innocent pair of equations, can have an extremely complicated behavior. However, the experience has also shown that a typical and very fruitful form is the one shown in Fig. 9.9. There we show—in the phase plane $u - v$—the form of the nullclines [that is, the curves $f(u,v) = 0$ and $g(u,v) = 0$], and the sign of the derivatives of the nonlinear reactive functions in each plane region.

We will discuss now how it is that such a simple system can model the sequence of events we have indicated at the beginning of this section. We recall that an excitable medium is a spatially distributed system composed of

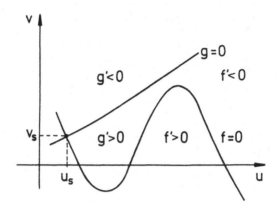

Fig. 9.9

excitable elements. The interaction between neighboring elements through a diffusive coupling makes it possible to produce excitation waves. If a local region of space is disturbed beyond a certain threshold value, then the auto-catalytic production of substance within the excited region causes to diffuse into the neighboring regions, driving those regions across the threshold and thus making the excitation spread spatially. This corresponds to a front propagation. In order to make a pictorial representation of this process, we refer to Fig. 9.11. There is a unique homogeneous steady state indicated by the point (u_{st}, v_{st}), that satisfies $f(u_{st}, v_{st}) = 0$ and $g(u_{st}, v_{st}) = 0$, and is locally stable but excitable: while the subthreshold disturbances are rapidly damped (perturbations in the region indicated by 1 in Fig. 9.10), larger disturbances (those driving the system beyond the point u_{th}) provoke an excitation cycle that is governed by the local reaction kinetics before the system returns to the steady state. This cycle is indicated in the figure through the sequence of numbers from 2 to 7, corresponding to different states of the system. In region 2, u increases by an autocatalytic process, until the phase trajectory reaches the curve $f(u, v) = 0$, where the density of v starts to increase, and the process moves following the nullcline as indicated by 3. After that the process reaches a maximum value of the density for v (4), and the process follows curve 5, where the density of decreases and after crossing the nullcline $g = 0$, region 6, the other branch of the nullcline $f = 0$ is reached and the system moves along this trajectory (indicated by 7) and reaches the steady state (u_{st}, v_{st}) again.

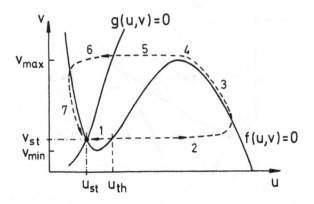

Fig. 9.10

The abrupt over-threshold autocatalytic production of u gives rise to the excitability of the system and the interaction between an causes the recovery from the excitation state. For this reason u is sometimes called the *trigger*, *propagator* or *activator* variable and v the *recovery* variable (also the *controller* or *inhibitor*) respectively. For instance, some examples of those variables in real systems are: membrane potential as *propagator* and ionic conductance as *inhibitor* in neuromuscular tissue; bromous acid as *propagator* and ferroin as *inhibitor* in the Belousov–Zhabotinskii reaction; infectious agent as *propagator* and level immunity as *inhibitor* in epidemics.

A typical form of the profile in a one dimensional media for the kind of waves that occur according to the previous discussion is shown in Fig. 9.11. The transition zone from the rest to the excited state is called the *front*, while the transition zone from the excited to the rest state is the *back* (or the *backfront*).

The process we have so far discussed is clearly not restricted to a one dimensional geometry. In fact, in two dimensional media the same line of argument leads to describe the so called *target structures*, that is perturbations that spread radially originating a sequence of growing rings, such as has been typically observed in the Belousov–Zhabotinskii reaction. When we look at such structures far from the point where they have been originated, the curvature has decreased and the structure acquires a one dimensional characteristic, i.e. in the direction of propagation it has the same profile as shown in Fig. 9.12, while it extends "infinitely" in the normal direction.

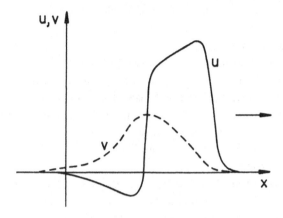

Fig. 9.11

A more quantitative analysis of pattern formation and propagation in *activator–inhibitor* systems requires some elaborate analytic or numerical techniques. For instance, it is possible to resort to some form of *singular perturbation* method, when one can assume that one of the diffusion constants (typically the one of the controller substance) is very small compared with the other. However, as mentioned earlier, we can mimic the behavior of the nullclines by making a piecewise linear approximation to them. Such an approach allows to proceed further at an analytical level and has been exploited by several authors.

We will not pursue such procedures here, instead we turn now to discuss, in a qualitative way, the origin of another very important type of structure that arise in two dimensional *activator–inhibitor* systems, the *spirals*.

9.6   The genesis of spirals

A common form of pattern in the reaction-diffusion description of two-dimensional excitable media is the rotating *spiral*. The interest in this kind of pattern arises from its occurrence in chemical (typically in the Belousov–Zhabotinskii reaction) as well as in biological (waves of electrical and neuromuscular activity in cardiac tissue, waves of spreading depression in the cerebral cortex) systems. From a topological point of view, spirals are related to dislocation type defects in striped patterns. A complete mathematical description of such structures is far from trivial. However, within the *activator–inhibitor* scheme, it is possible to understand intuitively the initial stage in the formation of an spiral wave. We will concentrate on providing such a simplified view.

We start considering a thought experiment with a solitary wave of the type discussed earlier. That means a propagating straight band, a two-dimensional wave with a profile in the direction of motion like the one shown in Fig. 9.11, and extending indefinitely in the normal direction. This is schematically indicated on the l.h.s. of Fig. 9.12, where "+" indicates the excited region, and "-" the unexcited ones. There $_f$ and $_b$ indicate the *front* and the *back-front* respectively. The thought experiment goes as follows. Consider that such a band is perturbed in some way (for instance by physically mixing the chemicals in a region overlapping with a finite segment of the band). Hence, the pulse-like structure is disturbed in that region, acquiring the form indicated on the r.h.s. of Fig. 9.12. It is clear that in both branches of the perturbed structure we will see that the points in the front or in the back-front will continue their motion. The only exception will be the point indicated by * (corresponding to the *tip* of the *spiral core*). This point is the boundary between the front and the back-front, and if we consider that the front velocity has to change continuously, it must have zero velocity.

The evolution will continue according to the following steps. We refer our argumentation to Fig. 9.13. On the left, we depicted the upper branch

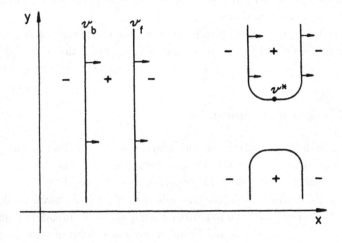

Fig. 9.12

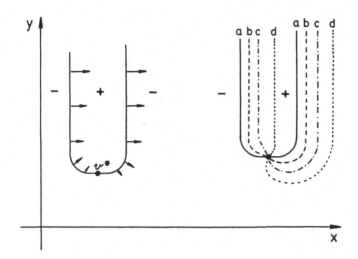

Fig. 9.13

of the perturbed band. The points on the front, far from *, move at the same original velocity, but when we come closer to *, their velocity reduces continuously. The same happens with points on the back-front. This initial situation is indicated by the curve labelled. After a short time has elapsed, the point * remains immobile, but all other points have moved into their new position indicated by the curve labelled. Clearly, the original form of the perturbed band is deformed. After another short time elapses, the same process is repeated and the curve changes to the one labelled; after another short time to, and so on. Carried to its logical extreme, this type of analysis would predict that the front would acquire a growing angular velocity and curvature, a process that finally produces a spiral. This is indicated in Fig. 9.14.

The experimental observation of spirals in chemically reactive media, particularly in the Belousov–Zhabotinskii reaction, shows that this kind of pattern always appears as pairs of symmetric, counter-rotating spirals. To understand this aspect within the same qualitative picture we return to the initial perturbed propagating band as indicated again in Fig. 9.15.

We have indicated on each branch the position of their respective "fixed" point *. The previous argumentation can, clearly, be applied to both branches. However, each one is the specular image of the other, implying that, if the motion in the neighborhood of the left one is a rotation in

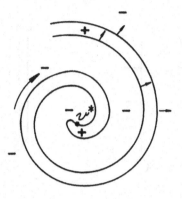

Fig. 9.14

the indicated direction, the motion around the one on the right will be a rotation in opposite direction. Hence, the same picture offers a description on the possible origin of spirals (that, as a matter of fact have been proved through numerical simulation of active media, i.e. by means of cellular automata), as well as their appearance in counter-rotating pairs.

The long time evolution of the spirals fast becomes a mathematically too complicated subject, and is clearly beyond the scope of this textbook. To close this section, we want to comment that, if we define the *vertex* of the spiral as the position of the front at the point *, it does not automatically follow that this point remains stationary. In fact, meandering has been observed in chemical systems. Such meandering is usually associated with instabilities at the vertex (or *core*) of the spiral due to the appearance of singularities.

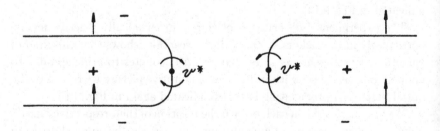

Fig. 9.15

9.7   Examples of NEP in extended systems

Our aim in this section is to present some examples of *nonequilibrium potential* for the kind of spatially extended systems we have studied in the previous sections. We will show a few examples for scalar as well as many components reaction–diffusion systems and, in the following section, exploit those results for the analysis of stochastic resonance in extended systems.

9.7.1   *Scalar reaction–diffusion model*

The specific model we shall focus here, with a known form of the nonequilibrium potential, corresponds to the same simple one-dimensional, one-component model of an electrothermal instability discussed earlier in this chapter. However, it can be considered as mimicking a broader class of bistable reaction–diffusion models. The particular, nondimensional form that we shall work with is

$$\partial_t T = \partial_{yy}^2 T - T + T_h \theta(T - T_c). \tag{9.58}$$

Here we analyze how the global stability of the patterns and the relative change in stability for this model were analyzed, as some b.c. parameter was changed and how those results depend on the threshold parameter.

For the sake of concreteness, we consider here a class of stationary structures $T(y)$ in the bounded domain $y \in (-y_L, y_L)$ with Dirichlet boundary conditions at both ends, $T(y = \pm y_L) = 0$. These are the spatially symmetric solutions to Eq. (9.58) already discussed in Sec. 9.3. Such structures can also be seen as a symmetrization of a set of stationary solutions to the ballast reaction–diffusion model in the interval $(0, y_L)$ with Neumann boundary condition at $y = 0$, namely, $dT/dy|_{y=0} = 0$.

The explicit forms of the stationary structures are given by Eq. (9.31a), and the double-valued coordinate y_c, at which $T = T_c$, is given by

$$y_c^{\pm} = \frac{1}{2} y_L - \frac{1}{2} \ln \left[z \cosh(y_L) \pm \sqrt{z^2 \cosh(y_L)^2 - 1} \right],$$

with $z = 1 - 2T_c/T_h$ $(-1 < z < 1)$.

When $y_c^{\pm}$ exists and $y_c^{\pm} < y_L$, the solution (9.31a) represents a structure with a central hot zone $(T > T_c)$ and two lateral cold regions $(T < T_c)$. For each parameter set there are two stationary solutions, given by the two values of y_c. In [Schat and Wio (1992)] it has been shown that the structure with the smallest hot region is unstable, whereas the other one is linearly stable. The trivial homogeneous solution $T = 0$ exists for any parameter set and is always linearly stable. These two linearly stable solutions are the only stable stationary structures under the chosen boundary conditions. Therefore, under suitable conditions, we have a bistable situation in which two stable solutions coexist, one of them corresponding to a cold–hot–cold (CHC) structure and the other one to the homogeneous trivial state. The unstable solution is always a CHC structure, with a relatively small hot region.

For the symmetric solution we are considering here, the nonequilibrium potential or Lyapunov functional (LF) reads [Izús *et al.* (1995)]

$$\mathcal{F}[T] = 2 \int_0^{y_L} \left\{ - \left(\int_0^T [-T' + T_h \theta(T' - T_c)]\, dT' \right) + \frac{1}{2}(\partial_y T)^2 \right\} dy.$$
(9.59)

Replacing (9.31a), we obtain the explicit expression

$$\mathcal{F}^\pm = -T_h^2 y_c^\pm z + T_h^2 \sinh(y_c^\pm) \frac{\cosh(y_L - y_c^\pm)}{\cosh(y_L)}.$$
(9.60)

For the homogeneous trivial solution $T(y) = 0$, instead, we have $\mathcal{F} = 0$.

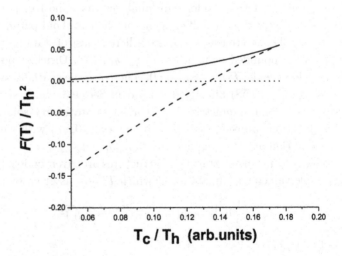

Fig. 9.16   Nonequilibrium potential $\mathcal{F}$, for the stationary patterns of the ballast resistor, as a function of T_c, for $L = 1$. The bottom curve corresponds to the stable pattern, while the top one corresponds to the unstable one. For the homogeneous pattern $T = 0$ we have $\mathcal{F} = 0$. The points T_c^*, correspond to the value at which the stable and the zero branch crosses.

In Fig. 9.16 we have plotted the LF $\mathcal{F}[T]$ as a function of $\phi_c = T_c/T_h$ for a fixed system size. The curves correspond to the inhomogeneous structures, $\mathcal{F}^{\pm}$, whereas the horizontal line stands for the LF of the trivial solution. The upper branch of each curve is the LF of the unstable structure, where $\mathcal{F}$ attains a maximum. At the lower branch and for $T = 0$, the LF has a local minimum. The curve exists up to a certain critical value of ϕ_c at which both branches collapse. It is possible to analyze the critical behavior, around this point (which we will not discuss here).

It is interesting to note that, since the LF for the unstable solution is always positive and, for the stable CHC structure, $\mathcal{F} < 0$ for $\phi_c \to 0$, and > 0 otherwise, the LF for this structure vanishes for an intermediate value of the parameter: $\phi_c = \phi_c^*$. At that point, the stable inhomogeneous structure and the trivial homogeneous solution interchange their relative stability. In fact, $T(y) = 0$ switches from being a metastable state, to being more stable than the inhomogeneous structure.

9.7.2   *Activator–inhibitor systems*

The importance of activator–inhibitor systems for applications in physics, chemistry and biology is by now very well established (see Sec. 9.5). Here we shall focus on a specific system belonging to this family of two-component models. We want to present an analysis of the global stability of stationary patterns exploiting the concept of nonequilibrium potential.

We start with a simplified (piecewise linear) version of the activator–inhibitor model sketched in Sec. 8.3, which preserves the essential features, and fix the parameters so as to allow for nontrivial solutions to exist. After scaling the fields, we get a dimensionless version of the model as:

$$\partial_t u(x,t) = D_u \partial_x^2 u - u + \theta[u - a] - v$$
$$\partial_t v(x,t) = D_v \partial_x^2 v + \beta u - \gamma v. \tag{9.61}$$

We confine the system to the interval $-L < x < L$ and impose Dirichlet boundary conditions on both extrema. According to the values of the parameters a, β and γ, we can have a monostable or a bistable situation. In the second case we have two homogeneous stationary (stable) solutions. One corresponds, in the (u,v) plane, to the point $(0,0)$ while the other is given by (u_0, v_0) with

$$u_0 = \frac{\gamma}{\beta + \gamma} \quad , \quad v_0 = \frac{\beta}{\beta + \gamma}$$

implying that the condition $\frac{\gamma}{\beta+\gamma} > a$ must be fulfilled. Without losing generality we may assume that $0 < a < 1/2$ and $u_0 < 2a$.

The inhomogeneous stationary patterns appear due to the nonlinearity of the system, and ought to have activated regions $(u > a)$ coexisting with non-activated regions $(u < a)$. This fact, together with the symmetry of the evolution equations and boundary conditions, implies the existence of symmetric inhomogeneous stationary solutions. We restrict ourselves to the simplest inhomogeneous, symmetric, stationary solutions. That is, a symmetric pattern consisting of a central region where the activator field is above a certain threshold $(u > a)$ and two lateral regions where it is below it $(u < a)$. As was already discussed, [Koga and Kuramoto (1980); Bouzat and Wio (1998)], different analytical forms (which are here linear combinations of hyperbolic functions) should be proposed for u and v depending on whether $u > a$ or $u < a$. These forms, as well as their first derivatives, need to be matched at the spatial location of the transition point, which we called x_c. Through that matching procedure and imposing boundary conditions we get the general solution for the stationary case. In order to identify the matching point x_c we have to solve the equation $u(x_c) = a$, resulting in general in a transcendental equation for x_c. In order to avoid the complications arising from the possible spatially oscillatory behavior of the solutions, we will work in a parameter range where the diffusion coefficient of the activator (D_u) is lower than some critical value (D_u^{osc}), [Koga and Kuramoto (1980); Ohta *et al.* (1989)], beyond which the solutions became spatially oscillatory. In particular there are up to four different solutions for x_c, and associated with each one we have different stationary solutions that we will indicate by u_{e1}, u_{e2}, u_{e3} and u_{e4}, with increasing values of the transition point x_c. A linear stability analysis of these solutions indicates that u_{e1} and u_{e3} are unstable while u_{e2} and u_{e4} are locally stable. The stable states will correspond to attractors (minima) of the functional while the unstable ones will be saddle points, defining the barrier height between attractors.

We now write the equations of our system specifying the time scale associated with each field. This allows us to perform an adiabatic approximation and obtain a particular form of the nonequilibrium potential for this system. Measuring the time variable on the characteristic time scale of the slow variable u (i.e.: τ_u), (9.61) adopt the form

$$\partial_t u(x,t) = D_u \partial_x^2 u(x,t) - u(x,t) + \Theta[u(x,t) - a] - v(x,t)$$

$$\eta \partial_t v(x,t) = D_v \partial_x^2 v(x,t) + \beta u(x,t) - \gamma v(x,t) \qquad (9.62)$$

where $\eta = \tau_v/\tau_u$. At this point we assume that the inhibitor is much faster than the activator (i.e. $\tau_v \ll \tau_u$). In the limit $\eta \to 0$, we can rewrite (9.62) as

$$\partial_t u(x,t) = D_u \partial_x^2 u(x,t) - u(x,t) + \Theta[u(x,t) - a] - v(x,t)$$
$$0 = D_v \partial_x^2 v(x,t) + \beta u(x,t) - \gamma v(x,t) \tag{9.63}$$

In the last pair of equations we can eliminate the inhibitor (now *slaved* to the activator) by solving the second equation using the Green function method

$$\{-D_v \partial_x^2 + \gamma\} G(x,x') = \delta(x - x')$$
$$v(x) = \beta \int dx' G(x,x') u(x'). \tag{9.64}$$

This slaving procedure reduces our system to a *nonlocal* equation for the activator only, having the form

$$\partial_t u(x,t) = D_u \partial_x^2 u(x,t) - u(x,t) + \Theta[u(x,t) - a]$$
$$-\beta \int dx' G(x,x') u(x'). \tag{9.65}$$

From this equation, and taking into account the symmetry of the Green function $G(x,x')$, we can obtain the Lyapunov functional for this system, which has the form

$$\partial_t u(x,t) = -\frac{\delta \mathcal{F}[u]}{\delta u}, \tag{9.66}$$

with

$$\mathcal{F}[u] = \int dx \left\{ \frac{D_u}{2} \{\partial_x u\}^2 + \frac{u^2}{2} - (u - a)\Theta[u - a] \right.$$
$$\left. +\frac{\beta}{2} \int dx' G(x,x') u(x') u(x) \right\}. \tag{9.67}$$

The spatial nonlocal term in the nonequilibrium potential takes into account the repulsion between activated zones. When two activated zones come near each other, the exponential tails of the inhibitor concentration overlap, increasing its concentration between both activated zones and creating an *effective* repulsion between them, the Green function playing the role of an exponential screening between the activated zones.

We can now exploit this LF in order to discuss the stability of the stationary solutions found earlier. According to the analysis done in [Drazer and Wio (1997)], we can see that, obtaining the "curvature" of the potential is equivalent to diagonalize the operator $\mathcal{F}_2[u_e]$ and finding its eigenvalues. Such an analysis is completely analogous to the linear stability one.

A complete stability analysis of the stationary solutions just found, as functions of the activator diffusivity, can be found in [Drazer and Wio (1997)].

9.7.3 *Three-component activator–inhibitor model*

Now we present an exact form of the nonequilibrium potential for a three-component system of the activator-inhibitor type (with one activator and two inhibitors) in a particular parameter region. Such a three-component system provides the adequate framework for a minimal model describing pattern formation in high-pressure or low-pressure chemical reactors [Bouzat and Wio (1998)]. The system we consider is described by the following set of equations

$$\frac{\partial u(x,t)}{\partial t} = D\nabla^2 u(x,t) + f(u(x,t)) - v(x,t) - w(x,t)$$

$$+ g_1^u \xi_1(x,t) + g_2^u \xi_2(x,t)$$

$$\epsilon_1 \frac{\partial v(x,t)}{\partial t} = \nu_1 \nabla^2 v(x,t) + \beta u(x,t) - \gamma v(x,t)$$

$$+ g_1^v \xi_1(x,t) + g_2^v \xi_2(x,t) \qquad (9.68)$$

$$\epsilon_2 \frac{\partial w(x,t)}{\partial t} = \nu_2 \nabla^2 w(x,t) + \beta' u(x,t) - \gamma' w(x,t)$$

$$+ g_1^w \xi_1(x,t) + g_2^w \xi_2(x,t),$$

where x represents an n-dimensional spatial coordinate. The $\xi_i(x,t)$'s are Gaussian white-noise sources of zero mean satisfying

$$\langle \xi_i(x,t)\xi_j(x',t')\rangle = \eta \delta_{ij}\delta(t-t')\delta(x-x'), \qquad (9.69)$$

where η is again a small parameter measuring the noise intensity. All the parameters and fields shall be considered as dimensionless (scaled) quantities. We shall only consider situations where the noise terms in the third of Eqs. (9.68) are negligible, and we set $g_1^w = g_2^w = 0$. Furthermore, we will analyze the system in the limit $\nu_1 = \epsilon_2 = 0$ with $\epsilon_1 = 1$ and $\nu_2 = \nu$, where for the now *temporally slaved inhibitor* w we have

$$w(x,t) = \int dx' G(x,x')u(x',t), \qquad (9.70)$$

where $G(x,x')$ is the Green function of the third of Eqs. (9.68) in the indicated limit [Lindenberg and Wio (2003)]. Hence the system can be reduced to an effective two-component system with a *nonlocal interaction* [Lindenberg and Wio (2003); Bouzat and Wio (1998)].

We may call $Q_u = (g_1^u)^2 + (g_2^u)^2$, $Q_v = (g_1^v)^2 + (g_2^v)^2$, and $Q_{uv} = g_1^u g_1^v + g_2^u g_2^v$. When the relation

$$\gamma = \frac{Q_u \beta + Q_v}{2Q_{uv}}, \tag{9.71}$$

holds, the equations of this effective two-component system adopt the form

$$\begin{pmatrix} \frac{\partial u(x,t)}{\partial t} \\ \frac{\partial v(x,t)}{\partial t} \end{pmatrix} = -\mathcal{T} \begin{pmatrix} \frac{\delta \Phi}{\delta u(x,t)} \\ \frac{\delta \Phi}{\delta v(x,t)} \end{pmatrix} + \begin{pmatrix} g_1^u \xi_1(x,t) + g_2^u \xi_2(x,t) \\ g_1^v \xi_1(x,t) + g_2^v \xi_2(x,t) \end{pmatrix}, \tag{9.72}$$

where the matrix $\mathcal{T}$ is given by

$$\mathcal{T} = \frac{1}{2} \begin{pmatrix} Q_u & 2Q_{uv} \\ 0 & Q_v \end{pmatrix} = \frac{1}{2} \begin{pmatrix} Q_u & Q_{uv} \\ Q_{uv} & Q_v \end{pmatrix} + \begin{pmatrix} 0 & \frac{Q_{uv}}{2} \\ -\frac{Q_{uv}}{2} & 0 \end{pmatrix} = \mathcal{S} + \mathcal{A}, \tag{9.73}$$

and the functional $\Phi[u(x,t), v(x,t)]$ by

$$\Phi[u(x,t), v(x,t)] = \int dx \tag{9.74}$$

$$\left[\frac{D}{Q_u} (\nabla u(x,t))^2 + V(u(x,t), v(x,t)) + \frac{1}{Q_u} \int dx' G(x,x') u(x,t) u(x',t) \right],$$

with

$$V(u,v) = -\frac{2}{Q_u} \int^u f(u') du' + \frac{2Q_{uv}\beta}{Q_u Q_v} u^2 + \frac{\gamma}{Q_v} v^2 - 2\frac{\beta}{Q_v} uv. \tag{9.75}$$

When the symmetric matrix S is positive definite (when $Q_u Q_v > Q_{uv}^2$ holds), the functional Φ, that in the associated deterministic system decreases monotonously with time, is the NEP of the system in Eqs. (9.72) satisfying the HJE, Eq. (7.44). Equation (9.71), which resembles a detailed balance condition, arises in this context as a mathematical constraint necessary for Φ, as defined by Eq. (9.74), to be the solution of the abovementioned HJE.

We limited the analysis to the parameter region where Eq. (9.71) is valid, the matrix S is positive definite, and hence the nonequilibrium potential is given by Eq. (9.74). Although these conditions impose restrictions on the range of application of our treatment, it is worth noting that, after choosing the values of the g_i^ν's satisfying the condition of positive definiteness, there is still a wide spectrum of interesting situations to analyze [Bouzat and Wio (1998)]. Furthermore the nonlinear function $f(u)$ is still arbitrary, opening a wealth of possibilities. When plotting Φ *vs.* ϕ_c, with Φ evaluated on the stationary patterns, we see a result similar to the one shown in Fig. 9.1.

Chapter 10

Noise-induced phenomena in extended systems

Whenever any action occurs in Nature, the quantity of action employed
by this change is the least possible.
Pierre M. de Maupertuis

10.1 Introduction

In this chapter we discuss some noise-induced phenomena that arise in spatially extended systems. Explicitly we will discuss stochastic resonance in extended systems and the so called *noise-induced phase transitions*. However, there are several other, and equally important, phenomena that we will not discuss here.

In the following sections we will show how to exploit the concept of *nonequilibrium potential* to analyze stochastic resonance in spatially extended systems, as well as a couple of (different) mechanisms that, due to the associated ergodicity breaking, lead us to a true nonequilibrium phase transitions.

10.2 Stochastic resonance in spatially extended media

10.2.1 *Scalar reaction–diffusion model*

Stochastic resonance in extended systems arises when considering, for instance, a large number of coupled resonant unities as those described in Sec. 8.4 [Gammaitoni *et al.* (1998)]. The description can be done adopting either a discrete or a continuous representation. Here we will work in a continuous version, more amenable to exploit the concept of *nonequilibrium potential* (NEP) as discussed in Secs. 7.4 and 9.7 [Wio *et al.* (2002); Wio and Deza (2007)].

In order to fix ideas, we will analyze a very simple continuous system where we have the expression of the NEP. It is the same model studied in Sec. 9.3, whose NEP was shown in Sec. 9.7. For sake of clarity we rewrite here the equations and some results.

The non dimensional version of the equation governing the system is

$$\frac{\partial}{\partial t}\phi(y,t) = \frac{\partial^2}{\partial y^2}\phi - \phi + \theta[\phi - \phi_c] + \xi(y,t), \tag{10.1}$$

where, respect to the form indicated in Sec. 9.3, we have included a spatial and temporal white noise source $\xi(y,t)$, with zero mean ($\langle \xi(y,t)\rangle = 0$) and correlation

$$\langle \xi(y,t)\xi(y',t')\rangle = 2\eta\delta(y-y')\delta(t-t').$$

In what follows, and for simplicity, we will assume Dirichlet boundary conditions in the interval $[-y_L, y_L]$.

For the symmetric solutions of such a system that we will consider here, the related form of the NEP is

$$\mathcal{F}[\phi] = 2\int_{-y_L}^{y_L}\left\{-\left(\int_0^\phi [-\psi' + \theta(\psi - \phi_c)]\,d\psi\right) + \frac{1}{2}(\partial_y\phi)^2\right\}dy. \tag{10.2}$$

In principle, in order to obtain information about the system's response to both, the noise and a modulation of a parameter, we need to evaluate the space-time correlation function $\langle \phi(y,t)\phi(y',t')\rangle$. However, the knowledge of the above indicated NEP allows us to follow a more simple and useful approach.

As was indicated in Chapter 9, there is a region of parameters where we find three solutions to the deterministic part of Eq. (10.1). Two of them correspond to stable patterns (as discussed there, ϕ_1 is the zero homogeneous state, while ϕ_2 is the non homogeneous one), and the third (ϕ_m, also non homogeneous) is unstable. The two stable patterns correspond to the two attractors in this high (infinity) dimensional space. We show in Fig. 10.1 the shape of such patterns.

The evaluation of the NEP in the stable and unstable patterns was also indicated in Chapter 9, but we reiterate such a picture in Fig. 10.2. In that figure it is apparent that the unstable pattern corresponds to a a kind of barrier or saddle point separating both attractors. Hence, we can exploit a simplified point of view, based on the two-state approach (see Sec. 8.4), which allows us to apply some known results almost directly. According to this, we can reduce the infinite dimensional system to a random system described by a discrete dynamical variable ϕ adopting two possible values: ϕ_1 and ϕ_2 (the two stable patterns), with probabilities $\Xi_{1,2}(t)$ respectively. Such probabilities satisfy the condition $\Xi_1(t) + \Xi_2(t) = 1$.

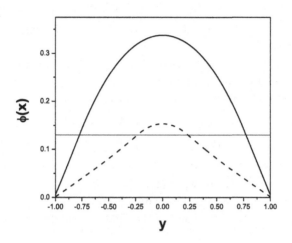

Fig. 10.1 Qualitative form of the patterns we are considering here, assuming Dirichlet boundary conditions.

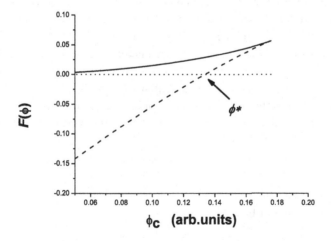

Fig. 10.2 Evaluation of the NEP on the patterns indicated in the previous figure, as a function of ϕ_c. The stable patterns are indicated by dotted (zero pattern) and dashed (non homogeneous pattern) lines while the continuous line indicates the NEP on the unstable (non homogeneous) pattern.

As indicated in Sec. 8.4, we can write the equation governing the evolution of $\Xi_1(t)$ (with a similar one for $\Xi_2(t) = 1 - \Xi_1(t)$) as

$$\frac{d\Xi_1}{dt} = -\frac{d\Xi_2}{dt} = W_2(t)\Xi_2(t) - W_1(t)\Xi_1(t)$$

$$= W_2(t) - [W_2(t) + W_1(t)]\Xi_1, \qquad (10.3)$$

where the $W_{1,2}(t)$ are the transition rates *out of* the spatially extended attractors $\phi = \phi_{1,2}$ states.

If the system is subject (through one of its parameters) to a time-dependent signal of the form $S(t) = A\cos(\omega_s t)$, up to first order in the amplitude (assumed to be small) the transition rates may be expanded as

$$W_1(t) \approx \mu_1 - \alpha_1 A\cos(\omega_s t)$$

$$W_2(t) \approx \mu_2 + \alpha_2 A\cos(\omega_s t), \qquad (10.4)$$

where the constants $\mu_{1,2}$ and $\alpha_{1,2}$ depend on the detailed structure of the system under study. Here we remark that the μ_i's, which are the (time independent) values of the W_i's without signal, are in general different from each other as a consequence of the different stability of the two states, and the same happens to the α_i's

$$\mu_i = W_i|_{S(t)=0} \quad ; \quad \alpha_i = -\left.\frac{dW_i}{dS(t)}\right|_{S(t)=0}. \qquad (10.5)$$

With the indicated modulation the system becomes non-stationary but we make an adiabatic assumption considering small signal frequencies that makes the NEP valid at each time for the corresponding value of the signal. Here we assume that the oscillating parameter is ϕ_c, and adopting the form $\phi_c(t) = \phi^* + A\cos(\omega_s t)$, where ϕ^* corresponds to the value at which the stability of both stable patterns is similar (indicated in the figure).

We can repeat the whole discussion of Sec. 8.4, and reach the expression for the signal-to-noise ratio (SNR), that up to the relevant (second) order in the signal amplitude A, results

$$\mathcal{SNR} = \frac{A^2\pi(\alpha_2\mu_1 + \alpha_1\mu_2)^2}{4\mu_1\mu_2(\mu_1 + \mu_2)}. \qquad (10.6)$$

In order to evaluate the transition rates between both states we consider a discrete space and field as

$$y \to y_i, \quad \phi(y) \to (\tilde{\phi}_1, \tilde{\phi}_2 \ldots, \tilde{\phi}_N),$$

and use the Kramers-like formula (see Sec. 8.2)

$$W_{\phi_i \to \phi_j} \equiv W_i = \frac{\lambda_+}{2\pi}\sqrt{\frac{\mathcal{F}_i''}{|\mathcal{F}_m''|}}\exp\left[-\frac{(\mathcal{F}_m - \mathcal{F}_i)}{\eta}\right], \qquad (10.7)$$

where η is the noise intensity, λ_+ is the unstable eigenvalue of the deterministic flux at the unstable state ϕ_m, $\mathcal{F}_i''$ and $\mathcal{F}_m''$ indicate the determinants of the matrix of second order derivatives of the NEP with respect to the discrete fields in the states ϕ_i and ϕ_m respectively, and $\mathcal{F}_i$ and $\mathcal{F}_m$ are the values of the NEP evaluated at the stationary states ϕ_i and ϕ_m, $i = 1, 2$. Finally, in order to compute the SNR as indicated above, we calculate the parameters μ_i and α_i numerically, accordingly to Eq. (10.5). The typical result is indicated in Fig. 10.3.

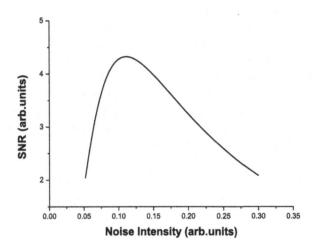

Fig. 10.3   Qualitative shape of the SNR as function of the noise intensity.

The whole analysis, as indicated in [Wio *et al.* (2002); Wio and Deza (2007)], gives a clear physical understanding to the numerical studies commented in [Gammaitoni *et al.* (1998)]. A comment worth to be made is that this kind of analysis is not restricted to one-dimensional systems, but is valid to any dimension where the dynamics consist in transitions among two stable patterns, through a saddle defined by the unstable pattern. In order not to make the presentation too cumbersome, in what follows we comment on the results of a few of possible extensions.

10.2.2   *Role of the NEP symmetry*

In [Wio *et al.* (2002); Wio and Deza (2007)] as well as in work cited therein, it was indicated that the maximum response, that is the maximum value of the maximum of the SNR curve, corresponds to the case where the stability of both extended stable states is similar. This fact was not only studied in a scalar system as discussed before, but also analyzing excitable systems like those discussed in the previous chapter.

The behavior of $SNR_{\max}$, that is the maximum of the curve SNR *vs.* noise intensity, as different model parameters are (not simultaneously) varied, show that SNR_{max} always increases with the symmetry of the potential. This fact led us to a very relevant result: the optimal values of the different model parameters (for instance diffusivity or threshold), as regards the maximization of $SNR_{\max}$, correspond to those making the potential more symmetric in each situation. Besides the analysis of the influence of symmetry on stochastic resonance, it is important to remark that the mere consideration of asymmetric situations has its own relevance. This is because such bistable asymmetric models provide, for example, the appropriate framework for describing SR in voltage–dependent ion channels, as proposed in some biological experiments. In those systems, the conducting state is associated to a higher-energy well than the non-conducting one ([Gammaitoni *et al.* (1998)] and references therein).

Hence, it is clear that in a technological application of this phenomenon, we cannot vary parameters at random in order to improve the system's response, but need to look for such a set of parameters that keeps the associated NEP more symmetrical.

10.2.3   *Enhancement due to selective coupling*

We can consider an extension of the one-component RD model previously discussed, but now assuming that the diffusive parameter depends on the field $\phi(x,t)$. As a matter of fact, since in the ballast resistor the thermal conductivity is a function of the energy density, the resulting equation for the temperature field includes a temperature-dependent diffusion coefficient in a natural way.

By adequate re-scaling of the field, space-time variables and parameters, we get a dimensionless time-evolution equation for the field $\phi(x,t)$

$$\partial_t \phi(x,t) = \partial_x \left(D(\phi)\partial_x\phi \right) + f(\phi) + \xi(x,t), \qquad (10.8)$$

where $\xi(x,t)$ is a white noise in space and time, and $f(\phi) = -\phi + \theta(\phi - \phi_c)$, $\theta(x)$ is the step function. All the effects of the parameters that keep the system away of equilibrium (such as the electric current in the electrothermal devices or some external reactant concentration in chemical models) are included in ϕ_c.

As was done before for the reaction term, a simple choice that retains however the qualitative features of the system is to consider the following dependence of the diffusion term on the field variable

$$D(\phi) = D_0(1 + h\,\theta[\phi - \phi_c]), \tag{10.9}$$

where h could be positive or negative. For simplicity, the same threshold ϕ_c for the reaction term and the diffusion coefficient could be chosen.

Let us assume that the system is limited to a bounded domain $x \in [-L, L]$ with Dirichlet boundary conditions at both ends, i.e. $\phi(\pm L, t) = 0$. As before, the piecewise-linear approximation of the reaction term in Eq. (10.8) was chosen in order to find analytical expressions for its stationary spatially-symmetric solutions. In addition to the trivial solution $\phi_0(x) = 0$ (which is linearly stable and exists for the whole range of parameters) we find another linearly stable nonhomogeneous structure $\phi_s(x)$—presenting an excited central zone (where $\phi_s(x) > \phi_c$) for $-x_c \leq x \leq x_c$—and a similar unstable structure $\phi_u(x)$, which exhibits a smaller excited central zone. The form of these patterns is analogous to what has been obtained previously. The difference is that in the present case $d\phi/dx|_{x_c}$ is discontinuous.

The indicated patterns are extrema of the NEP. In fact, the unstable pattern $\phi_u(x)$ is a *saddle-point* of this functional, separating the *attractors* $\phi_0(x)$ and $\phi_s(x)$. For the case of a field-dependent diffusion coefficient $D(\phi(x,t))$ as described by Eq. (10.8), the NEP reads [Wio *et al.* (2002); Wio and Deza (2007)]

$$\mathcal{F}[\phi] = \int_{-L}^{+L} \left\{ -\int_0^\phi D(\phi')f(\phi')\,d\phi' + \frac{1}{2}\left(D(\phi)\frac{\partial \phi}{\partial x}\right)^2 \right\} dx. \tag{10.10}$$

Given that $\partial_t \phi = -(1/D(\phi))\delta\mathcal{F}/\delta\phi$ one finds $d_t\mathcal{F} = -\int (\delta\mathcal{F}/\delta\phi)^2\,dx \leq 0$, thus warranting the LF property.

With the above indicated model and NEP we can proceed as before and study SR in this extended system. It was done (see [Wio *et al.* (2002); Wio and Deza (2007)]) and the results indicate that, when $h > 0$, such form of selective coupling could contribute to enhance the system's response. This is an very important aspect when designing some associated technological devices.

10.2.4 *Nonlocal coupling*

Another direction in which the present studies can be extended corresponds to the analysis of nonlocal couplings, as well as the competition between local and nonlocal couplings. Consider the same NEP described in the previous chapter, for the case of an activator–inhibitor system (u activator and v the inhibitor), but for the case of a the temporally slaved inhibitor. For this system the NEP $\mathcal{F}[u, v]$ has the form

$$\mathcal{F}[u(x,t)] = \int dx \left[\frac{D}{2} (\nabla u(x,t))^2 + C \int dx' G(x, x') u(x, t) u(x', t) \right],$$
(10.11)

that is, becomes only function of the activator field, and the coupling constant C, that in the usual activator-inhibitor system is positive (that is due to an inhibition interaction), could be assumed negative (that is an *excitatory interaction*)[Wio *et al.* (2002); Wio and Deza (2007)].

With the above indicated model we can repeat the same procedure as before in order to analyze the effect of a nonlocal coupling in the stochastic resonant response of a bistable extended system. However, we will not do that here, but only refer the reader to [Wio *et al.* (2002); Wio and Deza (2007)].

10.2.5 *Other cases*

More recently, other systems were treated within this formalism. Opposing to some claims for systems related to the kinetics of growing interfaces, to be discussed in a next chapter, like

> "*The KPZ equation is in fact a genuine kinetic equation describing a nonequilibrium process in the sense that the drift cannot be derived from an effective free energy. . .*"

in [Halpin-Healy and Zhang (1995)], such a nonequilibrium thermodynamic-like potential was shown to exist. Moreover, it was shown that several related systems are also included in such a description [Wio (2009)].

10.3　Noise-induced phase transitions

10.3.1 *Initial idea: Short-time instability*

In order to fix ideas about the phenomenon of *noise-induced phase transitions* (NIPT) we start discussing a simple model described by the following SDE

$$\dot{x} = -x + x\,\xi(t),$$
(10.12)

where $\xi(t)$ is a Gaussian white noise of intensity σ^2. This SDE should be interpreted in the sense of Stratonovich. Dividing by x, the equation adopts the form

$$\frac{\partial}{\partial t}[\ln x + t] = \xi(t).$$ (10.13)

The associated FPE, in terms of the variable $y = \ln x + t$, is

$$\frac{\partial}{\partial t} P(y, t) = \frac{\sigma^2}{2} \frac{\partial^2}{\partial y^2} P(y, t).$$ (10.14)

This diffusion-like equation can be exactly solved yielding, in terms of the original variable x,

$$P(x, t) = \frac{1}{x\sqrt{2\pi\sigma^2 t}} \exp\left[-\frac{(\ln x + t - \ln x_0)^2}{\sigma^2 t} \right].$$ (10.15)

This exact time dependent probability has the initial condition $P(x, t = 0) = \delta(x - x_0)$. In other words, x is log-normal distributed with $\langle \ln x \rangle = \ln x_0 - t$. Furthermore, for $t \to \infty$, one finds $P(x, t) \to 0$ for every finite value of x. We can conclude that $\lim_{t \to \infty} P(x, t) = \delta(x)$, in other words $x = 0$ is an *absorbing state*. This is in agreement with the formal solution of the steady state probability

$$P_{st}(x) \sim x^{-\frac{2}{\sigma^2} - 1}.$$ (10.16)

On the other hand, if we analyze the equation for the mean value we have

$$\langle \dot{x} \rangle = -[1 - \frac{2}{\sigma^2}]\langle x \rangle,$$ (10.17)

with the solution

$$\langle x \rangle = e^{-[1 - \frac{2}{\sigma^2}]t} \langle x(t = 0) \rangle.$$ (10.18)

This is an exact result that can also be obtained from Eq. (10.15). However, it is important to remark that $\langle x \rangle$ diverges for $t \to \infty$, whenever $\sigma^2 > 2$. This result seems to contradict the fact that the pdf converges to $\delta(x)$. The reason is that the average $\langle x \rangle$ involves an integration that is dominated by large but unlikely deviations. This divergence can be traced back to the opposite signs in the equations for $\langle x \rangle$ and for $P_{st}(x)$.

What here occurs as a mathematical peculiarity, acquires relevance when considering spatially distributed and coupled systems. However, it is the behavior of the first moment that explains the existence of noise induced phase transitions (NIPT). Let us formulate a spatially distributed version of Eq. (10.12). One of the simplest possibilities is to consider a lattice model, in each lattice point j with a stochastic dynamics given by one Eq. (10.12), and couple these elements by a "diffusion-like" coupling

$$\dot{x}_j = -x_j + x_j\, \xi_j(t) - \frac{D}{z} \sum_{k \in [j]} (x_j - x_k), \qquad (10.19)$$

where $k \in [j]$ indicates the z neighboring sites, D is the strength of the spatial coupling, and $\xi_j(t)$ are independent Gaussian white noises with intensity σ^2. In general there is no hope to find the general solution of such a set of equations, even for the stationary state. Hence, we focus in a *mean-field approximation* to solve this problem. It consist in replacing the values of x_j at neighboring sites by an averaged value j by $\langle x \rangle$, which has to be determined in a self-consistent way. With this Ansatz, the set of Eq. (10.19) is replaced by

$$\dot{x} = -x + x\, \xi(t) - D\,(x - \langle x \rangle), \qquad (10.20)$$

that has the stationary probability distribution

$$P_{st}(x) = N \exp\left(\int_0^x dy\, \frac{-y - \frac{\sigma^2}{2}y - D(y - \langle x \rangle)}{\frac{\sigma^2}{2}y^2} \right), \qquad (10.21)$$

that reduces to

$$P_{st}(x) = N x^{\frac{2+\sigma^2+2D}{\sigma^2}} \exp\left(-2\frac{D\langle x \rangle}{x\sigma^2} \right). \qquad (10.22)$$

Using this expression for $P_{st}(x)$ we obtain

$$\langle x \rangle = \langle x \rangle \frac{D}{1 + D - \frac{\sigma^2}{2}}. \qquad (10.23)$$

One concludes that $\langle x \rangle = 0$ is the unique solution for $\sigma^2 < 2$, while diverging solutions $\langle x \rangle = \pm\infty$ arise for $\sigma^2 > 2$. The surprising conclusion is that the instability of the first moment found in the scalar model for $\sigma^2 > 2$, produces a genuine phase transition (from an inactive $x_\nu = 0$ to an exploding state $x_\nu \to \infty$) for the spatially distributed version of the system, at least in a mean field approximation (however, this result is supported for numerical simulations in 2-d systems [Van den Broeck *et al.* (1997)]). Clearly, the addition of nonlinear terms (say x^3) will "temper" the transition leading to 1st or 2nd order transitions.

10.3.2 *NIPT due to a dynamical instability*

We present now a more elaborate model that, within the framework of both a mean-field approach and detailed numerical simulations, shows the phenomenon of NIPT. We briefly sketch here the approach from [Van den Broeck *et al.* (1994)].

The model under consideration, as introduced in [Van den Broeck *et al.* (1994, 1997); Mangioni *et al.* (1997)], is a d-dimensional extended system of typical linear size L is restricted to a hypercubic lattice of $N = L^d$ points, whereas time is still regarded as a continuous variable. The state of the system at time t is given by the set of stochastic variables $\{x_i(t)\}$ ($i = 1, \ldots, N$) defined at the sites $\mathbf{r}_i$ of this lattice, which obey a system of coupled ordinary SDE .

$$\dot{x}_i = f(x_i) + g(x_i)\eta_i + \frac{D}{2d} \sum_{j \in n(i)} (x_j - x_i) \qquad (10.24)$$

(along this section, the Stratonovich interpretation will be meant for the SDEs). Equation (10.24) is the discrete version of the *partial* SDE which in the continuum would determine the state of the extended system: we recognize in the first two terms the generalization of Langevin's equation for site i to the case of multiplicative noise (here η_i is a Gaussian white multiplicative noise acting on site x_i). The last term in Eq. (10.24) is nothing but the lattice version of the Laplacian $\nabla^2 x$ of the extended stochastic variable $x(\mathbf{r}, t)$ in a reaction–diffusion scheme. $n(i)$ stands for the set of $2d$ sites which form the immediate neighborhood of the site $\mathbf{r}_i$, and the coupling constant D between neighboring lattice sites is the diffusion coefficient.

The set of SDE in Eq. (10.24) have the following associated (multidimensional) FPE

$$\partial_t P([x_i], t) = \qquad (10.25)$$

$$\sum_i \frac{\partial}{\partial x_i} \left(-f(x_i) + \frac{D}{2d} \sum_{j \in n(x_i)} (x_i - x_j) + \frac{\sigma^2}{2} g(x_i) \frac{\partial}{\partial x_i} g(x_i) \right) P([x_i], t),$$

where $j \in n(x_i)$ indicate the neighbors to site x_i.

In order to solve this Eq. (10.25) (in an approximate mean-field approach) one repeats the approximate treatment done before. That is, we replace the values of x_j at neighboring sites by an averaged value $\langle x \rangle$, which has to be determined in a self-consistent way. Hence, Eq. (10.24) is replaced by

$$\dot{x} = f(x) + g(x)\eta + \frac{zD}{2d}(x - \langle x \rangle), \qquad (10.26)$$

with z the coordination number. The associated Fokker–Planck equation will adopt the reduced form

$$\partial_t P([x, \langle x \rangle], t) = \frac{\partial}{\partial x} \left(-f(x) + \frac{zD}{2d}(x - \langle x \rangle) + \frac{\sigma^2}{2} g(x) \frac{\partial}{\partial x} g(x) \right) P([x, \langle x \rangle], t),$$
(10.27)

with $\langle x \rangle$ defined through the self-consistency condition

$$\langle x \rangle = \int_{-\infty}^{+\infty} dx \; x \; P^{st}(x, \langle x \rangle) = F(\langle x \rangle). \tag{10.28}$$

Here $P^{st}(x, \langle x \rangle)$ is the stationary solution of Eq. (10.27) and is given by

$$P^{st}(x, \langle x \rangle) = Z^{-1} \exp \left[\int_0^x dy \frac{f(y) - \frac{2}{\sigma^2} g(y) g'(y) - D(y - \langle x \rangle)}{g(y^2)} \right], \tag{10.29}$$

being Z its norm.

For the specific example analyzed in [Van den Broeck *et al.* (1994)]

$$f(x) = -x(1 + x^2)^2,$$

and

$$g(x) = 1 + x^2.$$

In the case of $g(x) = 1$, that is additive noise, this model does not undergo a phase transition. However, it was shown that the simultaneous action of the multiplicative noise (with the choice of $g(x)$ indicated above) and the coupling induces the kind of short time instability discussed in the previous subsection, leading to a NIPT. In Fig. 10.4 we depict the phase diagram arising from the indicated mean-field scheme. A notable aspect is the reentrance phenomenon that is apparent when varying the noise intensity for a fixed value of the coupling.

In addition to the simple mean-field approach sketched here, a more elaborate scheme including correlations was done qualitatively in [Van den Broeck *et al.* (1994, 1997)], as well as numerical simulations confirming the existence of the NIPT (see the dashed line in Fig. 10.4). The simulations clearly show the evolution of the small clusters produced by the initial short time instability towards an ordered state.

It is worth here remarking that it is the simultaneous effect of multiplicative noise and coupling that yields the NIPT. If we have only coupling without fluctuations, or the opposite situation, that is fluctuations without coupling, there is no macroscopic effect: neither NIT nor NIPT.

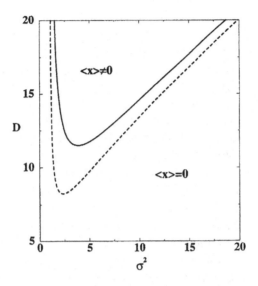

Fig. 10.4   Qualitative form of the phase diagram d vs σ^2 [taken from [Van den Broeck *et al.* (1994)]].

10.3.3   *Effect of colored noise*

The results shown before represent a major departure from what one can expect on the basis of equilibrium thermodynamics, according to which one should tend to think that as $D \to \infty$ an *ordered* situation is favored. Whereas that "intuitive" notion remains valid if the multiplicative noise that induces the nonequilibrium ordering phase transition is white [Van den Broeck *et al.* (1994, 1997)], it ceases to be so when the noise source has a finite correlation time $\tau > 0$.

Let us consider the same model as before but now η_i are *colored* multiplicative noises acting on site x_i. We assume a specific form for the noises $\{\eta_i\}$: Ornstein–Uhlenbeck noises, that is Gaussian distributed stochastic variables with zero mean and exponentially decaying correlations

$$\langle \eta_i(t)\eta_j(t')\rangle = \delta_{ij}(\sigma^2/2\tau)\exp(-|t-t'|/\tau). \tag{10.30}$$

They arise as solutions of an *uncoupled* set of Langevin SDE

$$\tau\dot{\eta}_i = -\eta_i + \sigma\xi_i \tag{10.31}$$

where the $\{\xi_i(t)\}$ are Gaussian white noises—with zero mean and δ-correlated: $\langle \xi_i(t)\xi_j(t')\rangle = \delta_{ij}\delta(t - t')$. For $\tau \to 0$, the Ornstein–Uhlenbeck noise $\eta_i(t)$ approaches the white-noise limit $\xi_i^W(t)$ with correlations $\langle \xi_i^W(t)\xi_j^W(t')\rangle = \sigma^2 \delta_{ij} \delta(t - t')$, which was the case studied in [Van den Broeck *et al.* (1994, 1997)].

The non-Markovian character of the process $\{x_i\}$ (due to the presence of the colored noises $\{\eta_i\}$) makes it difficult to study. However, some approximations have been devised which render an *effective Markovian* process that—whereas nicely simplifying the treatment—can still capture some of the essential features of the complete non-Markovian one. The UCNA is one of them [Jung and Hänggi (1987)].

In the white noise case, the results could be interpreted in terms of a "freezing" of the short-time behavior by a strong enough spatial coupling: for $D/\sigma^2 \to \infty$, the stationary probability distribution could be considered to be δ-like, just as the initial one. In the colored noise case, an analysis of the short-time behavior of the *order parameter* up to first order in τ reveals that the disordering effect of D can only be felt for *nonzero* self-correlation time. As τ increases, the minimum value of D required to stabilize the disordered phase decreases and the region in parameter space available to the ordered phase shrinks until it vanishes. Thus, the foregoing results can only be interpreted once we recall that the ordered phase arises from the cooperation of *two* factors: the presence of spatial coupling *and* the multiplicative character of the noise (which may eventually lead to "counterintuitive" results).

In short, the effect of colored noise is a shrinking of the ordered phase. We find a similar reentrance behavior as in the white noise case when varying noise intensity for a fixed value of the coupling, but now with a new reentrance for $\tau > 0$ when the noise intensity is fixed, and the coupling is varied (which is related to the reentrance found in [Castro *et al.* (1995)]).

10.3.4   *Another form of NIPT*

For some time, all arguments presented to account for the occurrence of NIPT have been dynamical ones (that is the existence of a short-time instability). However, in [Ibañes *et al.* (2001)] it was shown that, for some nonequilibrium model whose steady-state pdf and the associated effective potential are **exactly** known, it was possible to find a different mechanism leading to NIPT. We will not elaborate here on this new mechanism, but only briefly sketch the ideas.

Let us consider the following generic deterministic model for a real field $\phi(\vec{r}, t)$

$$\frac{\partial}{\partial t}\phi(\vec{r}, t) = -\Gamma[\phi(\vec{r}, t)]\frac{\delta V}{\delta\phi(\vec{r}, t)}, \qquad (10.32)$$

which corresponds to a relaxational flow [Wio (1997)] in a potential $V(\phi)$, with a field-dependent kinetic coefficient $\Gamma[\phi]$ [Hohenberg and Halperin (1977)].

When stochastic terms are introduced in such a description a fluctuation–dissipation relation is used [Ibañes *et al.* (2001)]. Following this approach, we define the Langevin equation (again in the Stratonovich interpretation) associated with the deterministic Eq. (10.32) as

$$\frac{\partial}{\partial t}\phi(\vec{r}, t) = -\Gamma[\phi(\vec{r}, t)]\frac{\delta V}{\delta\phi(\vec{r}, t)} + \Gamma[\phi(\vec{r}, t)]^{1/2}\,\xi(\vec{r}, t). \qquad (10.33)$$

We adopt that the noise is Gaussian with zero mean and correlation

$$\langle\xi(\vec{r}, t)\xi(\vec{r}', t')\rangle = 2\varepsilon\delta(\vec{r} - \vec{r}')\delta(t - t')$$

where ε is the noise intensity.

With the indicated conditions the stationary solution for $P^{\text{st}}([\phi])$, the pdf, has the Boltzmann's form

$$P^{st}([\phi]) \sim e^{-V_{\text{eff}}/\varepsilon}, \qquad (10.34)$$

where the effective potential has the form

$$V_{\text{eff}}([\phi]) \equiv V([\phi]) + \frac{\varepsilon_0}{2}\int d\vec{r}\,\ln\Gamma[\phi(\vec{r})]. \qquad (10.35)$$

Here ε_0 is a renormalized parameter, proportional to ε, which includes an ultraviolet cutoff (see [Ibañes *et al.* (2001)]).

Equation (10.35) defines the *exact nonequilibrium free energy* of the model in Eq. (10.33). We will now see how this class of systems can exhibit a NIPT. Let us focus in a particular system where V is described by a deterministic free energy potential with the usual Ginzburg–Landau form [Hohenberg and Halperin (1977)],

$$V([\phi]) = \int d\vec{r}\left(V_0[\phi(\vec{r})] + \frac{K}{2}|\nabla\phi(\vec{r})|^2\right), \qquad (10.36)$$

and (in order to stress the origin of the phenomenon) consider the monostable potential $V_0[\phi(\vec{r})] = \frac{a}{2}\phi(\vec{r})^2$, with $a > 0$. In order to simplify, we consider a system with a conserved order parameter, hence the kinetic coefficient Γ is only a function of the field and not of its derivatives [Hohenberg and Halperin (1977)]. In [Ibañes *et al.* (2001)] the authors choose

$$\Gamma[\phi] = [1 + c\phi(\vec{r})^2]^{-1}. \tag{10.37}$$

This corresponds to have larger fluctuations in dilute regions (lower ϕ) and smaller in dense ones (higher ϕ). For this particular case and according to Eqs. (10.34) and (10.35), the *local effective* potential (originally monostable) becomes bistable for $\varepsilon_0 > a/c$. Hence, we can expect that for K strong enough, a true NIPT towards an ordered state, controlled by ε, exists. This expectation can be checked performing a mean-field analysis.

Considering a spatially discrete version of the model, and following a similar procedure as used in the two previous sections, it is easy to find that the mean-field expression for the effective potential becomes

$$V_{\text{eff}}([\phi], M) = V_0([\phi]) + \frac{\varepsilon_0}{2} \ln \Gamma[\phi] + dK(\phi - M)^2, \tag{10.38}$$

where M is the averaged value of the order parameter to be determined self-consistently, and d the lattice dimension. As before, M is determined through

$$M = \frac{\int_{-\infty}^{\infty} \phi \, e^{-V_{\text{eff}}(\phi, M)/\varepsilon} d\phi}{\int_{-\infty}^{\infty} e^{-V_{\text{eff}}(\phi, M)/\varepsilon} d\phi}, \tag{10.39}$$

that requires a numerical solution. In Fig. 10.5 we show the solution of this equation for a two dimensional case. The results, indicated by a solid line, predicts a phase transition ($M = 0$ to $M \neq 0$) as the noise intensity ε is increased. This theoretical result was also confirmed with numerical simulations. Other aspects that we will not discuss here, like the divergence of the *generalized susceptibility*, or the fulfillment of some scaling laws are described in [Ibañes *et al.* (2001)].

As remarked in [Ibañes *et al.* (2001)], the dynamical system so introduced exhibits a nonequilibrium phase transition from a disordered to an ordered state as the intensity of the noise is increased. Hence, the system is disordered below a certain noise intensity and ordered above it, indicating that this is a possible scenario, within the context of SDE, for the appearance of a lower critical temperature in physical systems with an inverted phase diagram. Also, at variance with other NIPT previously reported, the physical mechanism inducing the transition is not a dynamical noise-induced short time instability. Rather, it is the the noise-induced bistability (NIT), and a balance between the deterministic forces and the stochastic ones that triggers the transition.

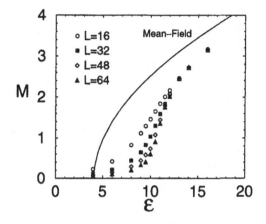

Fig. 10.5   Inverted phase diagram: order parameter M *vs.* noise intensity ε for $a = 1$, $c = 0.5$, $K = 1$. The continuous line corresponds to the mean-field result. Also indicated are the results of numerical simulations for different system sizes [taken from [Ibañes *et al.* (2001)]].

10.3.5   *Coupled ratchets: anomalous hysteresis*

The previously discussed short time mechanism for NIPT, was also exploited to analyze, through the symmetry breaking produced by such a transition, the possibility of the arising of a system of coupled motors [Reimann *et al.* (1999); Mangioni *et al.* (2001)]. This new twist to the problem of Brownian motors, implies to relax the requirement of a built-in bias: a system of periodically coupled nonlinear phase oscillators in a symmetric "pulsating" environment has been shown to undergo a NIPT, wherein the spontaneous symmetry breakdown of the stationary probability distribution gives rise to an *effective* ratchet-like potential [Reimann *et al.* (1999)]. The aforementioned mechanism and its striking consequences, such as the appearance of *negative zero-bias conductance* and *anomalous hysteresis*, have been introduced and illustrated through numerical simulations and explained by resort to the strong-coupling limit. By anomalous hysteresis we refer to the case where the cycle runs clockwise, in opposition to the normal one (as typified by a ferromagnet) that runs counterclockwise.

The indicated model (analogous to the one studied previously within NIPT) was addressed using an explicit mean-field approach, focusing on the relationship between the *shape* of the stationary probability distribution (as well as the *number of solutions* to the mean-field equations) and the transport properties in its different regions. In [Mangioni *et al.* (2001)], though a thorough exploration of the ordered phase, it was shown that within the ordered region there are subregions, with features related to the transition from anomalous to normal hysteresis in the behavior of the particle current as a function of the bias force. The main finding is that there exists a close relationship between the character of the hysteresis loop on one hand, and the shape of the stationary pdf as well as the number of "homogeneous" solutions on the other.

Even it being a very relevant aspect to the field of "nanomechanics" and more specifically to that of noise-induced transport or "Brownian motors", we will not extend further in this point but refer to the specific literature [Reimann *et al.* (1999); Mangioni *et al.* (2001)].

Chapter 11

Surface growth and kinetic roughening

What is the difference between theoretical physics and mathematical physics? Theoretical physics is done by physicists who lack the necessary skills to do real experiments; mathematical physics is done by mathematicians who lack the necessary skills to do real mathematics.

N. D. Mermin

11.1 Introduction

In recent years, an enormous amount of work has been devoted to the study of the dynamics of growing surfaces [Barabási and Stanley (1995); Krug and Spohn (1991); Meakin (1993); Halpin-Healy and Zhang (1995); Kardar (1998)]. Although in most cases the growth processes considered occur very far from equilibrium, it has been observed that the surface fluctuations exhibit scaling behavior similar to that found at second order transitions between equilibrium phases. To be precise, the resulting surface lacks any characteristic length or time scale (apart from those associated with the finite system size) and it is said to be a self-affine fractal object.

Typically, if $h(\mathbf{x}, t)$ denotes the surface height at a given position $\mathbf{x}$ we can rescale the lengths, $\mathbf{x} \to b\mathbf{x}$, and the surface height, $h \to b^\alpha h$, to obtain a surface that is statistically identical to the original one, i.e. $h \sim b^\alpha h(b\mathbf{x}, t)$. This can be also expressed more formally by saying that the surface is invariant under certain anisotropic scale transformation, or simply that we have a *scale-invariant* surface. The exponent α, for which scale invariance is recovered, is called the *roughness exponent* and provides a unique index to characterize the spatial profile of the surface. As we will see later on, the roughness exponent informs on how height–height correlations scale with distance and is also intimately linked to the intricate and fractal

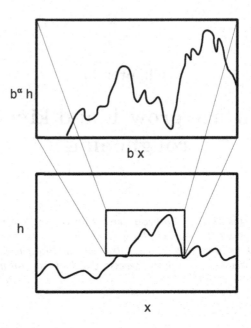

Fig. 11.1   Sketch of the scale invariance transform.

geometry of the height profile. Indeed, it can be shown that the surface fractal dimension is $D_f = d + 1 - \alpha$, where d is the substrate dimension.

A natural question to ask is how can a scale-invariant surface come into being from a random uncorrelated initial condition? We will see how the interplay between local interactions and random fluctuations, under certain generic conditions, is able to lead to the spread of correlations over the whole system and produce scale-invariant surfaces in a self-organized manner, i.e. without the fine tuning of external parameters. This phenomenon is known as kinetic roughening and occurs in a very generic way in a variety of systems including growth of thin-films, epitaxial growth, ion-beam sputtering, erosion, fracture, fluid-flow imbibition, and so on. Kinetic roughening can take place in physical systems that range at least 6 orders of magnitude in length scales. In this sense, it is a very universal phenomenon in nature.

11.2   Scaling of the surface width and critical exponents

In a typical situation an initially flat surface grows and roughens continuously as it is driven by some external noise. The noise term can be of thermal origin (like for instance fluctuations in the flux of particles in a deposition process), or a quenched disorder (like in the motion of driven interfaces through porous media). A rough surface can be characterized by the fluctuations of the height around its mean value. So, a basic quantity to look at is the *global* interface width, $W(L,t) = \langle \overline{[h(x,t) - \overline{h}]^2}\rangle^{1/2}$, where the overbar denotes average over all x in a system of size L and brackets denote average over different realizations. Rough surfaces then correspond to situations in which the stationary width $W(L, t \to \infty)$ grows with the system size.

Since there are no characteristic time or length scales in the system the surface width is expected grow in time as a power law $W(t,L) \sim t^\beta$. This transient regime will last until the asymptotic stationary regime is reached and the fluctuations of the surface will, therefore, saturate and become constant in time. At the stationary regime, since there is characteristic scale but the system size, the fluctuations should scale as $W(L) \sim L^\alpha$, where α is the roughness exponent introduced above. The path to the stationary regime is controlled by the correlation length $\xi(t) \sim t^{1/z}$, where z is the dynamical exponent. Akin to what we know from the dynamics near a critical phase transition, the correlation length informs about the spatial region of the system that has become correlated up to time t when started

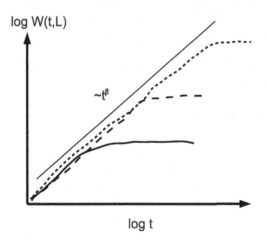

Fig. 11.2   Finite-size scaling of the global width $W(t, L)$ for increasing system sizes $L = L_0$, $2 \times L_0$, $4 \times L_0$. The time exponent $\beta = \alpha/z$ describes the transient regime before the fluctuations saturate. In the stationary state the width does not increase with time anymore and scales as $W_s(L) \sim L^\alpha$ with the system size.

from a random initial condition. Points separated at distances l below $\xi(t)$ are correlated, while regions that are far apart remain disconnected at time t. The stationary regime is then reached at time $t_s \sim L^z$, when the correlation length has reached the whole system size, $\xi(t_s) = L$.

This scaling behavior can be cast in the so-called Family–Vicsek dynamic scaling Ansatz [Family and Vicsek (1985); Barabási and Stanley (1995); Krug (1997)]

$$W(L,t) = t^{\alpha/z} f(L/\xi(t)). \tag{11.1}$$

The scaling function $f(u)$ behaves as

$$f(u) \sim \begin{cases} u^\alpha & \text{if } u \ll 1 \\ \text{const. if } u \gg 1 \end{cases}, \tag{11.2}$$

where α is the roughness exponent and characterizes the stationary regime, in which the horizontal correlation length $\xi(t) \sim t^{1/z}$ (z is the so called dynamic exponent) has reached a value larger than the system size L. The ratio $\beta = \alpha/z$ is called growth exponent and characterizes the short time behavior of the surface. As occurs in equilibrium critical phenomena, the corresponding critical exponents do not depend on microscopic details of the system under investigation. This has made it possible to divide growth processes into universality classes according to the values of these characteristic exponents [Barabási and Stanley (1995); Krug (1997)].

11.3 Scale-invariant surfaces, models, and universality classes

A very natural question for the theoretician is what is the model, ideally in the form of a stochastic partial differential equation, that describes a physical system in the thermodynamic/hydrodynamic limit. In particular, one can ask what are the continuous models that describe kinetic roughening of surfaces. Once we had a hydrodynamic description, we could ask what is the range of validity of the model or how the system states depend on the internal or external parameters. As we have seen along this book, in order to build continuous descriptions of a spatially extended system subject to fluctuations one can opt between two alternative and complementary approaches. On the one hand, the microscopic approach, where one builds the model theory from the basic interactions between the degrees of freedom at microscopic level (particle-particle or site-site). In roughening studies this corresponds to specifying the transition probability rates for the system to change from one microscopic state to other, say the probability rate for the system to increase the height at position i in one atom (or unit mass), $w(h_i \to h_i + 1 | \text{n.n})$, given the height states of nearest-neighbor (n.n) sites. This probability per unit time will depend on the details of the surface interactions (electromagnetic, mechanic,...) and the corresponding binding energies of the involved atoms. Needless to say that we use the term 'atom' here as a generalization of the elementary unit of our system at the level of description, being that grains, cells, elements of mass, or actual atoms, depending on the system we are studying. Once the probability rates are known from first principles we can write a master equation for the time evolution of the probability of the system state. From there, and using standard techniques we can obtain a Langevin equation for the surface evolution in time. This approach has several drawbacks. First, getting the final Langevin equation for $h(\mathbf{x}, t)$ involves difficult and long series of calculations that are only valid for the particular system under study. Second, several approximations are necessarily involved to cut the infinite series or to close the expansions in a self-consistent manner. The great advantage of this approach, however, is that the transport coefficients of the Langevin equation are obtained explicitly as functions of the microscopic parameters (binding energies, interaction potential and range, coupling constants, temperature,...). This is very useful to construct a model with predictive power if one is interested in quantitative values for crossover times or lengths, variation of transport coefficients with energy barriers, effect of different atomic species on roughness, and so on.

On the other hand, an alternative approach is possible in order to build a macroscopic Langevin description of kinetic roughening of surfaces with critical behavior. Since the system is critical we expect symmetries, conservation laws, and dimension to be sufficient to write down the basic model that describes the physics on the long wavelength limit, at hydrodynamic scales, and satisfies the symmetry requirements. Details of the local interactions are irrelevant provided the system exhibits critical behavior. The drawback in this case is of course that we do not know the dependence of macroscopic coefficients on the microscopic parameters. This is often called the Landau approach and it will be the one adopted here for our goal is to describe generic features of universality classes in kinetic roughening.

In the following subsections we introduce just a few important examples of stochastic partial differential equations that describe universality classes of scale-invariant surface growth process. The number of existing universality classes is larger and, for instance, other types of noise can also be needed to describe certain situations like quenched disordered systems, correlated thermal noise, power-law distributed (instead of Gaussian) noises, and so on. However, these models are out of the scope of the present introductory review and the interested reader can find further information in the references provided.

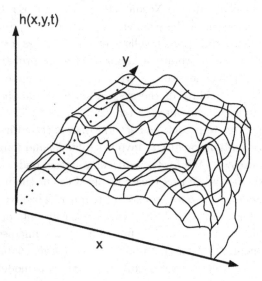

Fig. 11.3   Sketch of a random surface h in 2+1 dimensions.

Let us consider a surface in $d+1$ dimensions given by the height $h(\mathbf{x}, t)$ at position $\mathbf{x}$. The hydrodynamic description of the system is expected to be given by the Langevin-type of equation

$$\frac{\partial h}{\partial t} = \mathcal{G}(\mathbf{x}, h, \nabla h, \dots, t) + \eta(\mathbf{x}, t), \tag{11.3}$$

where the function $\mathcal{G}$ defines the particular model under consideration and incorporates the relevant symmetries and conservation laws. In particular, invariance under translation along the growth ($h \to h + h_0$) and substrate directions ($\mathbf{x} \to \mathbf{x} + \mathbf{x_0}$) as well as invariance in the choice of the time origin ($t \to t + t_0$) rule out an explicit dependence of $\mathcal{G}$ on h, $\mathbf{x}$ and t.

All these elementary symmetry considerations yield the most general model of growth as

$$\frac{\partial h}{\partial t} = \mathcal{G}(\nabla h, \cdots) + \eta(\mathbf{x}, t), \tag{11.4}$$

where the function $\mathcal{G}$ depends on ∇h, its derivatives, and combinations thereof. Moreover, the right–left symmetry must hold in general in isotropic systems with no net mass currents and this rules out terms that change sign with respect to the mirror transformation $\mathbf{x} \to -\mathbf{x}$. This, for instance, rules out terms like ∇h, $(\nabla h)^3$, and so on.

Growth is driven by an external noise $\eta(\mathbf{x}, t)$, which may represent for instance the influx of particles in a deposition processes. More generally the noise term represents all the fast-evolving small-scale degrees of freedom that are neglected in the hydrodynamic description, which focuses on the slowly varying long wavelengths. The noise is usually considered to be Gaussian, uncorrelated in space and time, and either non-conserved, $\langle \eta(\mathbf{x}, t)\eta(\mathbf{x}', t') \rangle = 2D\,\delta(\mathbf{x} - \mathbf{x}')\delta(t - t')$, or conserved, $\langle \eta_c(\mathbf{x}, t)\eta_c(\mathbf{x}', t') \rangle = -2D\,\nabla^2\delta(\mathbf{x} - \mathbf{x}')\delta(t - t')$, depending on the model under consideration.

11.3.1 *Random growth*

The simplest system one can think of is a surface growth model in which there are no interactions at all and every single site grows in an independent manner. This corresponds to Eq. (11.4) with $\mathcal{G} = 0$. The lack of lateral correlation leads to no roughening at all. The surface height at each site $\mathbf{x}$ grows randomly and the mean height is linear in time while the surface fluctuations scale as $W(t) \sim t^{1/2}$. No correlation means no sensing the finite system size and no stationary regime exist. This model is too trivial to describe any real complex process in nature.

11.3.2 *Edwards–Wilkinson equation*

The simplest nontrivial model of surface growth can be obtained by considering the first term in the expansion of $\mathcal{G}$ that is consistent with all the system's symmetries. Assuming no net currents are present, the lowest-order term is $\nabla^2 h$, which corresponds to surface diffusion. Physically, this term makes height differences of neighboring sites to be bounded—surface diffusion is a relaxation mechanism that smoothes out height fluctuations. So we arrive at the Edwards–Wilkinson (EW) [Barabási and Stanley (1995)] equation:

$$\frac{\partial h}{\partial t} = \nu\nabla^2 h + \eta(\mathbf{x}, t), \tag{11.5}$$

which appears in the context of thin-film growth by deposition of atoms, sandpiles, and the motion of depinned interfaces in disordered media. EW dynamics actually corresponds to the Langevin relaxation dynamics of the elastic energy $\mathcal{H} = \int d\mathbf{x}\sqrt{1 + (\nabla h)^2}$ at lowest order.

The EW model can be solved exactly by propagator methods simply by Fourier-transforming Eq. (11.5). However, it is instructive to see how can the critical exponents be obtained from a one-line calculation. This scaling approach shows the importance of scale-invariance in surface growth. If we rescale the variables as $\mathbf{x} \to b\mathbf{x}$, $h \to b^\alpha h$, and $t \to b^z t$ the EW equation becomes

$$\frac{\partial h}{\partial t} = \nu b^{z-2}\nabla^2 h + b^{-d/2+z/2-\alpha}\eta(\mathbf{x}, t). \qquad (11.6)$$

The existence of scale-invariant solutions for h implies that there is a pair of exponents α and z such that the equation remains invariant after rescaling. This leads to the scaling relations $z - 2 = 0$ and $-d/2 + z/2 - \alpha = 0$, which can be solved to arrive at $z = 2$ and $\alpha = (2 - d)/2$ for surfaces in dimension $d + 1$. These values, although obtained from nothing else than dimensional analysis, are exact because the model is linear. The model becomes mean-field (i.e. $\alpha = 0$) above a critical dimension $d_c = 2$.

11.3.3   *Kardar–Parisi–Zhang equation*

The EW equation is important as it appears as the lowest-order approximation of the elastic energy Hamiltonian in the small slopes limit $|\nabla h| \ll 1$. The first nonlinear correction is given by the most relevant term that would follow in the expansion in powers for $\mathcal{G}$. That would be the KPZ term $\lambda(\nabla h)^2$, which physically corresponds the surface growing in the direction locally normal to the surface at each point. The net result of this term is to produce shadowing effects and faster spread of correlations, thus changing the value of the dynamic exponent. The resulting dynamics is given by the well-known KPZ equation [Kardar *et al.* (1986)]:

$$\frac{\partial h}{\partial t} = \nu\nabla^2 h + \lambda(\nabla h)^2 + \eta(\mathbf{x}, t). \qquad (11.7)$$

KPZ is the simplest nonlinear equation that produces kinetic roughening. Even so, the dynamics of KPZ is terribly difficult to tackle analytically and a very great amount of effort has been devoted to finding exact or even approximate values for the critical exponents. Dimensional analysis fails in this case because the terms actually renormalize as the system size increases (so that the equation parameters scale in unknown fashion with system size). The flow of the parameters in the renormalization group

(RG) sense drives the system towards a nonperturbative strong-coupling fixed point, which makes it impossible the use of perturbative techniques.

In $d = 1$ KPZ exponents can be obtained exactly. On the one hand, KPZ dynamics satisfies Galilean invariance, $x \to x - \lambda t$, which leads to the scaling relation $z + \alpha = 2$ in any dimension. On the other hand, the Fokker–Planck equations associated with EW and KPZ are in $d = 1$ such that the stationary probability distribution of h for KPZ is the same as for EW, which is known exactly. This leads to the KPZ roughness exponent $\alpha = 1/2$ and, from the Galilean scaling relation, $z = 3/2$ in $d = 1$.

We do not know the critical exponents of KPZ in $d > 1$. There are, however, two popular conjectures that give estimates based solely on extensive numerical simulations. These conjectures, even if known to be ultimately incorrect, have been thought to be the closest approximation to the real values that we had available and, therefore, are worth to mention. On the one hand, Wolf and Kertész [Barabási and Stanley (1995)] proposed the expressions $\alpha = 1/(d + 1)$ and $z = (2d + 1)/(d + 1)$ based on simulation on the noise-reduced Eden model. On the other hand, Kim and Kosterlitz [Barabási and Stanley (1995)] conjectured $\alpha = 2/(d + 3)$ and $z = 2(d + 2)/(d + 3)$ based on numerical simulations of a discrete solid-on-solid growth algorithm. Even the existence of a critical dimension for KPZ is still under debate. The problem becomes further complicated since numerical simulations are affected by strong corrections to scaling, which mathematical form is still unknown.

The importance of the KPZ problem in the field of out-of-equilibrium statistical mechanics cannot be exaggerated. KPZ dynamics is of paramount importance to understand many dissipative spatially extended systems including solid-on-solid type atomic deposition growth, the Eden models, ballistic deposition models, slow combustion of paper, asymmetric exclusion process, polymers in random media, Burgers turbulence, the asymptotic behavior of the Kuramoto–Sivashinsky equation of chemical turbulence, dynamics of perturbations in space-time chaos, and so on (see [Barabási and Stanley (1995); Krug and Spohn (1991); Meakin (1993); Halpin-Healy and Zhang (1995)] and references therein).

11.3.4 *Mullins–Herring equation*

Under some circumstances surface diffusion becomes suppressed and the lowest-order term of $\mathcal{G}$ in the Langevin dynamics is not given by the Laplacian. This is particularly the case for ideal molecular-beam epitaxy (MBE) growth. Moreover, if equilibrium conditions apply, the next linear term in the expansion of $\mathcal{G}$ is expected to become the most relevant one. So we arrive at the so-called Mullins–Herring or linear MBE equation [Barabási and Stanley (1995)]:

$$\frac{\partial h}{\partial t} = -K\nabla^4 h + \eta(\mathbf{x}, t), \tag{11.8}$$

where K is a constant. Note the minus sign in front of the double Laplacian operator, which appears from the expansion of the elastic energy Hamiltonian (actually, a term $+K\nabla^4 h$ would be unstable). This relaxational dynamics can be seen as curvature diffusion in contrast to the EW dynamics that describes relaxation via surface diffusion. Being linear the model can be solved exactly and completely by Green propagator methods in Fourier space. If we are interested only in the critical exponents, those can be obtained by power-counting after rescaling as we did for the EW model and we get $\alpha = (3 - d)/2$ and $z = 4$, which are exact in any dimension.

11.3.5   Lai–Das Sarma–Villain equation

Again, the linear MBE model can be nonlinearly perturbed and the most relevant nonlinearity is given by $\nabla^2(\nabla h)^2$, which can be seen as a mass conserving KPZ term. The model reads:

$$\frac{\partial h}{\partial t} = -K\nabla^2 h + \lambda\nabla^2(\nabla h)^2 + \eta(\mathbf{x}, t). \tag{11.9}$$

This model has become important to describe nonlinear corrections to ideal MBE. Also here the nonlinear term leads to renormalization of the equation parameters as the system size is increased, so that power-counting does not produce the correct exponents. RG calculations up to one-loop lead to $\alpha = (4 - d)/3$ and $z = (8 + d)/3$, which due to symmetry are expected to be exact to all loops [Barabási and Stanley (1995)].

11.4   Anomalous roughening: Global versus local surface fluctuations

The existence of dynamic scaling as in (11.1), plus the assumption of the self-affine character of the interface—in the sense that there is no characteristic length scale in the surface besides the system size, and therefore all scales obey the same physics—allows one to obtain the exponents from the *local* width $w(l,t)$, which measures the surface height fluctuations over a window of size $l \ll L$. One may calculate other quantities related to correlations over a distance l as the height–height correlation function, $G(l,t) = \langle \overline{[h(x+l,t) - h(x,t)]^2} \rangle$, or the *local* width, $w(l,t) = \langle \langle [h(x,t) - \langle h \rangle_l]^2 \rangle_l \rangle^{1/2}$, where $\langle \cdots \rangle_l$ denotes an average over x in windows of size l. At times larger than $t_s(l) \sim l^z$ the local width is thus expected to saturate and

$$w(l, t \gg l^z) \sim l^{\alpha_{\mathrm{loc}}}, \tag{11.10}$$

where the *local* roughness exponent α_{loc} should equal the exponent obtained in (11.1), (11.2), $\alpha_{\mathrm{loc}} = \alpha$. For short times, $t \ll t_s(l)$, the local width scales with time as $w(l,t) \sim t^\beta$, which is the short scale analog of the scaling behavior for the global width given in (11.1). It is interesting to stress here that the equality $\alpha_{\mathrm{loc}} = \alpha$ in (11.10) is *not* guaranteed in general when Family–Vicsek scaling holds for the global width (11.2), since the self-affinity of the interface is an additional independent condition. This method to compute the critical exponents is very common in numerical simulations, but it is often the only practical way to measure the exponents in real experiments, where changing the system size is often impossible or at least unpractical.

Although the above scaling picture can be successfully applied to a great variety of models, which includes for instance the important universality classes associated to the EW and KPZ equations, it is not always valid. Many studies of discrete models and continuum equations show that in the case of the so-called *super-rough* surfaces, i.e. for surfaces with a global roughness exponent $\alpha > 1$, the usual assumption of the equivalence between the global and local descriptions of the surface is not valid. In these systems the local width (or equivalently the height–height correlation function) does not saturate as expected in the standard Family–Vicsek scaling, but crosses over to a new behavior in the intermediate time regime $l^z \ll t \ll L^z$, characterized by a different growth exponent κ, $w(l, t \gg l^z) \sim l\, t^\kappa$, where $\kappa = \beta - 1/z$. It is easy to see that now the local width saturates only when the global width does (this is, $t_s(l) = L^z$ for all l) and $w(l, t \gg L^z) \sim l L^{\alpha-1}$ in saturation, which, according to (11.10), yields a local exponent $\alpha_{\mathrm{loc}} = 1$ for super-roughening. This scaling behavior which is not encompassed by the Family–Vicsek Ansatz has been termed *anomalous scaling* in the literature [Barabási and Stanley (1995)] and later different models have been studied in which α_{loc} and κ take values different from 1 and $\beta - 1/z$, respectively.

Actually, growing surfaces can display anomalous scaling *no matter* what their value of the global roughness exponent is, either if $\alpha > 1$ or $\alpha < 1$. In the presence of anomalous scaling, not all length scales are equivalent in the system; the scaling behavior is different for short (local) and large length scales, hence α_{loc} and α differ.

This singular phenomenon was first noticed in numerical simulations of both continuous and discrete models of ideal molecular beam epitaxial growth. Anomalous roughening has later on been reported to occur in growth models in the presence of disorder, electrochemical deposition model, chemical vapor deposition, etc. Remarkably, anomalous roughening has also been reported in many experimental studies including molecular-beam epitaxy of Si/Si(111), sputter-deposition growth, electrodeposition, growth of superlattices, propagating fracture cracks, and fluid flow in porous media, among other systems.

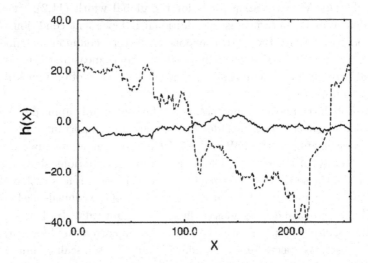

Fig. 11.4   Example of two fractal curves with the same local roughness exponent but different global one. One interface (solid line) has $\alpha = \alpha_{\mathrm{loc}} = 1/2$ and is a realization of the linear EW model in Eq. (11.5), and it is thus a true self-affine interface. The other curve (dashed line) has $\alpha_{\mathrm{loc}} = 1/2$ but $\alpha = 3/4$ and is a typical front of the random diffusion growth process exhibiting anomalous roughening [see [J.M. López and M.A. Rodríguez, Phys. Rev. E **54**, R2189 (1996)]].

The physical origin of anomalous roughening will be treated in detail in next Section. Now we will analyze the mathematical form of correlations and the corresponding scaling functions in general situations where α can be different from α_{loc}. The mechanisms that lead to anomalous scaling behavior can be separated into two classes, according to the behavior of the structure factor or power spectrum of the surface, $S(k, t)$, to be defined below. One of the mechanisms is super-roughening, even if $S(k, t)$ behaves as in the Family–Vicsek case. The other mechanism corresponds to what is called *intrinsic* anomalous scaling, where an independent exponent appears which measures the difference between the short and large length scale power laws, namely, the difference between the local and global roughness exponents α_{loc} and α. We will show how to identify the corresponding anomalous scaling by extracting the independent critical exponents from the correlation functions. For simplicity we restrict our further analysis to growth models in one dimension, but the conclusions are valid in arbitrary dimensions. Since a clear picture of the different scaling behaviors which can appear is crucial to correctly identify them in experiments and models, we believe a somewhat detailed presentation is in order.

As discussed above, in order to study a problem of kinetic roughening the height–height correlation function

$$G(l, t) = \langle \overline{[h(l + x, t) - h(x, t)]^2} \rangle, \qquad (11.11)$$

is frequently used. The average is performed over x running the whole system. This correlation function scales in the same way as the square of the local width, $G(l, t) \sim w^2(l, t)$, and provides an alternative method to determine the critical exponents.

However, as we will see in the following, a more transparent understanding of the complete dynamic scaling can be obtained by studying the Fourier transform of the interface height in a system of linear size L,

$$\widehat{h}(k, t) = L^{-1/2} \sum_x [h(x, t) - \overline{h}(t)] \exp(ikx), \qquad (11.12)$$

where the spatial average of the height has been subtracted. In this representation, the properties of the surface can be investigated by calculating the structure factor or power spectrum

$$S(k, t) = \langle \widehat{h}(k, t) \widehat{h}(-k, t) \rangle, \qquad (11.13)$$

which contains the same information on the system as the height–height correlation function $G(l, t)$ defined in (11.11), both of them being related by

$$G(l,t) = \frac{4}{L} \sum_{2\pi/L \leq k \leq \pi/a} [1 - \cos(kl)] S(k,t) \tag{11.14}$$

$$\propto \int_{2\pi/L}^{\pi/a} \frac{dk}{2\pi} [1 - \cos(kl)] S(k,t).$$

11.4.1　*Family–Vicsek scaling*

Family–Vicsek scaling reads, when expressed in terms of the structure factor,

$$S(k,t) = k^{-(2\alpha+1)} s_{FV}(kt^{1/z}), \tag{11.15}$$

with s_{FV} the following scaling function

$$s_{FV}(u) \sim \begin{cases} \text{const. if } u \gg 1 \\ u^{2\alpha+1} \text{ if } u \ll 1 \end{cases}, \tag{11.16}$$

Indeed, Eqs. (11.15), (11.16) can easily be seen to be equivalent to Eqs. (11.1), (11.2), by noting that the global width is nothing but the integral of $S(k,t)$, i.e.

$$W^2(L,t) = \frac{1}{L} \sum_k S(k,t) = \int \frac{dk}{2\pi} S(k,t), \tag{11.17}$$

where the momentum integral is limited to $2\pi/L \leq k \leq \pi/a$ and represents a continuum approximation to the sum over the discrete set of modes. In a discrete growth model, a is identified with the lattice spacing. Note as well that Eq. (11.16) implies that, for $kt^{1/z} \gg 1$, the spectrum does *not* depend on time, and hence at saturation ($t^{1/z} \gg L$), $S(k,t)$ is a pure power law independent of system size.

Going back to real space, and having assumed a Family–Vicsek scaling for $S(k,t)$, Eq. (11.16), one obtains, using Eq. (11.14),

$$G(l,t) = \int_{\frac{2\pi}{L}}^{\frac{\pi}{a}} [1 - \cos(kl)] \frac{s_{FV}(kt^{1/z})}{k^{2\alpha+1}} dk$$

$$= l^{2\alpha} \int_{\frac{2\pi l}{L}}^{\frac{\pi l}{a}} [1 - \cos(u)] \frac{s_{FV}(\rho u)}{u^{2\alpha+1}} du, \tag{11.18}$$

where we have introduced the ratio $\rho = t^{1/z}/l$. The limiting behaviors of (11.18) are very simple to study when $\alpha < 1$. For instance, at saturation

$t^{1/z} \gg L$, the scaling function in the integrand of Eq. (11.18) can be substituted for its (constant) limiting behavior at large arguments (see (11.16)), and one is left with $G(l,t) \sim l^{2\alpha}$, the limits $a \to 0$, and $L \to \infty$ having been taken. Analogously one can arrive at the local scaling given by

$$G(l,t) \sim \begin{cases} t^{2\alpha/z} & \text{if } t^{1/z} \ll l \\ l^{2\alpha} & \text{if } t^{1/z} \gg l \end{cases} = l^{2\alpha} \, g_{FV}(l/t^{1/z}), \qquad (11.19)$$

where the scaling function $g_{FV}(u)$ is *constant* for $u \ll 1$ and $g_{FV}(u) \sim u^{-2\alpha}$ for $u \gg 1$. This is the usual situation in which $\alpha < 1$ and the local and global roughness exponents are equal ($\alpha = \alpha_{\text{loc}}$ and $\kappa \equiv 0$).

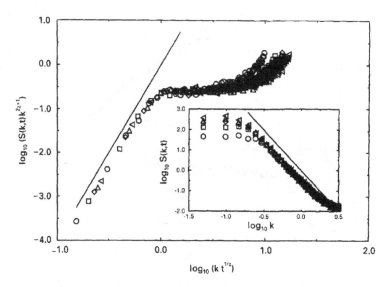

Fig. 11.5 Example of super-roughening scaling in Fourier space. The inset shows the structure factor for the linear MBE equation, Eq. (11.8), calculated in a system of total size $L = 128$ for times $t = 10^2$, 3×10^2, 5×10^2, 7×10^2, 9×10^2. The solid straight line has slope -4 and is plotted to guide the eye. In the main panel the data are collapsed using exponents $\alpha = 3/2$ and $z = 4$. The solid line has slope 4, showing that the scaling function of the structure factor is consistent with Eqs. (11.15) and (11.16).

11.4.2 *Super-roughening*

In the case of growth models generating super-rough surfaces ($\alpha > 1$), but with a *structure factor fulfilling Family–Vicsek scaling*, the integrals in (11.18) are divergent in the limit $l \ll t^{1/z}$ ($\rho \gg 1$) for $L \to \infty$, given the strong singularity at the origin of integration. Taking the limit $l \ll t^{1/z}$ first for fixed a, L, one obtains

$$G(l,t) \sim \begin{cases} l^2 \, t^{2(\alpha-1)/z} & \text{if } l \ll t^{1/z} \ll L \\ l^2 L^{2(\alpha-1)} & \text{if } l \ll L \ll t^{1/z} \end{cases}, \qquad (11.20)$$

so that $\alpha_{\text{loc}} = 1$ and $\kappa = \beta - 1/z$. In the early time regime $t^{1/z} \ll l \ll L$, $G(l,t) \sim t^{2\alpha/z}$. The fact that α_{loc} cannot exceed 1 for super-rough surfaces ($\alpha > 1$) is a purely geometric property which follows from definition (11.11).

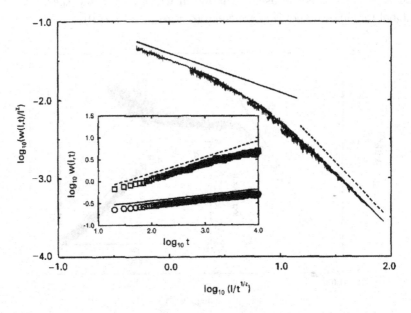

Fig. 11.6 Numerical results for the local width of the linear MBE equation in a system of size $L = 1000$. In the inset we plot $w(l,t)$ *vs.* time for two different window sizes $l = 5$ (o) and $l = 150$ ($\square$). The straight lines have slopes 0.13 (solid) and 0.37 (dashed), corresponding to κ and β respectively. Data are collapsed in the main panel for $l = 5$, up to $l = 150$ using exponents $\alpha = 3/2$ and $z = 4$. The straight lines correspond to $u^{-\alpha}$ (dashed) and $u^{-(\alpha-\alpha_{\text{loc}})}$ (solid) and have slopes $-3/2$ and $-1/2$ respectively. This leads to $\alpha_{\text{loc}} = 1$.

11.4.3 *Intrinsic anomalous roughening*

Apart from the super-roughening case, there is another important situation which leads to anomalous scaling. In some growth models the structure factor may present an unconventional scaling not described by Family–

Vicsek equation (11.16). Let us consider a dynamic scaling form for $S(k,t)$ as in Eq. (11.15) but with the scaling function $s_{FV}(u)$ replaced by

$$s_A(u) \sim \begin{cases} u^{2\theta} & \text{if } u \gg 1 \\ u^{2\alpha+1} & \text{if } u \ll 1 \end{cases}, \qquad (11.21)$$

where the label A denotes intrinsic anomalous spectrum. Here θ is a *new* exponent which "measures" the anomaly in the spectrum. In a system of size L, Eqs. (11.15), (11.21) hold only up to the saturation time $t_s(L) = L^z$, after which the system size L replaces the correlation length $t^{1/z}$ in all expressions. Thus in particular, at saturation the structure factor depends on the size of the system as $S(k,t) \sim L^{\theta} k^{\theta-(2\alpha+1)}$. As a consequence, the stationary spectrum shows severe finite size effects, to the extent that it is not defined in the thermodynamic limit $L \to \infty$.

A scaling behavior such as Eqs. (11.15), (11.21) for the structure factor does not affect the behavior of the *global* width, which preserves its Family–Vicsek form, $W(L,t) \sim t^{\beta}$ for $t \ll L^z$ and $W(L, t \gg L^z) \sim L^{\alpha}$. On the contrary, the *local* properties of the surface change dramatically if $S(k,t)$ scales as in (11.21). That can be seen by computing the height–height correlation function from Eq. (11.14),

$$G(l,t) = l^{2\alpha} \int_{\frac{2\pi l}{L}}^{\frac{\pi l}{a}} [1 - \cos(u)] \, \frac{s_A(\rho u)}{u^{2\alpha+1}} \, du \qquad (11.22)$$

which as before gives $G(l,t) \sim t^{2\beta}$ for times $t \ll l^z$. However, for intermediate times $l^z \ll t \ll L^z$ ($\rho \gg 1$) the integral (11.22) picks up a non-trivial contribution from the behavior of $s_A(u)$ at large arguments, so that

$$G(l,t) \sim l^{2\alpha} \rho^{2\theta} = l^{2\alpha-2\theta} t^{2\theta/z}$$

Thus, the complete scaling of the height difference correlation function (or, equivalently, the square of the local width) can be written as

$$G(l,t) \sim \begin{cases} t^{2\beta} & \text{if } t^{1/z} \ll l \ll L \\ l^{2\alpha_{\text{loc}}} t^{2\kappa} & \text{if } l \ll t^{1/z} \ll L \\ l^{2\alpha_{\text{loc}}} L^{2\theta} & \text{if } l \ll L \ll t^{1/z} \end{cases} = l^{2\alpha} g_A(l/t^{1/z}), \qquad (11.23)$$

where the local roughness exponent is $\alpha_{\text{loc}} = \alpha - \theta$, $\kappa = \theta/z = \beta - \alpha_{\text{loc}}/z$, and the scaling function $g_A(u)$ is *not* constant anymore for small arguments, but behaves as

$$g_A(u) \sim \begin{cases} u^{-2(\alpha-\alpha_{\text{loc}})} & \text{if } u \ll 1 \\ u^{-2\alpha} & \text{if } u \gg 1 \end{cases}. \qquad (11.24)$$

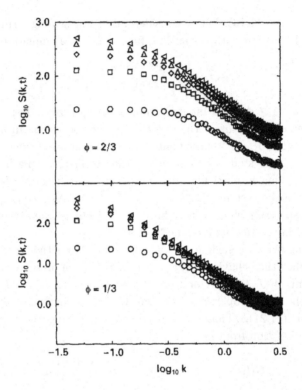

Fig. 11.7   Example of intrinsic anomalous scaling in Fourier space. Structure factor of the random diffusion model [see [J.M. López, M.A. Rodríguez, and R. Cuerno, Physica A **246**, 329 (1997)]] for two different degrees of the disorder and times $t = 10^2$, 3×10^2, 5×10^2, 7×10^2, 9×10^2. Upper panel, results for disorder parameter $\phi = 2/3$ (leads to $\alpha = 3/2$). Lower panel, data for $\phi = 1/3$ (leads to $\alpha = 3/4$). The shift in time is the fingerprint of the intrinsic anomalous character of the scaling. See [J.M. López and M.A. Rodríguez, Phys. Rev. E **54**, R2189 (1996)] for further details.

The fact that $\theta \neq 0$ in Eq. (11.21) yields a local roughness exponent $\alpha_{\mathrm{loc}} \neq \alpha$ and an anomalous growth exponent $\kappa \neq 0$. Therefore, there are now *three* independent exponents describing the scaling properties of the surface, whereas for Family–Vicsek scaling (even in the presence of super-roughening) there are only two. Which triplet of independent exponents one considers depends mainly on the kind of scaling properties one is measuring. In experiments local properties of the surface (local width, etc.) are normally measured, and one might naturally consider as independent exponents e.g. α_{loc}, z, and κ. On the other hand, the focus of numerical simulations is usually on finite-size effects, thus probing global properties of the surface such as the global width or power spectrum. In the latter cases a common choice is, e.g. α, θ, and β. We should stress that, although the scaling relations for exponents are exactly the same as for the super-roughening case discussed above, the present type of anomalous scaling is due to $\theta \neq 0$ and may and does occur in growth models with $\alpha < 1$. The scaling behavior (11.23) of the local width is formally equivalent to that of super-roughening (for which case $\alpha_{loc} = 1$ and $\kappa = \beta - 1/z$), a fact that produced some confusion in the earliest works where both anomalies were mistakenly identified somehow in early studies.

In view of the local scaling, Eq. (11.23), the structure factor can be rewritten conveniently as

$$
S(k,t) \sim \begin{cases} t^{(2\alpha+1)/z} & \text{if } kt^{1/z} \ll 1 \\ k^{-(2\alpha_{loc}+1)}t^{2(\alpha-\alpha_{loc})/z} & \text{if } kt^{1/z} \gg 1 \end{cases}, \tag{11.25}
$$

where we can observe two interesting facts that characterize what we call an *intrinsic anomalous scaling*. First, $S(k, t^{1/z} \gg k^{-1})$ decays as $k^{-(2\alpha_{loc}+1)}$, and *not* following the $k^{-(2\alpha+1)}$ law characteristic of Family–Vicsek scaling (super-roughening case included). Second, there is an unconventional dependence of $S(k, t^{1/z} \gg k^{-1})$ on time which leads to a non stationary structure factor for $t^{1/z}k \gg 1$. The combination of these two facts allows to distinguish between anomalous scaling due to super-roughening ($\alpha > 1$) and intrinsic anomalous scaling and has shown to be very useful in experimental studies.

11.5 Generic scaling Ansatz

Ramasco, López and Rodríguez [Ramasco *et al.* (2000)] introduced the most complete and general scaling description of scale invariance in surface growth. This theory includes all scaling forms presented above under a common scaling framework as follows. In order to explore the most general form that kinetic roughening can take, we study the scaling behavior of surfaces satisfying what Ramasco *et al.* called a *generic* dynamic scaling form of the correlation functions. They argued that a growing surface satisfies a generic dynamic scaling when there exists a correlation length $\xi(t)$, i.e. the distance over which correlations have propagated up to time t, and $\xi(t) \sim t^{1/z}$, being z the dynamic exponent. If no characteristic scale exists but ξ and the system size L, then power-law behavior in space and time is expected and the growth saturates when $\xi \sim L$ and the correlations

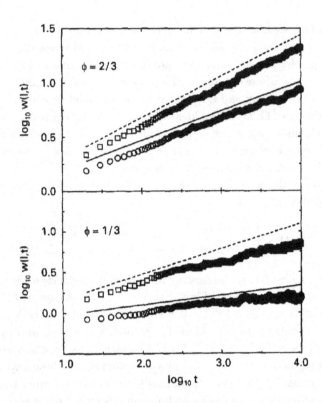

Fig. 11.8   Example of intrinsic anomalous scaling. Local width *vs.* time for the random diffusion model [see [J.M. López, M.A. Rodríguez, and R. Cuerno, Physica A **246**, 329 (1997)]] calculated over windows of size $l = 5$ (o) and $l = 150$ (□) for two different degrees of disorder. In the upper panel (for disorder parameter $\phi = 2/3$), the straight lines have slopes 0.27 (solid) and 0.38 (dashed), to be compared with the exact values $\kappa = 1/4$ and $\beta = 3/8$, respectively. In the lower panel ($\phi = 1/3$), the lines have slopes 0.12 (solid) and 0.31 (dashed), to be compared with $\kappa = 1/10$ and $\beta = 3/10$, respectively. Thus, the scaling is anomalous, i.e. $\kappa > 0$, no matter if the global roughness exponent is larger or smaller than one [see also [J.M. López and M.A. Rodríguez, Phys. Rev. E **54**, R2189 (1996)]].

(and from Eq. (11.14) also the structure factor) become time-independent. The global roughness exponent α can now be calculated in this regime from $G(l = L, t \gg L^z) \sim L^{2\alpha}$ (or $W(L, t \gg L^z) \sim L^{\alpha}$). In general, as we will see below, the scaling function that enters the dynamic scaling of the local width (or the height–height correlation) takes different forms depending on further restrictions and/or bounds for the roughness exponent values. These kind of restrictions are very often assumed and not valid for every growth model. For instance, only if the surface were *self-affine* saturation of the correlation function $G(l,t)$ would also occur for intermediate scales l at times $t \sim l^z$ and with the very same roughness exponent. However, the latter does not hold when anomalous roughening takes place as can be seen from Eq. (11.23).

Following Ramasco *et al.* [Ramasco *et al.* (2000)] the goal is to investigate *all* the possible forms that the scaling functions can exhibit when solely the existence of generic scaling is assumed. So, if the roughening process under consideration shows generic dynamic scaling (in the sense above explained), and no further assumptions (like for instance surface self-affinity or implicit bounds for the exponents values) are imposed, then Ramasco *et al.* proposed that the structure factor is given by

$$S(k,t) = k^{-(2\alpha+1)} s(kt^{1/z}), \tag{11.26}$$

where the scaling function has the general form

$$s(u) \sim \begin{cases} u^{2(\alpha-\alpha_s)} & \text{if } u \gg 1 \\ u^{2\alpha+1} & \text{if } u \ll 1 \end{cases}, \tag{11.27}$$

and the exponent α_s is called *spectral* roughness exponent. This scaling Ansatz is a natural generalization of the scaling proposed for the structure factor in the previous section for anomalous scaling.

In the case of the global width, one can make use of

$$W^2(L,t) = \frac{1}{L} \sum_k S(k,t) = \int \frac{dk}{2\pi} S(k,t), \tag{11.28}$$

to prove easily that the global width scales as in Eqs. (11.1) and (11.2), independently of the value of the exponents α and α_s.

However, the scaling of the local width is much more involved. The existence of a generic scaling behavior like (11.27) for the structure factor always leads to a dynamic scaling behavior,

$$w(l,t) \sim \sqrt{G(l,t)} = t^\beta g(l/\xi), \tag{11.29}$$

of the height–height correlation (and local width), but the corresponding scaling function $g(u)$ is not unique. When substituting Eqs. (11.27) and (11.26) into (11.14), one can see that the various limits involved ($a \to 0$, $\xi(t)/L \to \infty$ and $L \to \infty$) do not commute. This results in a different scaling behavior of $g(u)$ depending on the value of the exponent α_s.

Let us now summarize how all scaling behaviors reported in the literature are obtained from the generic dynamic scaling Ansatz (11.27). We shall also show how a new roughening dynamics naturally appears in this scaling theory. Two major cases can be distinguished, namely $\alpha_s < 1$ and $\alpha_s > 1$. On the one hand, for $\alpha_s < 1$ the integral in Eq. (11.14) can be easily computed and one gets

$$g_{\alpha_s < 1}(u) \sim \begin{cases} u^{\alpha_s} & \text{if } u \ll 1 \\ \text{const if } u \gg 1 \end{cases}. \tag{11.30}$$

So, the corresponding scaling function is $g_{\alpha_s < 1} \sim f_A$ and $\alpha_s = \alpha_{\text{loc}}$, i.e. the intrinsic anomalous scaling function in Eq. (11.23). Moreover, in this case the interface would satisfy a Family–Vicsek scaling (for the local as well as the global width) only if $\alpha = \alpha_s$ were satisfied for the particular growth model under study. Thus, the standard Family–Vicsek scaling turns out to be just one of the possible scaling forms compatible with generic scaling invariant growth, but it is not the unique solution compatible with *generic* scale invariance.

On the other hand, a new anomalous dynamics shows up for growth models in which $\alpha_s > 1$. In this case, one finds that, in the thermodynamic limit $L \to \infty$, the integral Eq. (11.14) has a divergence coming from the lower integration limit. To avoid the divergence one has to compute the integral keeping L fixed. We then obtain the scaling function

$$g_{\alpha_s > 1}(u) \sim \begin{cases} u & \text{if } u \ll 1 \\ \text{const if } u \gg 1 \end{cases}. \tag{11.31}$$

So that in this case one always gets $\alpha_{\text{loc}} = 1$ for any $\alpha_s > 1$. Thus, for growth models in which $\alpha = \alpha_s$ one recovers the super-rough scaling behavior.

However, it is worth noting that neither the spectral exponent α_s or the global exponent α are fixed by the scaling in Eqs. (11.27) and (11.31) and, in principle, they could be different. Therefore, growth models in which $\alpha_s > 1$ but $\alpha \neq \alpha_s$ could also be possible and represent a new type of dynamics with anomalous scaling. The main feature of this new type of anomalous roughening is that it can be detected only by determining the scaling of the structure factor. Whenever such a scaling takes place in the problem under investigation the new exponent α_s will only show up when analyzing the scaling behavior of $S(k, t)$ and will not be detectable in either $W(L, t)$, $w(l, t)$ or $G(l, t)$. In fact, as we have shown, the stationary regime of a surface exhibiting this kind of anomalous scaling will be characterized by $W(L) \sim L^\alpha$ and $w(l, L) \sim \sqrt{G(l, L)} \sim lL^{\alpha-1}$, however, the structure factor scales as $S(k, L) \sim k^{-(2\alpha_s+1)}L^{2(\alpha-\alpha_s)}$ where the spectral roughness

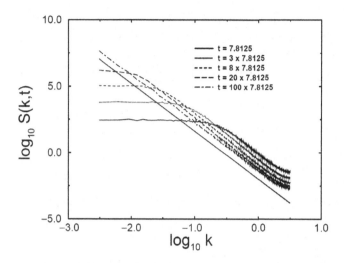

Fig. 11.9 Example of anomalous scaling for faceted growth. Structure factor of the Sneppen model for interface depinning at different times [J.J. Ramasco, J.M. López, M.A. Rodríguez, Phys. Rev. Lett. **84**, 2199 (2000)]. The continuous straight line is a guide to the eye and has a slope -3.7. Note the anomalous downwards shift of the curves for increasing times as compared with the intrisic anomalous case in Fig. 11.7.

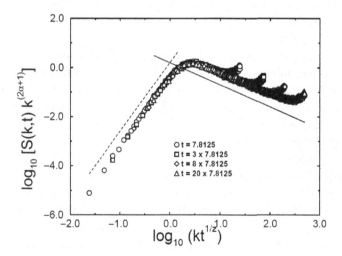

Fig. 11.10 Data collapse of the graphs in Fig. 11.9. The exponents used for the collapse are $\alpha = 1.0$ and $z = 1.0$. The straight lines have slopes -0.7 (solid) and 3.0 (dashed) and are a guide to the eye. The scaling function is given by Eq. (11.27) with a spectral roughness exponent $\alpha_s = 1.35 \pm 0.03$. The deviations from the scaling for large values of the argument $kt^{1/z}$ are due to the finite lattice spacing.

exponent α_s is a new and *independent* exponent. Ramasco *et al.* also showed [Ramasco *et al.* (2000)] that this new scaling class is associated with faceted surfaces.

One can then summarize generic scaling analytical results in the following useful table:

$$\begin{cases} \text{if } \alpha_s < 1 \Rightarrow \alpha_{\text{loc}} = \alpha_s \begin{cases} \alpha_s = \alpha \Rightarrow \text{Family–Vicsek} \\ \alpha_s \neq \alpha \Rightarrow \text{Intrinsic} \end{cases} \\ \text{if } \alpha_s > 1 \Rightarrow \alpha_{\text{loc}} = 1 \begin{cases} \alpha_s = \alpha \Rightarrow \text{Super-rough} \\ \alpha_s \neq \alpha \Rightarrow \text{Faceted growth} \end{cases} \end{cases} \tag{11.32}$$

This theoretical scheme has been often used to classify experiments as well as simulations.

11.6 Dynamics of the surface slope and the origin of anomalous scaling

A general analytical method to determine anomalous exponents from the continuum growth equations is still lacking. Therefore, most efforts have been focused on determining the anomalous critical exponents using simulation of discrete models or direct numerical solutions of the Langevin-type equations of growth.

In this section, we study the scaling approach proposed by López [López (1999)] to determine the scaling exponents of the local surface fluctuations in continuum growth models exhibiting anomalous kinetic roughening. This theory was an important step forward to solve the problem of anomalous kinetic roughening since it allows to understand the physical origin of the anomaly and predict when a particular growth model is expected to have anomalous properties ($\alpha \neq \alpha_{\text{loc}}$). In principle, it also allows one to give estimates of the local exponents. We illustrate the method by studying several growth equations with and without anomalous scaling.

The key observation of this theory is that anomalous scaling stems from the fact that the mean-square local slope has a non-trivial dynamics. A standard (Family–Vicsek) self-affine scaling of the local interface fluctuations, i.e. $\alpha = \alpha_{\text{loc}}$, implies that the square local slope $G(l = a, t) = \langle [h(\mathbf{x} + \mathbf{a}, t) - h(\mathbf{x}, t)]^2 \rangle$, where a is the lattice spacing, becomes $G(a, t) \sim$ constant very rapidly in time. One can also see that this constant must go to zero as $a \to 0$. However, as can be easily seen from Eq. (11.23), in growth models exhibiting anomalous scaling, the local slopes scale as

$$G(a,t) \sim t^{2\kappa}, \qquad (11.33)$$

where the exponent $\kappa = \beta - \alpha_{loc}/z > 0$. The existence of such a diverging mean local slope introduces a new correlation length in the growth direction, which enters the scaling of the local fluctuations of the height. Therefore, in any growth model, the existence (or absence) of anomalous scaling of the local height fluctuations can be investigated by computing the mean local slope $G(a,t)$. In the continuous limit, $a \to 0$, $G(a,t)$ can be written as $G(a,t) \simeq s(t)a^2$ where

$$s(t) = \langle \overline{(\nabla h)^2} \rangle. \qquad (11.34)$$

This corresponds to calculating the mean square local derivative of the interface height. In general, this quantity scales as a power law $s(t) \sim t^{2\kappa}$. Negative values of the exponent κ will result in a normal Family–Vicsek scaling of the local fluctuations with the same roughness exponent. On the contrary, for $\kappa > 0$ the correlation length $s(t)$ diverges in time and becomes a relevant length scale in the problem that changes the local scaling, as discussed above. In this case, anomalous scaling with a local roughness exponent

$$\alpha_{loc} = \alpha - z\kappa, \qquad (11.35)$$

is expected to occur. This scaling relation is a fundamental result of the theory since it tells us that the local roughness exponent α_{loc} is different from the global one α when the slopes have a diverging dynamics in time with exponent κ. This scaling relation has been tested many times in both experiments and numerical simulations. As we will see in the following, this simple observation allow us to find a general method to compute anomalous critical exponents.

Now, let us see how the scaling behavior of the mean local slope $s(t)$ can be obtained from the continuum growth equation. By applying the operator ∇ to both sides of the growth equation (11.4) one gets to an equation of motion for the local derivative $\Upsilon(\mathbf{x}, t) = \nabla h$ of the form

$$\frac{\partial \Upsilon}{\partial t} = \frac{\delta \mathcal{G}}{\delta \Upsilon} \nabla \Upsilon + \nabla \eta. \qquad (11.36)$$

In general, this Langevin equation may contain nonlinear terms that break the translational symmetry $h \to h + h_0$ in the growth direction. This implies that the resulting interface $\Upsilon(\mathbf{x}, t)$ may not be rough ($\alpha < 0$).

The global width of the interface $\Upsilon(\mathbf{x}, t)$ is given by

$$W_\Upsilon(t) = \langle \overline{\Upsilon(\mathbf{x}, t)^2} \rangle^{1/2} = \langle \overline{(\nabla h)^2} \rangle^{1/2} \tag{11.37}$$

where $\langle \overline{\nabla h} \rangle = 0$ has been used. This leads to the general result that $s(t) = W_\Upsilon^2(t)$ and the exponent κ of the average local interface slope $s(t) \sim t^{2\kappa}$ is given by the time exponent of the global width of the local derivative $\Upsilon(\mathbf{x}, t)$,

$$W_\Upsilon(t) \sim t^\kappa. \tag{11.38}$$

It then becomes clear that one could obtain the anomalous exponents of the interface $h(\mathbf{x}, t)$ by finding the time scaling behavior of the fluctuations of $\Upsilon = \nabla h$.

In the following we investigate the existence of anomalous scaling in several continuum growth models. In the examples that we analyze here, a Flory-type approach introduced by Hentschel and Family [Hentschel and Family (1991)] suffices to obtain the exponent κ from the corresponding Eq. (11.36) for every model in good agreement with existing simulation results. The Flory approach can be seen as a stability analysis of the equation of motion (11.36) for the corresponding local derivative $\Upsilon(\mathbf{x}, t)$ of the interface.

11.6.1   *Kardar–Parisi–Zhang equation*

The first system we examine is the noise-driven interface growth model given by the Kardar–Parisi–Zhang (KPZ) equation [Kardar *et al.* (1986)]

$$\frac{\partial h}{\partial t} = \nu \nabla^2 h + \lambda (\nabla h)^2 + \eta(\mathbf{x}, t). \tag{11.39}$$

This equation was originally introduced to describe ballistic deposition growth far from equilibrium. In $1 + 1$ dimensions the critical exponents $\alpha = 1/2$ and $z = 3/2$ can be calculated exactly. On the basis of much numerical work, it is well established that the KPZ equation does not exhibit anomalous roughening. Let us see how this result can be derived in dimension $1 + 1$ by use of López approach.

In this case, the growth equation for the local derivative, Eq. (11.36), reads

$$\frac{\partial \Upsilon}{\partial t} = \nu \frac{\partial^2 \Upsilon}{\partial x^2} + \lambda \frac{\partial}{\partial x} \Upsilon^2 + \frac{\partial \eta}{\partial x}, \tag{11.40}$$

in $1 + 1$ dimensions. From this growth equation the scaling behavior of the fluctuations of the interface $\Upsilon(x, t)$ can be obtained. We find that the width scales as $W_\Upsilon(t) \sim t^{-1/5}$, where the exponent can be obtained by a Flory-type approach [Hentschel and Family (1991)] as follows.

In the spirit of [Hentschel and Family (1991)], we assume that at long times $t \gg t_l$, and averaged over length scales l, the typical magnitude of the fluctuations in $\Upsilon(x, t)$ scale as Υ_l, and that these fluctuations last for times of the order t_l. The idea now is to estimate the magnitude of the individual terms. Basically for any equation such as Eq. (11.40) to show time scaling each separate term, when coarse-grained over length scales l, must be of the same order of magnitude or negligible. Only under these circumstances can scaling behavior arise. The various terms in Eq. (11.40) may be estimated as $\langle |\partial \Upsilon / \partial t| \rangle_l \sim \Upsilon_l / t_l$, $\langle |\partial^2 \Upsilon / \partial x^2| \rangle_l \sim \Upsilon_l / l^2$, $\langle |\partial \Upsilon^2 / \partial x| \rangle_l \sim \Upsilon_l^2 / l$. As for the noise, one can estimate its fluctuations on length scales l and time scales t_l as $\langle |\partial \eta / \partial x| \rangle_l \sim (l^3 t_l)^{-1/2}$ for smooth surfaces, while for rough surfaces $\langle |\partial \eta / \partial x| \rangle_l \sim (\Upsilon_l l^2 t_l)^{-1/2}$ (see [Hentschel and Family (1991)] for details). Note that the correct scaling of the noise term depends on the particular equation under study.

At sufficiently large length scales the nonlinear term in Eq. (11.40) will dominate the diffusion term, and so, equating the $\langle |\partial \Upsilon / \partial t| \rangle_l$ term with the nonlinear term we obtain that a typical fluctuation scales as $\Upsilon_l \sim l/t_l$. To proceed further we now equate our estimate for the noise fluctuation $(\Upsilon_l l^2 t_l)^{-1/2}$ to the inertial term, which gives $\Upsilon_l \sim (t_l/l^2)^{1/3}$. So, we can estimate that a fluctuation of Υ scales in time as $\Upsilon_l \sim t_l^{-1/5}$.

So, in the case of the KPZ equation in $1 + 1$ dimensions a negative exponent $\kappa = -1/5$ is found indicating a standard scaling as expected. A similar Flory-type computation also shows that, in fact, the KPZ model exhibits a self-affine scaling in $d + 1$ dimensions and $\kappa = -1/(4 + d)$.

A particular case of the KPZ equation is the linear interface growth model ($\lambda = 0$), first studied by Edwards and Wilkinson [Barabási and Stanley (1995)]. For this model Eq. (11.40) can be solved exactly and we obtain the exponent $\kappa = -1/4$, as it corresponds to a standard Family–Vicsek scaling.

11.6.2 *Surface growth with conservation law*

The KPZ equation does not conserve the total volume of the interface. The conserved version of KPZ was studied by Sun, Guo and Grant [Barabási and Stanley (1995)] and is given by

$$\frac{\partial h}{\partial t} = -K \nabla^4 h + \lambda \nabla^2 (\nabla h)^2 + \eta_c(\mathbf{x}, t), \tag{11.41}$$

where

$$\langle \eta_c(\mathbf{x}, t)\eta_c(\mathbf{x}', t')\rangle = -2D\nabla^2\delta(\mathbf{x} - \mathbf{x}')\delta(t - t'). \qquad (11.42)$$

Here the exponents are known with a good approximation in any dimension [Barabási and Stanley (1995)], in particular $\alpha = 1/3$ and $z = 11/3$ for $d = 1$ are exact. From the corresponding equation for the local derivative Eq. (11.36) and the noise fluctuations $\sim (l^3 t_l)^{-1/2}$, we obtain the scaling behavior $W_\Upsilon(t) \sim t^{-2/11}$ for the fluctuations of $\Upsilon(x, t)$. This result shows that the SGG equation has also a normal scaling of the local fluctuations of the height.

11.6.3   *Linear MBE model*

As a simple example of anomalous roughening, we now consider the linear model for MBE growth [Barabási and Stanley (1995)] given by

$$\frac{\partial h}{\partial t} = -K\nabla^4 h + \eta(\mathbf{x}, t), \qquad (11.43)$$

Despite its simplicity, this equation has played an important role in the theory of MBE. The critical exponents are easily calculated in any dimension, and in particular one has $\alpha = 3/2$ and $z = 4$ in dimension $d = 1$. The model exhibits super-rough interfaces ($\alpha > 1$). This leads to anomalous (super-rough) scaling. In fact, the local scaling is given by Eq. (11.23) with a local roughness exponent $\alpha_{\text{loc}} \simeq 1$.

In this case the growth equation for the local derivative

$$\frac{\partial \Upsilon}{\partial t} = -K\frac{\partial^4 \Upsilon}{\partial x^4} + \frac{\partial}{\partial x}\eta(x, t), \qquad (11.44)$$

is linear and can be easily solved. The Flory approach now gives the exact exponent κ. We can estimate the curvature diffusion term as $\langle |\partial^4\Upsilon/\partial x^4|\rangle_l \sim \Upsilon_l/l^4$, being the estimate for the noise term $\langle |\partial\eta/\partial x|\rangle_l \sim (l^3 t_l)^{-1/2}$ in this case. Equating fluctuations of each of the two terms on the right-hand side of Eq. (11.44) with the inertial term, we obtain the scaling behavior $W_\Upsilon(t) \sim t^{1/8}$ for the width of the local interface derivative. The positive value of the exponent κ means that the local slope $s(t)$ becomes a relevant correlation length in the problem and anomalous roughening is to be expected. The scaling relation (11.35) gives us an exact determination of the local roughness exponent $\alpha_{\text{loc}} = 1$ for this model in dimension $d = 1$. This is in good agreement with existing numerical results.

11.6.4 Lai–Das Sarma–Villain equation

The last example we study is the LDV [Barabási and Stanley (1995)] equation for MBE growth

$$\frac{\partial h}{\partial t} = -K\nabla^4 h + \lambda\nabla^2 (\nabla h)^2 + \eta(\mathbf{x}, t). \tag{11.45}$$

This equation is expected to describe the behavior of the long-wavelength fluctuations of the interface height in several discrete models of MBE growth [Barabási and Stanley (1995)]. Note that this equation differs from the SGG equation discussed above in the non-conserved character of the noise in this case.

According to a dynamical RG analysis of Eq. (11.45) the global exponents $\alpha = (4-d)/3$ and $z = (8+d)/3$ are expected to be exact to all loops. However, numerical simulations of the LDV equation and several discrete growth models in the same universality class have shown that the model exhibits anomalous scaling with a local roughness exponent $\alpha_{\mathrm{loc}} \simeq 0.7$ in $1+1$ dimensions.

Again, the existence of anomalous scaling in this model in $1+1$ dimensions can be investigated by use of the growth equation for the local derivative of the interface. In this case Eq. (11.36) reads

$$\frac{\partial \Upsilon}{\partial t} = -K\frac{\partial^4 \Upsilon}{\partial x^4} + \lambda\frac{\partial^3}{\partial x^3}(\Upsilon^2) + \frac{\partial}{\partial x}\eta(x, t). \tag{11.46}$$

A Flory approach can also be applied to this case to determine the scaling of $W_\Upsilon \sim t^\kappa$. As we did for the KPZ equation, we can estimate the terms in Eq. (11.46) as $\langle|\partial\Upsilon/\partial t|\rangle_l \sim \Upsilon_l/t_l$, $\langle|\partial^4\Upsilon/\partial x^4|\rangle_l \sim \Upsilon_l/l^4$, $\langle|\partial^3\Upsilon^2/\partial x^3|\rangle_l \sim \Upsilon_l^2/l^3$, and the noise being $\langle|\partial\eta/\partial x|\rangle_l \sim (\Upsilon_l l^2 t_l)^{-1/2}$ as in the KPZ case. Assuming that the nonlinear term dominates the curvature diffusion term at large scales, we obtain $W_\Upsilon \sim t^{1/11}$.

According to this the exponent $\kappa = 1/11$ is positive and, as a consequence, the LDV equation in $1+1$ dimensions is expected to display anomalous roughening. The local roughness exponent $\alpha_{\mathrm{loc}} = 8/11 \simeq 0.73$ is given by the scaling relation (11.35) with $\alpha = 1$ and $z = 3$ for $d = 1$. The estimate for α_{loc} is in excellent agreement with the numerical result $\alpha_{\mathrm{loc}} \simeq 0.73 \pm 0.04$ [Krug (1994)].

Chapter 12

Final Comments

And so, I missed my chance with one of the Lords of Life.
And I have something to expiate;
A pettiness.
D. H. Lawrence

The field of nonequilibrium phenomena has received many different names depending on the author or the school: *Synergetics* by H. Haken and collaborators [Haken (1978)], *Self-Organization Systems* by I. Prigogine and the *Brussels School* [Nicolis and Prigogine (1977)], while several other people know it as *Complex Systems*. Irrespective of the name one wants to use, it covers such a wide spectrum that it was clearly impossible to cover them all in this textbook and it was necessary to leave out some very interesting and important subjects. Our aim in this presentation was, as in the first edition, to introduce a minimum set of ideas and techniques that could provide a view of the theoretical framework used for the description of far from equilibrium phenomena, as well as to try to make obvious its relevance and its many applications in the most diverse areas of physics, chemistry and biology.

In this revised edition we have rearranged and enlarged the original material in four parts. A first one devoted to stochastic processes, where we have commented on the many reasons to justify the increasing interest in the study of fluctuations, indicating that fluctuations might be used as a source of additional information about the dynamics of the system and that they are at the origin of some macroscopic effects such as the appearance of *spatio-temporal patterns* in physical, chemical, or biological systems. There we presented a brief introduction to the techniques and methods of stochastic processes, because this is the most adequate framework to describe the temporal behavior of fluctuations.

The second part was devoted to study thermodynamics and kinetics near equilibrium. However we have started presenting a scheme leading to irreversible kinetic equations by discussing the *BBGKY* hierarchy. Afterwards we have introduced the by now common ideas and methods for treating systems away from equilibrium, although restricting ourselves to a small class of phenomena, essentially described by linear transport theory. We refer specifically to *Onsager's* ideas about *regression to equilibrium* and *linear response theory*, respectively. Within such a context we have discussed: the concept of the *departure from thermodynamic equilibrium*, as a general order parameter; the role played by steady states out of equilibrium by discussing the *minimum entropy production theorem*; and briefly discussed the so called *fluctuation theorem*. We followed by analyzing the properties of time dependent correlation functions and their role within the framework of *linear response theory*, and presented the *fluctuation–dissipation theorem*.

The third and fourth parts were devoted to introduce a mesoscopic approach appropriate for discussing systems *far from thermodynamic equilibrium*, that is systems, that far from being isolated, are submitted to strong external constraints such as energy or chemical reactive fluxes.

The third part corresponds to systems without spatial extension, where we have briefly discussed aspects of dynamical systems, introducing the idea of *nonequilibrium potential*. We have also discussed some examples of the so called *noise induced phenomena* (noise induced transitions, stochastic resonance, Brownian motors).

The fourth part starts with a discussion of instabilities and pattern formation, were are several aspects not only advanced but basic as well, that have not been touched upon. We have chosen to discuss some of the basic features of pattern formation within a reaction-diffusion approach, and to leave out of the present discussion some of the basic features of pattern formation in fluids that originate the well known Rayleigh-Bénard convection or Taylor-Couette flow. Instead, we have discussed a few examples regarding the application of the *nonequilibrium potential* to some typical reaction-diffusion systems. We continue with a discussion of *noise induced phenomena* but now in extended systems (stochastic resonance –exploiting the concept of nonequilibrium potential–, noise induced phase transitions, coupled ratchets). Finally, we have included a discussion on surface growth, kinetic roughening and associated scaling relations.

As commented before, the field is so vast that is impossible to cover it all in a single volume. Hence, several topics have not been included. For instance, regarding the subject of pattern formation we have not considered the theoretical framework that, near the onset of linear instability for weak nonlinearities, transforms the complicated nonlinear equations into some *universal forms* leading to *amplitude* or *phase equations*, making it possible to describe different forms of instabilities such as those of *Eckhaus, Oscillatory, Zig-Zag*, etc. Also, we have made no reference to the derivation and application of higher order equations, neither have we discussed the effect of boundaries on pattern selection and propagation, nor the description of defects due to perturbations or boundary effects.

Among several other subjects we have left out is the whole area of *chaos*, and the very interesting problem of diffusion in disordered media, and its description within the framework of the *Continuous-Time Random-Walk*, that allows to make explicit the interconnection between spatial and time disorder, as well as problems of *population settling-down in fluctuating media.*

The list is very long, but, as we commented earlier, the size of a textbook intended for a one semester course makes it impossible to include them all. We can say that the subjects we have discussed are in some sense the *tip of the iceberg* representing the large body of subjects conforming this field.

The subjects so far studied as well as the kind of phenomena they are intended to describe, illustrates the *hierarchical structure* lying in statistical physics as a whole, which is common feature in human sciences, but not usual in physical sciences. Depending on the scale in which we are interested, the laws of physics differ: on a microscopic scale we have electrons and nuclei interacting through Coulomb forces, while on a macroscopic scale we meet, for instance in a gas or a solid, an order which is not apparent from the microscopic laws. For instance, the discovery that matter far from equilibrium acquires new properties, typical of such non-equilibrium situations, came as a surprise. However, the existence of such a hierarchy does not imply that it is enough to apply the most *fundamental* concepts to build up the others. Each level has its own conceptual structures, laws and methodology. In relation with this, let us recall a phrase from Norbert Wiener:

> *"One of the most interesting aspects of the world is that it can be considered to be made up of patterns. A pattern is essentially an arrangement. It is characterized by the order of the elements of which it is made rather than by the intrinsic nature of these elements."*

Nonetheless, we hope that both the material included and the way of presenting it will offer a feeling of and attract attention to, the many interesting aspects of this field. If this textbook could awake the curiosity of only a few students and induce them to delve deeper into one or another of the different aspects of the field, or to devote themselves to do research in one of its many facets, its writing will be justified.

Bibliography suggested in the first edition

the idea still exists that a book should not reveal things; a book should,
simply, help us discover those things.
Jorge Luis Borges

The bibliography of this second edition contains the references (most of them nonexistent by the time of the first edition). Instead, most of the bibliography suggested in the original version was not referred to in the text. It aimed at indicating relevant, pedagogic and helpful bibliography for the different topics introduced in the text. This included

- a compilation of very general textbooks in statistical physics (mainly those forming the core of standard courses in *equilibrium* statistical thermodynamics) as a prerequisite for the present material,
- specialized textbooks that cover partially the subjects presented here, and delve deeper into those subjects or complement them,
- tutorial review articles covering the individual themes,
- articles from physics journals that stress the pedagogical side, and introduce interesting problems and examples.

Since we believe it still plays the role for which it was originally designed, we have decided to keep most of the original bibliography apart (we have just erased duplicated items). As declared in the first edition, "*The aim was not to be complete, but to indicate what, in the author's opinion, seems to be the most pedagogical bibliography.*" We must, however, apologize to all those authors whose work was (unintentionally) omitted.

a) GENERAL TEXTBOOKS:

Balescu, R. (1975). *Equilibrium and Nonequilibrium Statistical Mechanics* (Wiley, New York).

Balian, R. (1991). *From Microphysics to Macrophysics*, vol.I and II (Springer-Verlag, Berlin).

Chandler, D. (1987). *Introduction to Modern Statistical Mechanics* (Oxford U.P., Oxford, UK).

Feynman, R. P. (1972). *Statistical Mechanics* (Benjamin, Reading, MA).

Goodstein, D. (1985). *States of Matter* (Dover Pub. Co., New York).

Huang, K. (1963). *Statistical Mechanics* (Wiley, New York).

Isihara, A. (1971). *Statistical Physics* (Acad. Press, New York).

Klimontovich, Yu. L. (1986). *Statistical Physics*, (Harwood A.P., New York).

Lifshitz, E. M. and Pitaevskii, L. P. (1980). *Statistical Physics*, Landau and Lifshitz Course of Theoretical Physics, Vol.9 (Pergamon, Oxford, UK).

Ma, S. K. (1982). *Statistical Mechanics* (World Scientific, Singapore).

Reichl, L. E. (1980). *A Modern Course in Statistical Physics* (Univ. Texas Press, Austin, TX).

Reif, F. (1965). *Fundamentals of Statistical and Thermal Physics* (McGraw–Hill, New York).

Toda, M., Kubo, R. and Saito, N. (1985). *Statistical Physics I* (Springer-Verlag, Berlin).

b) SPECIALIZED TEXTBOOKS:

Forster, D. (1975). *Hydrodynamic Fluctuations, Broken Symmetry and Correlation Functions* (W. A. Benjamin, New York).

de Groot, S. R. and Mazur, P. (1962). *Non-Equilibrium Thermodynamics* (North-Holland, Amsterdam).

Haken, H. (1978). *Synergetics, an Introduction*, 2nd Ed. (Springer-Verlag, New York).

Keizer, J. (1987). *Statistical Thermodynamics of Nonequilibrium Processes* (Springer-Verlag, Berlin).

Kreuzer, H. J. (1984). *Nonequilibrium Thermodynamics and its Statistical Foundations* (Clarendon P., Oxford, UK).

Kubo, R., Toda M. and Hashitsume, N. (1985). *Statistical Physics II* (Springer-Verlag, Berlin).

Lifshitz, E. M. and Pitaevskii, L. P. (1981). *Physical Kinetics*, Landau and Lifshitz Course of Theoretical Physics, Vol.10, (Pergamon, Oxford, UK).

Nicolis, G. and Prigogine, I. (1977). *Self-Organization in Nonequilibrium Systems* (Wiley, New York).

Pippard, A. B. (1985). *Response and Stability* (Cambrige U.P., Cambridge, UK).

van Kampen, N. G. (1982). *Stochastic Processes in Physics and Chemistry* (North-Holland, Amsterdam).

Risken, H. (1983). *The Fokker–Planck Equation* (Springer-Verlag, Berlin). See also the Series: *Studies in Statistical Mechanics*, published by North-Holland, Amsterdam, since 1959, and edited by J. de Boer and G. R. Uhlenbeck in its origin, and more recently by J. L. Lebowitz and E. W. Montroll.

c) BOOKS, REVIEWS & DIDACTIC ARTICLES:

Prigogine, I. (1980). *From Being to Becoming* (W.H. Freeman, San Francisco).

Zeldovich, Ya. B., Ruzmaikin, A. A. and Sokoloff, D. D. (1990). *The Almighty Chance* (World Scientific, Singapore).

Hänggi, P. and Thomas, H. (1982). Stochastic Processes, Time Evolution Symmetries and Linear Response, *Phys. Rep.* **88**, 207

Hänggi, P., Talkner, P. and Borkovec, M. (1990). Reaction-Rate Theory: Fifty Years After Kramers, *Rev. Mod. Phys.* **62**, 251.

Kac, M. and Logan, J. (1979). Fluctuations, in *Studies in Statistical Mechanics*, vol.VII, Eds. E. W. Montroll and J. L. Lebowitz (North-Holland, Amsterdam).

Montroll, E. W. and West, B. J. (1979). On an Enriched Collection of Stochastic Processes, in *Fluctuation Phenomena*, Eds. E. W. Montroll and J. L. Lebowitz, (North-Holland, Amsterdam).

Van den Broeck, C. (1985). The master equation and some applications in physics, in *Stochastic Processes Applied to Physics*, Eds. L. Pesquera and M. A. Rodríguez (World Scientific, Singapore).

Weiss, G. H. and Rubin, R. J. (1983). Random Walks: Theory and Selected Applications, *Adv. Chem. Phys.* (Eds. I. Prigogine and S. A. Rice), **52**, 363.

Bernardini, C. (1989). Statistical treatment of effusion of a dilute gas, *Am. J. Phys.* **57**, 1116.

Doering, Ch. (1991). Modelling Complex Systems : Stochastic Processes, Stochastic Differential Equations and Fokker–Planck Equations, in *Lectures in Complex Systems*, vol.III, Eds. L. Nadel and D. Stein, (Addison–Wesley, New York).

Güemez, J., Velasco, S. and Calvo Hernández, A. (1989). A generalization of the Ehrenfest urn model, *Am. J. Phys.* **57**, 828.

de la Peña, L. (1980). Time evolution of the dynamical variables of a stochastic system, *Am. J. Phys.* **48**, 1080.

Raposo, E. P., Oliveira, S. M., Nemirovsky, A. M. and Coutinho-Filho, M. D. (1991). Random walks: a pedestrian approach to polymers, critical phenomena and field theory, *Am. J. Phys.* **59**, 633.

Cercignani, C. (1975). *Theory and Applications of the Boltzmann Equation* (Scottish Ac. Press, Edinburgh).

Chapman, S. and Cowling, T. G. (1970). *The Mathematical Theory of Non-Uniform Gases*, (Cambridge U.P., Cambridge, UK).

Resibois, P. and de Leener, M. (1977). *Classical Kinetic Theory of Fluids* (Wiley, New York).

Syros, C. (1978). The Linear Boltzmann Equation: Properties and Solutions, *Phys. Rep.* **45**, 211.

Ter Haar, D. (1961). Theory and Applications of the Density Matrix, *Rep. Prog. Phys.* **24**, 304.

Baker, G. L. (1986). A simple model of irreversibility, *Am. J. Phys.* **54**, 704.

Eggarter, T. P. (1973). A comment on Boltzmann's H-theorem and time reversal, *Am. J. Phys.* **41**, 874.

Hansen, J.-P. (1978). Correlation functions and their relationship with experiments, in *Microscopic Structure and Dynamics of Liquids*, Eds. J. Dupuy and A. J. Dianoux (Plenum, New York).

Chung, S. H. and Stevens, J. R. (1991). Time-dependent correlation functions and the evaluation of the strtched exponential or Kohlrausch–Watts function, *Am. J. Phys.* **59**, 1024.

Berne, B. J. and Pecora, R. (1976). *Dynamic Light Scattering* (Wiley, New York).

Lenk, R. (1977). *Brownian Motion and Spin Relaxation* (Elsevier, Amsterdam).

Zubarev, D. N. (1974). *Nonequilibrium Statistical Thermodynamics* (Consultant Bureau, New York).

Li, Ke-Hsue (1986). Physics of open systems, *Phys. Rep.* **134**, 1.

Stinchcombe, R. B. (1978). Kubo and Zubarev formulations of response theory, in *Correlation Functions and Quasiparticle Interactions in Condensed Matter*, Ed. J. Woods Halley (Plenum, New York).

Cooper, C. A. and Davies, H. T. (1972). Momentum autocorrelation function of noninteracting particles in a box, *Am. J. Phys.* **40**, 972.

Lottici, P. P. (1978). Momentum autocorrelation function from classical Green's functions, *Am. J. Phys.* **46**, 507.

Matthews, J. and Nicolet, M. A. (1976). Current correlation function derived from a model based on Brownian motion, *Am. J. Phys.* **44**, 448.

Hu, B. Yu-Kuang (1993). Simple derivation of a general relationship between imaginary- and real-time Green's correlation functions, *Am. J. Phys.* **61**, 457.

Lefever, R. (1985). Noise induced transitions in nonequilibrium systems, in *Stochastic Processes Applied to Physics*, Eds. L. Pesquera and M. A. Rodríguez (World Scientific, Singapore).

Nicolis, G. and Van den Broeck, C. (1984). Stochastic Theory of Transition Phenomena in Nonequilibrium Systems, in *Nonequilibrium Cooperative Phenomena in Physics and Related Fields*, Ed. M. G. Velarde (Plenum, New York).

Suzuki, M. (1981). Passage from an Initial Unstable State to a Final Stable State, *Adv. Chem. Phys.* **46**, 195.

Thomas, H. (1978). Instabilities and Fluctuations in Systems far from Thermodynamic Equilibrium, in *Noise in Physical Systems*, Ed. D. Wolff (Springer-Verlag, New York).

Hänggi, P. and Riseborough, P. (1982). Dynamics of nonlinear dissipative oscillators, *Am. J. Phys.* **51**, 347.

Mielczarek, E. V., Turner, J. S., Leiter, D. and Davis, L. (1983). Chemical clocks: experimental and theoretical models of nonlinear behavior, *Am. J. Phys.* **51**, 32.

van Kampen, N. G. (1986). Stochastic behavior in nonequilibrium systems, *Helv. Phys. Acta* **59**, 896.

Weaver, D. L. (1982). Exact analytical stochastic model for first-order nonequilibrium phase transitions, *Am. J. Phys.* **50**, 1038.

Fife, P. C. (1979). *Mathematical Aspects of Reacting and Diffusing Systems* (Springer-Verlag, Berlin).

Kuramoto, Y. (1984). *Chemical Oscillations, Waves, Turbulence* (Springer-Verlag, Berlin).

Mikhailov, A. S. and Loskutov, A. Yu. (1992). *Foundations of Synergetics II*, (Springer-Verlag, Berlin).

Murray, J. D. (1989). *Mathematical Biology* (Springer-Verlag, Berlin).

Walgraef, D. (1988). *Structures Spatiales loin de l'equilibre* (Mason, Paris).

Cross, P. C. (1988). Theoretical Methods in Pattern Formation in Physics, Chemistry and Biology, in *Far From Equilibrium Phase Transitions*, Ed. L. Garrido (Springer-Verlag, Berlin).

Fife, P. C. (1984). Current Topics in Reaction-Diffusion Systems, in *Nonequilibrium Cooperative Phenomena in Physics and Related Fields*, Ed. M. G. Velarde (Plenum, New York).

Meinhardt, H. (1992). Pattern Formation in Biology: A Comparison of Models and Experiments, *Rep. Prog. Phys.* **55**, 797.

Mikhailov, A. S. (1989). Selected Topics in Fluctuational Kinetics, *Phys. Rep.* **184**, 307-374.

Newell, A. C. (1989). The Dynamics and Analysis of Patterns, in *Complex Systems*, Ed. D. Stein, (Addison-Wesley, New York).

Normand, Ch., Pomeau, Y. and Velarde, M. G. (1977). Convective Instability: A Physicist's Approach, *Rev. Mod. Phys.* **49**, 581.

Velarde, M. G. (1982). Dissipative Structures and Oscillations in Reaction-Diffusion Models with or without Time-Delay, in *Stability of Thermodynamic Systems*, Ed. J. Casas-Vázquez and G. Lebon, (Springer-Verlag, Berlin).

Albano, A. M., Abraham, N. B., Chyba, D. E. and Martelli, M. (1984). Bifurcations, propagating solutions, and phase transitions in a nonlinear chemical reaction with diffusion, *Am. J. Phys.* **52**, 161.

Titus, W. J. (1989). A one-dimensional realization of a general model of cluster-cluster aggregation, *Am. J. Phys.* **57**, 1131.

Bibliography

Barabási, A.-L. and Stanley, H. E. (1995). *Fractal Concepts in Surface Growth* (Cambridge U.P., Cambridge, UK).

Boksenbojm, E., Wynants, B. and Jarzynski, C. (2010). Nonequilibrium thermodynamics at the microscale: Work relations and the second law, *Physica A* **389**, pp. 4406–4417.

Bouzat, S. and Wio, H. S. (1998). Nonequilibrium potential and pattern formation in a three component reaction-diffusion system, *Phys. Lett. A* **247**, p. 297.

Bulsara, A. R., Chilleni, S., Kiss, L., McClintock, P. V. E., Mannella, R., Marchesoni, F. V., Nicolis, G. and Wiesenfeld, K. (eds.) (1995). *Nuovo Cim. D*, Vol. 17.

Bustamante, C., Liphardt, J. and Ritort, F. (2005). The nonequilibrium thermodynamics of small systems, *Physics Today* **58**, p. 43.

Castellano, C., Fortunato, S. and Loreto, V. (2009). Statistical physics of social dynamics, *Rev. Mod. Phys.* **81**, pp. 591–646.

Castro, F., Sánchez, A. and Wio, H. S. (1995). Reentrance phenomena in noise induced transitions, *Phys. Rev. Lett.* **75**, pp. 1691–.

Crooks, G. E. (1998). Nonequilibrium measurements of free energy differences for microscopically reversible markovian systems, *J. Stat. Phys.* **90**, p. 1481.

Cross, P. C. and Hohenberg, P. C. (1993). Pattern formation outside of equilibrium, *Rev. Mod. Phys.* **65**, pp. 851–1112.

de la Lama, M. S., Szendro, I. G., Iglesias, J. R. and Wio, H. S. (2006). Van kampen's expansion approach in an opinion formation model, *Eur. Phys. J. B* **51**, pp. 435–442.

de la Lama, M. S., Szendro, I. G., Iglesias, J. R. and Wio, H. S. (2007). Erratum: Van kampen's expansion approach in an opinion formation model [eur. phys. j. b **51**, 435–442 (2006)], *Eur. Phys. J. B* **58**, p. 221.

Drazer, G. and Wio, H. S. (1997). Nonequilibrium potential approach: Local and global stability of stationary patterns in an activator–inhibitor system with fast inhibition, *Physica A* **240**, pp. 571–.

Duderstadt, J. J. and Martin, W. R. (1979). *Transport Theory* (Wiley, New York).

Esposito, M. and Van den Broeck, C. (2010). Three detailed fluctuation theorems, *Phys. Rev. Lett.* **104**, p. 090601.

Evans, D. J. and Searles, D. (2002). The fluctuation theorem, *Adv. Phys.* **51**, pp. 1529–1585.

Family, F. and Vicsek, T. (1985). Scaling of the active zone in the eden process on percolation networks and the ballistic deposition model, *J. Phys. A: Math. Gen.* **18**, pp. L75–L81.

Forster, D. (1975). *Hydrodynamic Fluctuations, Broken Symmetry and Correlation Functions* (Benjamin, New York).

Frank, T. D. (2005). *Nonlinear Fokker–Planck Equations* (Springer, Berlin).

Gammaitoni, L., Hänggi, P., Jung, P. and Marchesoni, F. (1998). Stochastic resonance, *Rev. Mod. Phys.* **70**, pp. 223–287.

Gardiner, C. W. (1985). *Handbook of Stochastic Methods*, 2nd edn. (Springer-Verlag, Berlin).

Gingl, Z., Kiss, L. B. and Moss, F. (1993). Non-dynamical stochastic resonance: Theory and experiments with white and various coloured noises, *J. Stat. Phys.* **70**, pp. 795–802.

Gingl, Z., Kiss, L. B. and Moss, F. (1995). Non-dynamical stochastic resonance: Theory and experiments with white and various coloured noises, *Nuovo Cim. D* **17**, pp. 795–802.

Graham, R. (1978). *Lecture Notes in Physics*, Vol. 84 (Springer-Verlag, Berlin).

Graham, R. (1987). *Instabilities and Nonequilibrium Structures*, Vol. 1 (D. Reidel, Dordrecht).

Gunton, J. D. and Droz, M. (1983). *Introduction to the theory of metastable and unstable states, Lecture Notes in Physics*, Vol. 183 (Springer-Verlag, Berlin).

Haken, H. (1978). *Synergetics: An Introduction*, 2nd edn. (Springer-Verlag, New York).

Halpin-Healy, T. and Zhang, Y.-C. (1995). Kinetic roughening phenomena, stochastic growth, directed polymers and all that, *Phys. Rep.* **254**, pp. 215–414.

Hentschel, H. and Family, F. (1991). Scaling in open dissipative systems, *Phys. Rev. Lett.* **66**, pp. 1982–1985.

Hohenberg, P. C. and Halperin, B. I. (1977). Theory of dynamical critial phenomena, *Rev. Mod. Phys.* **49**, pp. 435–479.

Horsthemke, W. and Lefever, R. (1984). *Noise-Induced Transitions* (Springer-Verlag, Berlin).

Ibañes, M., García-Ojalvo, J., Toral, R. and Sancho, J. M. (2001). Noise-induced scenario for inverted phase diagrams, *Phys. Rev. Lett.* **87**, p. 020601.

Izús, G., Ramírez, O., Deza, R., Wio, H. S., Zanette, D. and Borzi, C. (1995). Global stability of stationary patterns in reaction–diffusion systems, *Phys. Rev. E* **52**, pp. 129–136.

Jarzynski, C. (1997). Nonequilibrium equality for free energy differences, *Phys. Rev. Lett.* **78**, p. 2690.

Jung, P. (1993). Periodically driven stochastic systems, *Phys. Rep* **234**, pp. 175–295.

Jung, P. and Hänggi, P. (1987). Dynamical systems: A unified colored-noise approximation, *Phys. Rev. A* **35**, pp. 4464–.

Kadanoff, L. P. and Martin, P. (1963). Hydrodynamic equations and correlation functions, *Ann. Phys.* **24**, pp. 419–469.

Kardar, M. (1998). Nonequilibrium dynamics of interfaces and lines, *Phys. Rep.* **301**, pp. 85–112.

Kardar, M., Parisi, G. and Zhang, Y.-C. (1986). Dynamic scaling of growing interfaces, *Phys. Rev. Lett.* **56**, pp. 889–892.

Kirkaldy, J. S. (1992). Spontaneous evolution of spatiotemporal patterns in materials, *Rep. Prog. Phys.* **55**, pp. 723–795.

Koga, S. and Kuramoto, Y. (1980). *Prog. Theor. Phys.* **63**, p. 106.

Kreuzer, H. J. (1984). *Nonequilibrium Thermodynamics and its Statistical Foundations* (Clarendon, Oxford).

Krug, J. (1994). Turbulent interfaces, *Phys. Rev. Lett.* **72**, pp. 2907–2910.

Krug, J. (1997). Origins of scale invariance in growth processes, *Adv. Phys.* **46**, pp. 139–282.

Krug, J. and Spohn, H. (1991). in *Solids Far From Equilibrium,* C. Godréche (ed.) (Cambridge U.P., Cambridge, UK).

Langer, J. S. (1987). *Lectures in the Theory of Pattern Formation,* Chance and Matter (North-Holland, Amsterdam).

Langouche, F., Roekaerts, D. and Tirapegui, E. (1982). *Functional Integration and Semiclassical Expansions* (Reidel, Dordrecht).

Lindenberg, K. and Wio, H. S. (2003). in *Proc. Pan-American Advanced Studies Institute: New Challenges in Statistical Physics,* Eds. K.Lindenberg and V. Kenkre (AIP, Melville, NY).

Lindner, J. F., Meadows, B. K., Ditto, W. L., Inchiosa, M. E. and Bulsara, A. (1995). Array enhanced stochastic resonance and spatiotemporal synchronization, *Phys. Rev. Lett.* **75**, pp. 3–6.

Lindner, J. F., Meadows, B. K., Ditto, W. L., Inchiosa, M. E. and Bulsara, A. (1996). Scaling laws for spatiotemporal synchronization and array enhanced stochastic resonance, *Phys. Rev. E* **53**, pp. 2081–2086.

López, J. M. (1999). Scaling approach to calculate critical exponents in anomalous surface roughening, *Phys. Rev. Lett.* **83**, pp. 4594–4597.

Malchow, H. and Schimansky-Geier, L. (1985). *Noise and Diffusion in Bistable Nonequilibrium Systems* (Teubner, Berlin).

Mangioni, S., Deza, R., Wio, H. S. and Toral, R. (1997). Disordering effect of color in nonequilibrium phase transition induced by multiplicative noise, *Phys. Rev. Lett.* **79**, pp. 2389–.

Mangioni, S. E., Deza, R. R. and Wio, H. S. (2001). Transition from anomalous to normal hysteresis in a system of coupled brownian motors: A mean-field approach, *Phys. Rev. E* **63**, p. 041115.

Manoliu, A. and Kittel, C. (1979). Correlation in the Langevin theory of Brownian motion, *Am. J. Phys.* **47**, pp. 678–680.

Maruyama, K., Nori, F. and Vedral, V. (2002). Brownian motors: noisy transport far from equilibrium, *Phys. Rep.* **361**, pp. 57–265.

Maruyama, K., Nori, F. and Vedral, V. (2009). *Colloquium*: The physics of maxwells demon and information, *Rev. Mod. Phys.* **81**, pp. 1–23.

McNamara, B. and Wiesenfeld, K. (1989). Theory of stochastic resonance, *Phys. Rev. A* **39**, pp. 4854–4869.

Meakin, P. (1993). The growth of rough surfaces and interfaces, *Phys. Rep.* **235**, pp. 189–289.

Mikhailov, A. S. (1990). *Foundations of Synergetics I* (Springer-Verlag, Berlin).

Moss, F. (1992) (SIAM, Philadelphia).

Nicolis, G. (1989). *Physics of Far-from-Equilibrium Systems and Self-Organization* (Cambridge U.P., Cambridge, UK).

Nicolis, G. (1995). *Introduction to Nonlinear Science* (Cambridge U.P., Cambridge, U.K.).

Nicolis, G. and Prigogine, I. (1977). *Self-Organization in Nonequilibrium Systems* (Wiley, New York).

Ohta, T., Mimura, M. and Kobayashi, R. (1989). *Physica D* **34**, p. 115.

Prigogine, I. (1980). *From Being to Becoming* (Freeman, San Francisco).

Ramasco, J. J., López, J. M. and Rodríguez, M. A. (2000). Generic dynamic scaling in kinetic roughening, *Phys. Rev. Lett.* **84**, pp. 2199–2202.

Reichl, L. E. (1980). *A Modern Course in Statistical Physics* (U. of Texas Press, Austin).

Reimann, P., Kawai, R., Van den Broeck, C. and Hänggi, P. (1999). Coupled brownian motors: Anomalous hysteresis and zero-bias negative conductance, *Europhys. Lett.* **45**, pp. 545–551.

San Miguel, M. (1985). *Stochastic Methods and Models in the Dynamics of Phase Transitions*, Stochastic Processes Applied to Physics (World Scientific, Singapore).

San Miguel, M. and Toral, R. (2000). in *Instabilities and Nonequilibrium Structures VI*, E. Tirapegui, J. Martínez-Mardones, and R. Tiemann, eds. (Kluwer Academic Publishers), pp. 35–130.

Schat, C. and Wio, H. S. (1992). An electrothermal instability: Influence of albedo boundary conditions on the stationary states of an exactly solvable model, *Physica A* **180**, pp. 295–.

Schulman, L. S. (1981). *Techniques and Applications of Path Integration* (Wiley, New York).

Stauffer, D., Moss de Oliveira, S., de Oliveira, P. M. C. and Sa Martins, J. S. (2006). *Biology, Sociology, Geology by Computational Physicists* (Elsevier, Amsterdam).

Van den Broeck, C. (2010). The many faces of the second law, *J. Stat. Mech.: Theory and Experiment* , p. P10009.

Van den Broeck, C., Parrondo, J. M. and Toral, R. (1994). Noise-induced nonequilibrium phase transition, *Phys. Rev. Lett.* **73**, pp. 3395–.

Van den Broeck, C., Parrondo, J. M., Toral, R. and Kawai, R. (1997). Nonequilibrium phase transitions induced by multiplicative noise, *Phys. Rev. E* **55**, pp. 4084–.

van Kampen, N. (1982). *Stochastic Processes in Physics and Chemistry* (North Holland, Amsterdam).

Wang, G. M., Sevick, E. M., Mittag, E., Searles, D. J. and Evans, D. J. (2002). Experimental demonstration of violations of the second law of thermodynamics for small systems and short time scales, *Phys. Rev. Lett.* **89**, p. 050601.

Weidlich, W. (2002). *Sociodynamics-A systematic approach to mathematical modelling in social sciences* (Taylor & Francis, London).

Wiegel, F. (1986). *Introduction to Path Integral Methods in Physics and Polymer Science* (World Scientific, Singapore).

Wio, H. S. (1990). *Introducción a las Integrales de Camino* (Univ. Illes Balears, Palma de Mallorca).

Wio, H. S. (1994). *An Introduction to Stochastic Processes and Nonequilibrium Statistical Physics*, 1st edn. (World Scientific, Singapore).

Wio, H. S. (1997). *Nonequilibrium Potential in Reaction–Diffusion Systems*, Lecture Notes in Physics, Vol. 493 (Springer-Verlag, Berlin), pp. 135–195.

Wio, H. S. (2009). Variational formulation for the KPZ and related kinetic equations, *Int. J. Bif. Chaos* **19**, pp. 2813–2821.

Wio, H. S. and Deza, R. R. (2007). Aspects of stochastic resonance in reaction–diffusion systems: The nonequilibrium-potential approach, *Eur. Phys. J. Special Topics* **146**, pp. 111–126.

Wio, H. S., Von Haeften, B. and Bouzat, S. (2002). Stochastic resonance in spatially extended systems: The role of far from equilibrium potentials, *Physica A* **306**, pp. 140–156, proc. STATPHYS21.

Index

Γ-space, 68, 74

affinities, 90, 109
analysis
  linear stability, 156
anomalous scaling, 275, 276, 283, 286,
    288, 290, 293
  intrinsic, 277, 280, 283, 286
approximation
  adiabatic elimination, 162, 165
  Born, 135
  coarse graining, 23
  relaxation time, 84
attraction basin, 158
attractor, 154, 158

bifurcation, 154, 163, 166
  Hopf, 164, 216
  subcritical, 164
  supercritical, 164
  Turing, 216
bifurcation diagram, 166
bistability, 176
Boltzmann
  $\mathcal{H}$–theorem, 82
  kinetic equation, 71–73
Boyle Gas, 105
Brownian
  motion, 23, 26
  motor, 199
  particle, 202

condition
  boundary, 212
    albedo, 212, 225
    Dirichlet, 212, 225
    Neumann, 212, 225
  detailed balance, 94
  self-consistency, 258
critical
  opalescence, 182
  slowing-down, 182

decay processes, 13
dynamic scaling Ansatz
  Family–Vicsek, 268
  generic, 283
dynamical exponent, 267

effect
  Dufour, 105
  Gunn, 153
  Ludwig–Soret, 105
  ratchet, 203
effusion, 37
equation
  balance, 74
    energy, 77
    entropy, 78
    mass, 75, 109
    momentum, 77
    probability density, 64
  Chapman–Kolmogorov, 10, 12, 18,
      29, 31

Clausius–Clapeyron, 104
Fokker–Planck, 22, 27
kinetic, 70
  Boltzmann, 71–73
  Vlasov, 71
Langevin, 24, 27
Liouville, 66
master, 12
expansion
  Ω–expansion, 37, 40, 44, 47, 57
    limitations, 56
  Kramers–Moyal, 21, 22, 26, 48
exponent
  dynamical, 267
  roughness, 265, 267
    global, 275–277, 279, 286
    local, 276, 277, 279, 281, 289,
      292, 293
    spectral, 285–288

Fermi
  Golden Rule, 135
fixed point, 155
  center, 157
  focus, 157
  node, 157
fluctuations, 154
formula
  Green–Kubo, 97
fractal
  self-affine, 265
function
  characteristic, 5
  self-correlation, 11
functional
  Onsager–Machlup, 34

hierarchy
  BBGKY, 62, 67, 70–72
hypothesis
  Onsager's, 91
hysteresis, 154

instability
  Bénard, 153

integral
  path-, 30, 34
  Wiener, 32

kinetic roughening, 265, 268, 269, 272
  anomalous, 288
  universality classes, 269
kinetics
  Ising model, 15
Knudsen Gas, 105

law
  Arrhenius, 178
limit cycle, 154

master equation, 12
model
  pulsating coupled ratchets, 263
  activator–inhibitor, 241
    three-component, 244
  ballast resistor, 217, 228, 229, 252
  Brusselator, 216
  dog–flea, 16
  Ehrenfest urn, 16
  Fisher, 228
  fluctuation–dissipation compliant,
    261
  genetic, 191
  hot-spot, 217
  Ising
    kinetics, 15
  Malthus–Verhulst, 184
  metal-wire thermocouple, 105, 107
  opinion formation, 48
  Oregonator, 217
  ratchet
    minimal, 202, 203
    on–off, 203
    temperature, 203, 204
  reaction–diffusion, 210, 211
    scalar, 239, 247
  Schlögl, 176, 182, 212, 213, 215,
    220
  short-time instability, 254
  surface growth
    Edwards–Wilkinson, 271

Kardar–Parisi–Zhang, 272
Lai–Das Sarma–Villain, 274
linear MBE, 273
Mullins–Herring, 273
nonlinear MBE, 274
random deposition, 271
Van der Pol oscillator, 161
molecular-beam epitaxy (MBE), 274

noise
white, 11
noise-induced
phase transitions, 254
transitions, 184

Onsager
hypothesis, 91
operator
density, 78
reduced, 80
osmotic pressure, 105

parameter
control, 154, 163
order, 260, 262
conserved, 262
path
-integral, 30, 34
most probable, 35
reference, 35
space
measure, 32
phase
curve, 155
plane, 155
portrait, 158
space, 158
trajectory, 155
probability distribution
conditional, 7
Gaussian, 7
joint, 6, 7

reaction
Belousov–Zhabotinskii, 153, 209,
217, 234, 235, 237

relation
Onsager, 100
roughness exponent, 265, 267
rule
golden, 135

saddle point, 157
scale invariance, 265
generic, 283, 286
scaling
anomalous, 275, 276, 283, 286, 288,
290, 293
intrinsic, 277, 280, 283, 286
relations, 272
scheme
activator–inhibitor, 236
stability
global, 165
linear, 156
stochastic
differential equation, 24, 27
process, 5
Gaussian, 27
independent, 8
Langevin, 11
Markovian, 9
Ornstein–Uhlenbeck, 10
Poisson, 11
white noise, 11
Wiener–Lévy, 10, 11
resonance, 195, 247
variable, 4
characteristic function, 5
cumulants, 6
moments, 5
probability distribution, 4
variance, 6
structures
dissipative, 153
super-roughening, 275, 277, 279, 280,
282, 283, 286, 292
surface growth, 265
scale-invariant, 265
universality classes, 268, 269
symmetry breaking, 154, 165

system
  activator–inhibitor, 226, 235, 241,
    254

technique
  singular perturbation, 161
theorem
  $\mathcal{H}$–theorem, 82
  central limit, 40
  Doob, 11
  equipartition, 25
  fluctuation–dissipation, 25
  Kolmogorov, 7
  Pawula, 48
  Wiener–Khintchine, 47
  Wiener–Kintchine, 198

transport coefficient, 84
  conductivity
    electrical, 84
    thermal, 87

variable
  recovery, 234
  trigger, 234

Wiener
  –Khintchine theorem, 47
  integral, 32
  measure, 32

About the Authors

Horacio S. WIO has been researcher at Centro Atómico Bariloche, Argentina where he was Head of the Theoretical Physics Division, Leader of the Statistical Physics Group; and Professor at Instituto Balseiro. He was also Visiting Researcher or Professor at several institutions in Argentina, Brazil, Chile, Mexico, USA, Spain, Italy, France, Germany, Belgium, Associated Researcher at ICTP, Trieste, and member of the Editorial Board of several professional journals. He was awarded a *Marie Curie Chair* by the European Commission during the period 2004-2007, and is currently Professor at the Universidad de Cantabria, Spain.

Roberto R. DEZA is full Professor of Physics since 1987 at the National University of Mar del Plata, Argentina—School of Exact and Natural Sciences (FCEyN), of which he has been the vice Dean from 1992 to 1996. He leads since 1993 the Mathematical & Computational Physics Group at the Physics Department, FCEyN and is presently the Head of the Complex Systems Laboratory at the Mar del Plata Institute for Physical Research (IFIMAR). He has taught at Instituto Balseiro and several other universities in Argentina, and has been a Visiting Researcher at UCSB (USA), ICTP (Italy) and UCS & UC (Spain).

Juan M. LÓPEZ is Scientific Researcher at Consejo Superior de Investigaciones Científicas (CSIC) in Spain and currently affiliated with the Instituto de Física de Cantabria (Spain), where he is devoted to full-time research. Previously, he held research positions at Imperial College London (UK), Università di Roma La Sapienza (Italy).